Upconverting Nanomaterials

Upconverting Nanomaterials

Editor

Mudiwa Afolayan

scitus
academics

Upconverting Nanomaterials
Edited by **Mudiwa Afolayan**

ISBN: 978-1-68117-715-1
Library of Congress Control Number: 2016958824

SCITUS Academics LLC,
616, Corporate Way, Suite 2, 4766,
Valley Cottage, NY 10989

www.scitusacademics.com

Notice

Printed in United States of America on Acid Free Paper

Preface

Upconversion is an optical process that involves the conversion of lower energy photons into higher energy photons. It has been extensively studied since the mid-1960s and has been widely used in optical instruments. In the past, high-quality rare earth doped upconversion nanoparticles have been synthesized successfully with the rapid development of nanotechnology and are becoming more visible in natural sciences. The methods of synthesis are typically phase-based processes such as thermal decomposition, hydrothermal reaction and synthesis based on ionic liquids. Rare earth-doped upconversion nanoparticles that can turn long wavelengths near infrared radiation into short visible wavelength radiation through a non-linear optical process emerge as a brand new class of fluorescent biolables. They can overcome a number of conventional biolables, including quantum dots and organic colors.

This compendium highlights recent advances in the synthesis, surface modification and biological applications of upconversion nanoparticles. It can serve as an effective tool to explore state-of-the-art research on upgrading nanosystems while providing the necessary basic background to non-specialist readers. The various aspects of converting materials are approached both from a theoretical point of view, especially from a converting point of view and from a useful point of view. By introducing synthetic strategies, functionality, the production of core shell structures and nanocomposites, this book provides doctoral students, practitioners and researchers with a wealth of ideas that can be applied to different research fields. The advances in functional strategies that allow the broad use of upconversion nanocrystals for photonics, volumetric displays, cancer therapy and multimodal imaging are mainly emphasized.

Table of Contents

Chapter 1

Upconversion Nanomaterials: Synthesis, Mechanism, and Applications in Sensing

Jiao Chen and Julia Xiaojun Zhao *

Department of Chemistry, University of North Dakota, Grand Forks, ND 58202, USA

ABSTRACT

Upconversion is an optical process that involves the conversion of lower-energy photons into higher-energy photons. It has been extensively studied since mid-1960s and widely applied in optical devices. Over the past decade, high-quality rare earth-doped upconversion nanoparticles have been successfully synthesized with the rapid development of nanotechnology and are becoming more prominent in biological sciences. The synthesis methods are usually phase-based processes, such as thermal decomposition, hydrothermal reaction, and ionic liquids-based synthesis. The main difference between upconversion nanoparticles and other nanomaterials is that they can emit visible light under near infrared irradiation. The near infrared irradiation leads to low autofluorescence, less scattering and absorption, and deep penetration in biological samples. In this review, the synthesis of upconversion nanoparticles and the mechanisms of upconversion process will be discussed, followed by their applications in different areas, especially in the biological field for biosensing.

Keywords: upconversion nanoparticles; biosensing; NIR-radiation

1. INTRODUCTION

The upconversion phenomenon has been observed in transition metals, actinides, but mainly in the rare earth (RE) elements, which contain the lanthanide (Ln) series, yttrium, and scandium [1]. Ln^{3+} ions have special $4f^n\ 5d^{0-1}$ inner shell configurations that are well-shielded by outer shells and have abundant and unique energy level structures. These Ln^{3+} ions can exhibit sharp luminescence emissions via intra-4f or 4f-5d transitions. Their remarkable luminescence properties, such as narrow bandwidth, long-time emission, and anti-Stokes emission, have been widely applied in lasers, solar cells, analytical sensors, optical imaging, photodynamic therapy, and so on. At present, the low luminescence efficiency is one of the main limiting factors. Therefore, to obtain the highest upconversion luminescence efficiency, it is critical to choose an appropriate host material with lower phonon energy (high phonon frequencies of the host lattice lead to nonradiative relaxation). To date, host materials, including fluoride, chloride and bromide, have been shown to enhance upconversion luminescence intensity. Most of chlorides and bromides are sensitive to moisture, and thus are not suitable for labeling biomolecules (used mostly in aqueous solutions) [2]. RE fluorides, mainly REF_3 and $AREF_4$ (A = alkali), have been considered as an excellent host material due to their high refractive index and high transparency arising from low-energy phonons. These two advantages further lead to low probability of nonradiative decay and increased luminescence quantum yield.

Most fluorescent materials, including dye molecules, quantum dots, and dye-doped silica/gold nanomaterials, emit light by the downconversion process (emitting lower-energy photons under higher-energy irradiation). Although the uses of a conventional organic dye molecule or quantum dot (QD) based biomarker have achieved significant progress in real-time detection and bioimaging, they still have drawbacks. These fluorescent materials are generally excited by ultraviolet (UV) or visible light, which may induce autofluorescence and photodamage to biological samples, resulting in low signal-to-noise ratio and limited sensitivity. These limitations prompted the development of a new type of high-quality and well-shaped nanomaterials known as upconversion nanomaterials (UCNs). These UCNs usually consist of an inorganic host that is doped with Ln^{3+} ions. They show good biocompatibility and generally low cytotoxicity, and are in fact non-cytotoxic to a broad range of cell lines [3,4]. Furthermore, surface modification by ligand engineering [5–7], ligand attraction [8], surface polymerization [3,9–13], self-assembly [14–16] or layer-by-layer assembly technology [17], broadens their application fields. Surface modification by a silica shell is by far the most popular, common, and practical approach [3,11,18–20]. Proteins, DNA, biological macromolecules or other desirable targets can be easily linked to UCNs. In particular, the UCNs' unique property of emitting visible light under NIR irradiation makes them a suitable candidate both for *in vivo* and *in vitro* bioimaging [21–23].

Theoretically, most lanthanide ions can undergo the NIR-to-visible upconversion process; however, relatively efficient upconversion is only possible with a few trivalent lanthanide ions (e.g., Er^{3+} and Tm^{3+}) under low pump densities (980 nm excitation). Up to now, the most often-used upconversion nanoparticles are Yb^{3+}-Er^{3+} or Yb^{3+}-Tm^{3+} co-doped $NaYF_4$ nanomaterials ($NaYF_4$:Yb^{3+}, Er^{3+} or $NaYF_4$: Yb^{3+}, Tm^{3+} UCNs). In this review, we will briefly discuss the mechanisms of upconversion phenomenon, but mainly focus on the recent progress in UCNs' chemical syntheses and their applications in different areas, especially in the biological field.

2. MECHANISM

Different upconversion luminescence mechanisms have been recognized either alone or in combination. Three basic mechanisms—excited state absorption (ESA), photon avalanche (PA), and energy transfer upconverion (ETU) (also known as *addition de photon par transferts d'energies*, APTE effect) will be discussed here.

2.1. Excited State Absorption (ESA)

ESA involves multistep excitation by sequentially absorbing one or more photons from the ground state to intermediate reservoir stage, and finally populates at excited state, from which upconverison luminescence occurs (Figure 1).

2.2. Photon Avalanche (PA)

PA process is a more complex process, which can be characterized by three distinct nonlinear behaviors: transmission, emission, and rise time on the pump power intensity with generally the existence of a critical pump threshold. Take a four-energy system as an example (Figure 2), in which E_0, E_1 and E_2, and E represents ground state, intermediate states, and upper excited states, respectively.

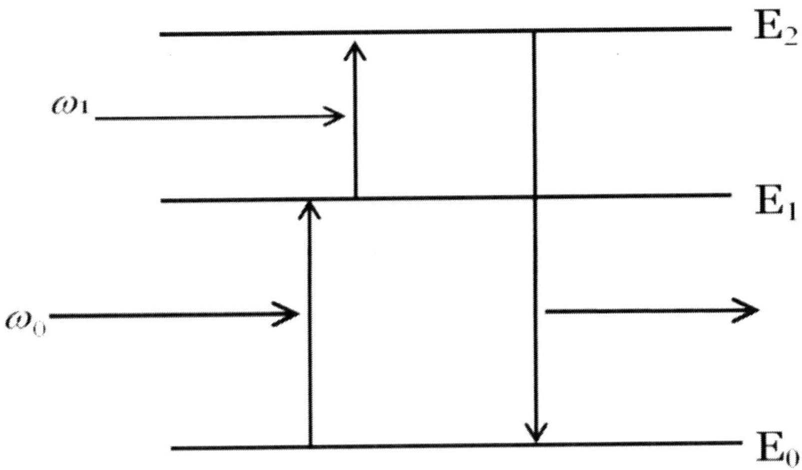

Figure 1. Schematic diagram of excited-state absorption (w' > w_1, w_0). E_0, E_1, and E_2 represent ground state, intermediate, and excited state, respectively. When one ion or electron in the E_0 state absorbs one photon, it is first excited to the E_1 state; after sequentially absorbing the second photon, it can jump to the excited state E_2 and emit higher energy photons when coming back to the ground state.

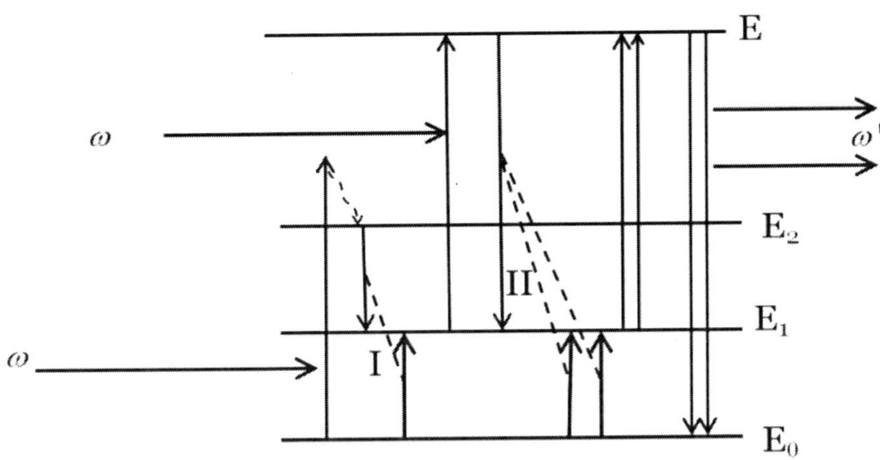

Figure 2. Schematic diagram of photon avalanche.

An electron or ion is excited when it absorbs the excitation radiation. The excitation radiation is usually not resonant with the absorption transition from the ground state to the intermediate states, but a little higher than E_2. Through the cross relaxation, it goes down to E_2 state. Energy transfer occurs between the E_2 state electron and the E_0 state electron, resulting in the formation of two electrons in the E_1 state. One of them absorbs the excitation radiation and is excited to the E state, in which it interacts with E_0 state electrons and energy transfer II occurs to form three E_1 electrons. Here, the excitation radiation is resonant with the absorption transition from E_1 to E. By repeating the whole steps again and again, the number of electrons in the E state increases dramatically. When the electrons go back to the E_0state, high-energy photons are emitted. In summary, the PA process basically involves resonant excited-state absorption, efficient cross relaxation and substantial population of the reservoir level, which leads to strong upconversion emission. Although PA is efficient, it still suffers from some disadvantages, such as limited maximum-output due to the weak ground–state absorption, and high pump powers needed to reach the threshold condition.

2.3. Energy Transfer Upconversion (ETU)

ETU is by far the most efficient upconversion process in RE doped nanomaterials and it is independent of the pump power. In the process of ETU, two situations, resonant non-radiative transfer and phonon-assisted non-radiative transfer in two-ion-involved system, will be mainly discussed (Figure 3) [1]. When the excited energies of sensitizer (S) and activator (A) are nearly equal and the distance between them is near enough, energy can be transferred from S to A, exciting A from its ground state to excited state before S emits photons. In phonon-assisted non-radiative transfer, an energy mismatch exists between S and A ions, so phonon assistance is necessary to have the energy transfer process.

For example, in the system of $NaYF_4$:Yb^{3+} (Figure 4) Er^{3+} UCNs, red, blue, and green light can be emitted through this process. The Yb^{3+} ion, with its excited state $^2F_{5/2}$, has an energy comparable to $^4I_{11/2}$ (Er^{3+}), which can act as a sensitizer for Er^{3+} and transfer its energy to an unexcited Er^{3+} ion through the energy transfer process: $^2F_{5/2}$ (Yb^{3+}) + $^4I_{15/2}$ (Er^{3+}) → $^2F_{7/2}$ (Yb^{3+}) + $^4I_{11/2}$ (Er^{3+}). By further cross relaxation and phonon-assisted process, red emission (~654 nm) can be emitted from $^4F_{9/2}$ [$^2F_{5/2}$(Yb^{3+}) + $^4I_{13/2}$ (Er^{3+}) → $^2F_{7/2}$ (Yb^{3+}) + $^4F_{9/2}$ (Er^{3+})]. The blue (~408 nm) and green luminescence emissions (~526 nm and ~533 nm) are emitted from $^2H_{9/2}$ – $^4I_{15/2}$ and $^2H_{11/2}$ – $^4I_{15/2}$ (also $^4S_{3/2}$ – $^4I_{15/2}$) via similar ways, respectively.

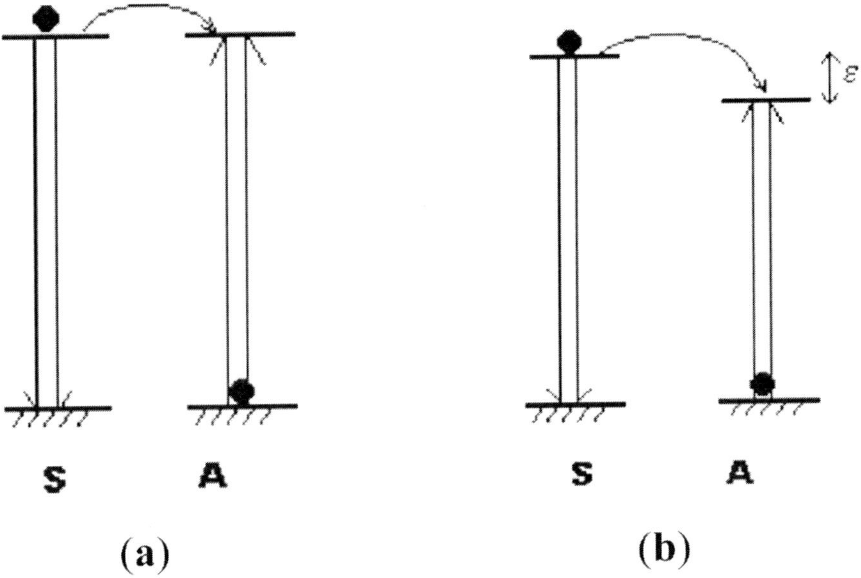

Figure 3. Energy transfer processes between two ions: (**a**) resonant non-radiative transfer; (**b**) phonon-assisted non-radiative transfer. (S: sensitizer ions, A: activator ions) [1].

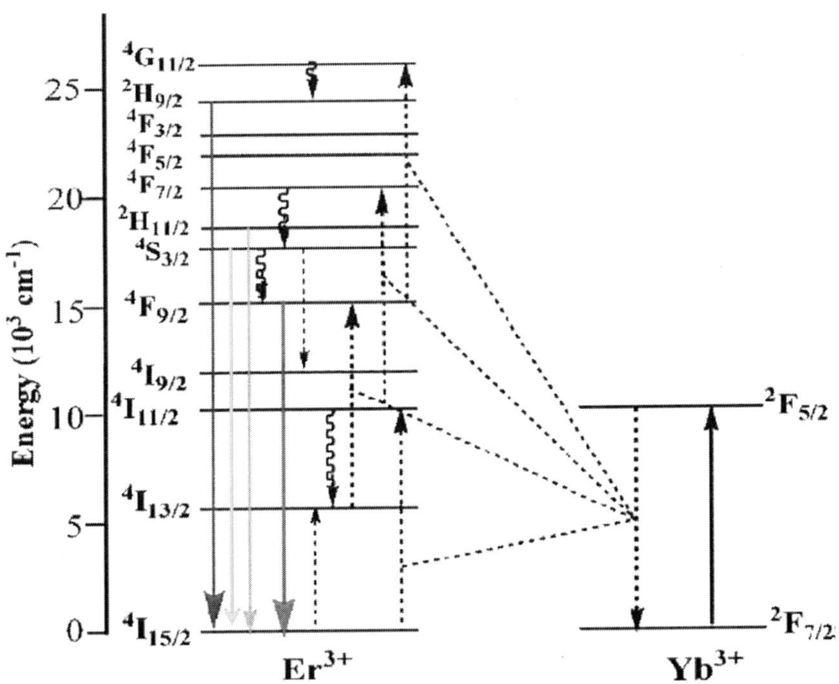

Figure 4. Energy diagram of the Er^{3+}/Yb^{3+} codoped materials [24].

3. SYNTHESIS OF RARE EARTH FLUORIDE NANOMATERIALS

In order to obtain high luminescence efficiency, synthesizing high-quality UCNs is very critical. So far, there are three common methods to synthesize UCNs, including thermal decompostition [13,25–30], hydrothermal synthesis [31–33], and ionic liquids-based synthesis [34].

3.1. Thermal Decomposition Method

Thermal decomposition, which gives well shaped particles, with good size control, after a relatively short reaction time, is one of the most popular methods. It usually involves dissolving organic precursors in high-boiling organic solvents with the assistance of surfactants. The commonly used organic precursors are trifluoroacetate compounds, and the surfactants typically have polar capping groups and long hydrocarbon chains, such as oleic acid (OA), omeylamine (OM), and 1-octadecence (ODE). Mai *et al.* [27] systematically investigated the growth mechanism of nanocrystals and pointed out that the trifluoroaetate precursors in hot surfactant solutions went through a unique delayed nucleation pathway. The synthesis reaction was separated as four stages including nucleation in a delayed time, particle growth by monomer supply, size shrinkage by dissolution, and aggregation. Figure 5 illustrates the synthesis steps of α-$NaYF_4$:Yb^{3+}, Er^{3+} UCNs and by varying the reaction time, concentration of reagents, and reaction temperature, various sizes and shapes of $NaYF_4$:Yb^{3+}, Er^{3+} UCNs can be obtained.

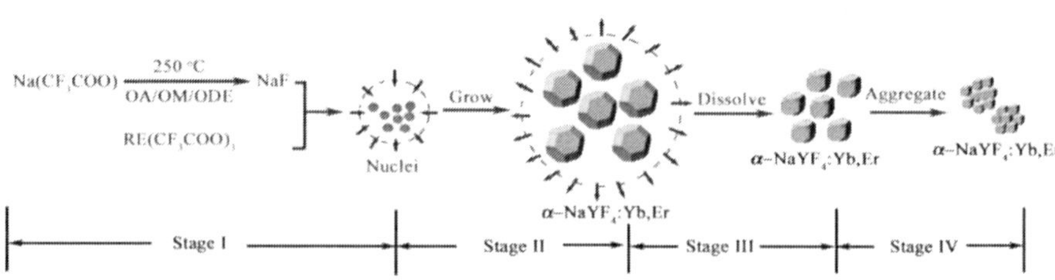

Figure 5. Schematic illustration of the growth stages of α-$NaYF_4$:Yb^{3+}, Er^{3+} nanocrystals via a delayed nucleation pathway [27].

Figure 6 shows a few examples of RE fluoride nanomaterials synthesized by thermal decomposition method, which are well-shaped and monodispersed. Although it gives a narrow size distribution and good shape control, the thermal decomposition method still suffers some drawbacks. It normally requires high reaction temperature (250–330 °C), organic solvents, and an oxygen-free with inert gas protection. In addition, most of the synthesized nanomaterials are stabilized by surfactant, which brings difficulties in the biological applications. Further surface modification is also required.

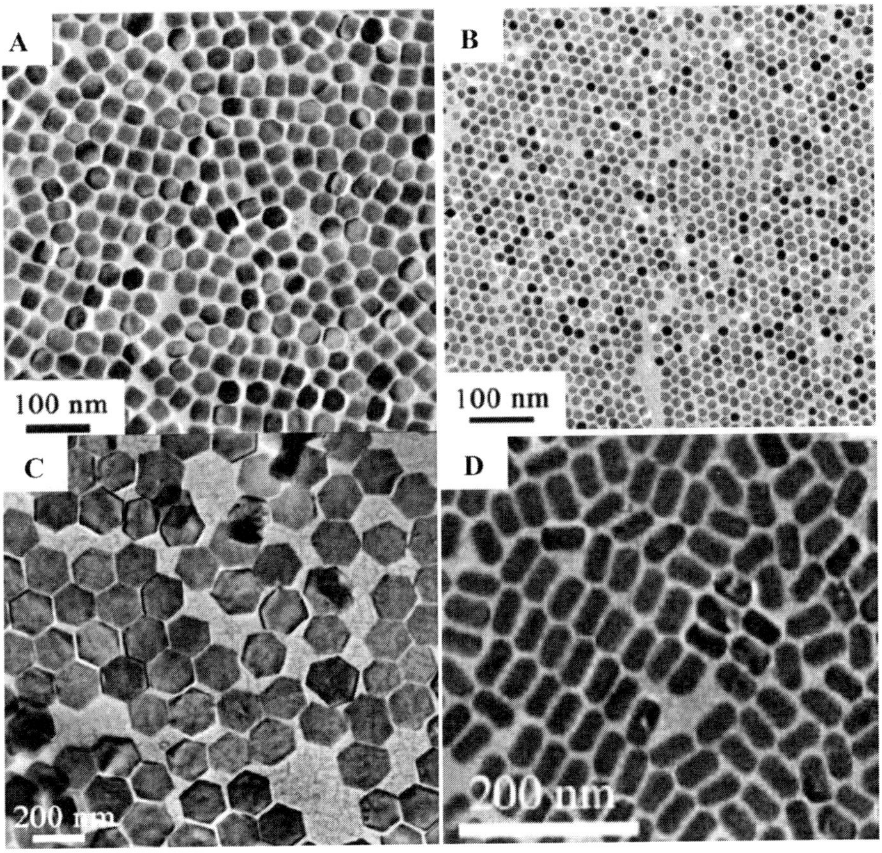

Figure 6. Representative shapes of various RE fluoride nanomaterials by thermal decomposition method [27,28].

3.2. Hydrothermal Synthesis Method

Hydrothermal synthesis is a typical solution-based approach, which is usually employed under high temperatures and pressures [5,31,35,36]. Unlike the thermal decomposition method, which can only use an organic compound as a solvent, hydrothermal synthesis can occur in a water-based system and at a lower reaction temperature (160–220 °C) in a relatively environmentally friendly approach. It is an effective and convenient process in preparing inorganic materials with diverse controllable morphologies and architectures. For example, various shapes of hexagonal $NaYF_4$ crystals, such as prism, disk, tube, rod, and octadecahedral shapes were synthesized by applying this method [37,38]. Recently, in Lin's group, the mechanism of synthesizing different shapes of RE fluoride nano-/microcrystals was systematically investigated [33]. It was reported that the organic additive trisodium citrate, the fluoride source, and pH value have great effects on the shapes (Figure 7).

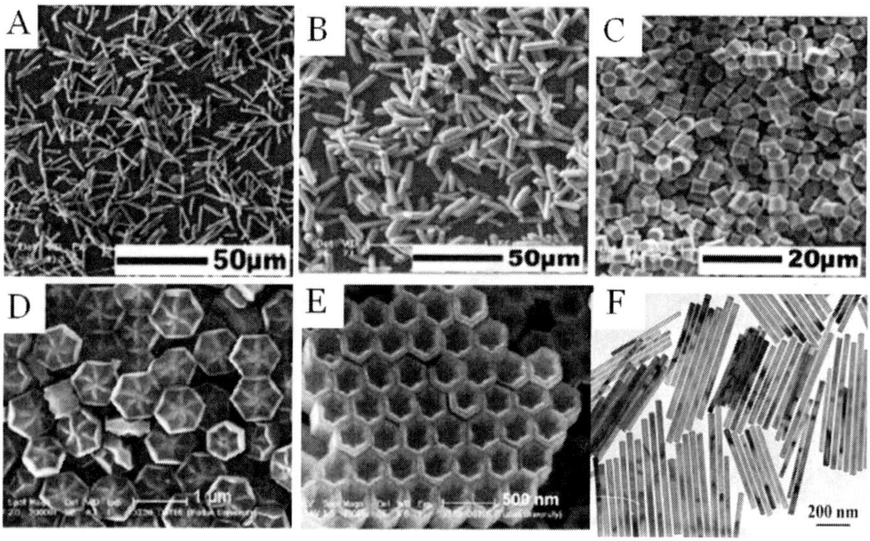

Figure 7. Different shapes of RE fluoride nano-/microcrystals by hydrothermal method [37–39].

3.3. Ionic Liquids-Based Synthesis Method

Compared to the other two methods, ionic liquids-based synthesis is a relatively green method due to not needing organic solvents, low reaction temperature, and a short reaction time. Nanomaterials are synthesized in ionic liquid media, which are known as "green solvents" because of their chemical stability, low vapor pressure, and non-flammability. However, the synthesized nanomaterials are of a lower quality with a broader size distribution, lower monodispersity, and less uniformity when compared with other two methods [40], which limits their applications.

Therefore, the thermal decomposition and hydrothermal synthesis methods are still the most widely used and best methods for producing well-controlled shapes and sizes of RE fluoride nanomaterials.

4. PROPERTIES AND APPLICATIONS

There has been a growing interest in studying the upconveriosn phenomenon after it was first recognized in the mid-1960s. It has been widely used in solid-state lasers (especially blue-light-emitting lasers) [41], solar cells [42,43], and waveguide amplifiers [44,45]. It was not considered to be used in the biological field until the tremendous advances in nanotechnology over the last decade. The successful synthesis of high-quality, lanthanide-doped upconversion nanomaterials (UCN) have broadened their applications. Nondestructive optical memory [46] and rewritable optical storage [47] applications based on UCNs have been developed recently. Furthermore, these UCNs with their controllable sizes and unique luminescence properties have become prominent in the biological field as promising alternatives to conventional organic dyes and quantum dots (QDs). Compared to organic dyes and QDs, UCNs have high quantum yields, long lifetimes, high photostability, a narrow emission peak, and most importantly, low optical background noise due to the absence of autofluorescence under NIR radiation. Multiple analytes can be detected simultaneously from different UCNs excited by the same IR laser [48]. Furthermore, their applications in biolabeling [6,10,25,49], homogeneous assay [11,50,51], and as reporters for DNA microarrays [52] have been extensively studied. Since these UCNs need to be used in cells or animals, the toxicity study is also very important. Therefore, their viability will now be discussed, followed with their applications in the biological field, including photodynamic therapy, imaging, and others. The recent advances of these UCNs in other fields will be discussed as well.

4.1. Toxicity

Due to the rapid progress in developing UCNs' applications in the biological field, the safety and toxicity of UCNs are a growing concern and extremely important. The most common methods to evaluate the toxicity are through cellular morphology and mitochondrial function, such as methylthiazolyl tetrazolium (MTT) and 3-(4,5-dimethylthiazol-2-yl)-5-(3-carboxymethoxyphenyl)-2-(4-sulfophenyl)-2H-tetrazolium (MTS) assays. Studies have shown that UCNs are non-/low-toxicity to a broad range of cell lines [4,6,10,18,53,54]. The cells were usually treated with different concentrations of UCNs for various time range, and by measuring the cell viability percentage, the UCNs toxicity was determined. For example, in Zhang's group, they studied both polyethyleneimine (PEI) [55] and silica-coated [18] UCNs' cell toxicity. In the case of silica-coated UCNs, although the cell viability decreased as a function of both concentration and time, 93.4% of skeletal myoblast cells and 93.2% of BMSCs cells were still alive under 24 h incubation at a concentration of 1 μg/mL, which indicated great biocompatibility [18]. Recently, Li and co-workers investigated the UCNs' long-term *in vivo* distribution and toxicity [56]. Their findings show that the mice survived for 115 days after the intravenously injection 15 mg/kg of UCNs with no apparent adverse effects observed to their health, indicating the possibility for long-term targeted imaging and therapy studies *in vivo*. Hu *et al.* treated the KB cells with different concentrations of PEG-modificed UCNs (Figure 8) [6]. Even at a high concentration of 250 μg/mL, the cell viability still remained above 80%, showing the low cytotoxicity of the nanoparticles.

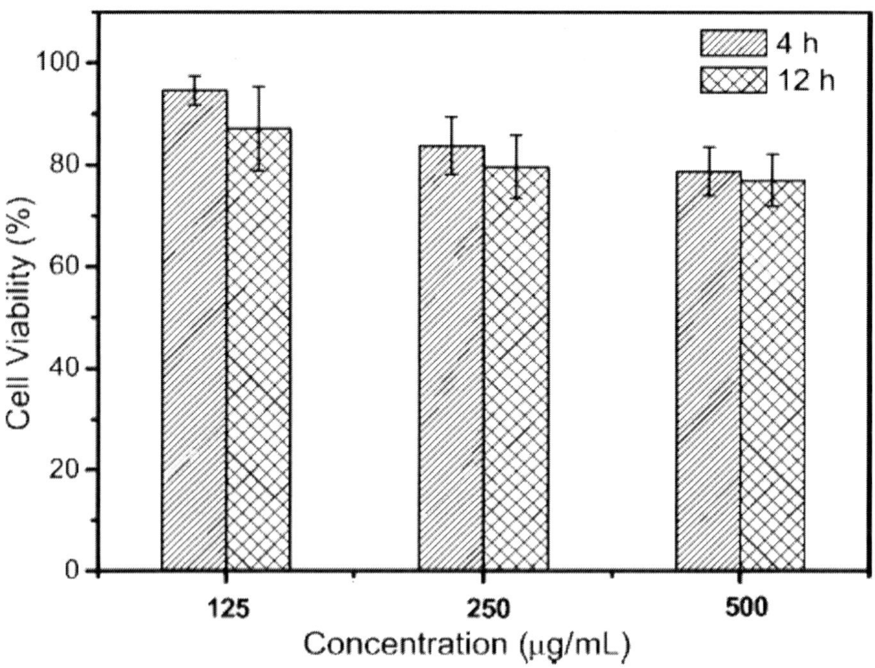

Figure 8. *In vivo* cell viability of KB cells incubated with mPEG-UCNPs at different concentrations for 4–12 h [6].

4.2. Applications in Cancer Therapy

Light has been used to treat various diseases for more than 3,000 years in many countries, such as Ancient Egypt, India, and China [57,58]. More than 100 years ago, Oscar Raab, a German medical student, accidently observed a combination of light and acridine could induce cell death. This was later named as a "photodynamic action" [59]. Photodynamic therapy (PDT) was not approved until 1993 in Canada, where Photofrin was used as a photosensitizer for bladder cancer treatment. A typical PDT process requires three elements: light, photosensitizer, and oxygen. When the photosensitizer is irradiated under certain wavelength, it will be excited from the ground state to the excited state. The energy is released when it returns to the ground state, which is transferred to the nearby oxygen to generate reactive oxygen species (ROS, like singlet oxygen, $^1O_2$). ROS can cause oxidative damage to nearby cells and ultimately kill the cells [60,61]. PDT, as a cancer therapy, has become a popular and acceptable technique in recent years [62,63] due to a better selectivity for the tumor and a lower systemic toxicity for fewer side effects compared to radiation therapy and chemotherapy [64,65]. Some nanomaterials, such as gold nanomaterials [66–68], QDs [69–73], and polymers [74], have been employed in PDT. However, most of them are through the downconversion luminescence, which needs high energy to activate the photosensitizers. Furthermore, the light needed to activate photosensitizers only can penetrate about a centimeter and can cause normal cell death as well. Their inherent cytoxicity (like QDs), downconversion luminescence property and limited penetration depth all cause hindrances in the further applications in biomedical field. Therefore, it is important to find a better way with high PDT efficiency, deeper penetration, and less side effects.

UCN satisfies all the requirements due to its deeper penetration (capable of converting NIR light into visible light), lower toxicity, higher stability, and easier surface modification. UCNs used in PDT are usually pre-coated with a shell, which has the functions of: (1) doping matrix for photosensitizers; (2) specific target on tumor cells; and (3) UCNs stabilization. The $NaYF_4$:Yb/Er UCN is one of the most common used UCN in PDT due to its high UC efficiency [19,75–77]. Zhang and co-workers first demonstrated a novel design for PDT based on UCNs for the treatment of bladder cancer cells (Figure 9) [78]. UCNs were coated with a layer of mesoporous silica shell, in which the photosensitizers were doped. An antibody, which had specific antigens that were expressed on the target cell's surface, was covalently attached on the surface of the silica shell. Due to the spectra overlap between photosensitizers' absorbance and UCNs' emission, the photosensitizer-doped UCNs can generate $^1O_2$under NIR irradiation and further kill the target cells. The generation of $^1O_2$ is usually indicated by the quenching phenomenon of ABDA or ADPA.

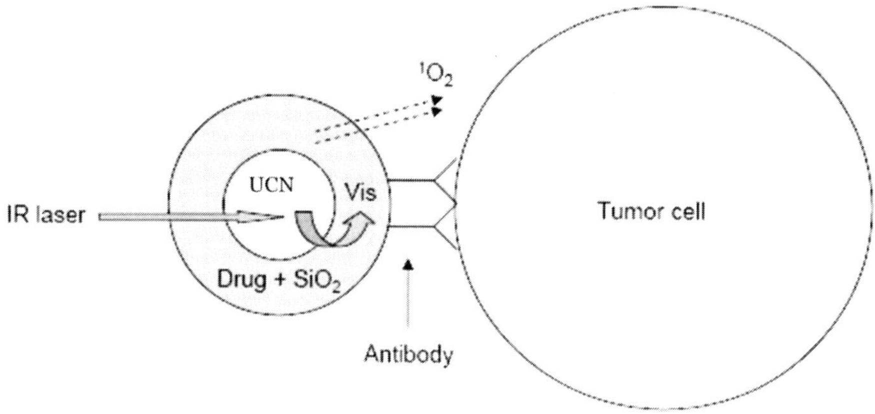

Figure 9. Schematic of the design of the versatile photosensitizer based on UCNs [78].

Zhang *et al.* applied similar principle to study PDT effect on cancer cells [77]. One thin layer of silica was coated on UCNs, followed by second layer of mesoporous silica, which was further doped with photosensitizers. The photosensitizers doped in the silica matrix can be protected from degradation in the harsh biological environment but the formed singlet oxygen can still be released. The obtained fluorescence image showed the uptake of mesoporous silica coated UCNs by murine bladder cancer cells under the excitation of 980 nm. The significance cell viability difference between photosensitizer-doped UCNs and no-photosensitizer UCNs was observed when both were exposed to NIR laser only for 5 min, indicating the production of singlet oxygen.

4.3. Applications in Optical Imaging

Optical imaging has become an essential tool in biological research and clinical application due to its ability of visualizing morphological details in tissue. Although traditional optical imaging has been widely used in *in vivo* and *in vitro* imaging with a combination of different functional nanoparticles, it suffers some inherent limitations such as significant auto-fluorescence from biological tissues, short penetration depth ability, and causing DNA damage or cell death [33,79–83]. These limitations occur because most of them are based on conventional downconversion phenomena, emitting low-energy fluorescence when excited by a high-energy light (usually UV or visible light). Therefore, finding an appropriate alternative, which has low background signal but also deep penetration ability, is very important.

The Ln^{3+}-doped materials were first used in tissue imaging in 1999, where Zijlmans *et al.* observed a low autofluorescence signal and no bleaching even after continuous exposure to high excitation energy levels [48]. However, the size of the applied particles was in the level of submicron size, which limited the applications. In recent years, with the rapid development of synthesis techniques, smaller size but with high-quality UCNs are easily obtained. The UCNs-based imaging technique has been explored and widely used in cell, tissue, and animal imaging [12,48,84–89]. Wu and co-workers [90] have recently shown that UCNs can be used for single-molecule imaging and the individual UCNs were bright enough to be imaged with a modest-power CW laser. The high photostability of the UCNs was also observed even after more than 1 h of continuous laser illumination, implying the extraordinary ability of UCNs for long-period observation of cells. Another bioimaging advantage of UCNs is their low background fluorescence and high signal-to-noise ratio.

Lim *et al.* [22,53] performed *in vivo* and scanning electron microscopy imaging of UCNs in *C. elegans*due to its short life cycle, rapid growth, and appropriate size to optical microscopy. In their study, *C. elegans* were fed with UCNs and imaged at fixed time intervals in order to track the movement of the phosphors through their digestive system. No significant change in the phosphors was monitored up to 24 h and after feeding the worms with food (Figure 10), the phosphors were secreted in under 2 h. It demonstrated that UCNs were nonbleaching, biocompatible, and nontoxic, which make them ideal candidates in the biological system.

Since Chatterjee *et al.* [55] first studied the *in vivo* imaging of UCNs in small mammals which showed much higher fluorescence compared to QDs, it has made significant progress and attracted great interest. In Li's group [4], they found a high relaxivity of 5.60 s^{-1} $(mM)^{-1}$ of UCNs, and the UCNs were successfully applied as contrast agents form magnetic resonance imaging (MRI) *in vivo*. The concept of upconversion and magnetic resonance dual-modality imaging *in vivo* of whole-body animals using UCNs with magnetic resonance properties has showed great promise to serve as a platform technology for the next-generation of probes for bioimaging *in vivo*. Fluorescence targeted imaging *in vivo* has proven very useful in tumor recognition and drug delivery. Xiong *et al.* [23] developed a high contrast upconversion imaging protocol based on UCNs as luminescent labels for targeted imaging of tumors both in *in vivo* and *in vitro*. No autofluorescence

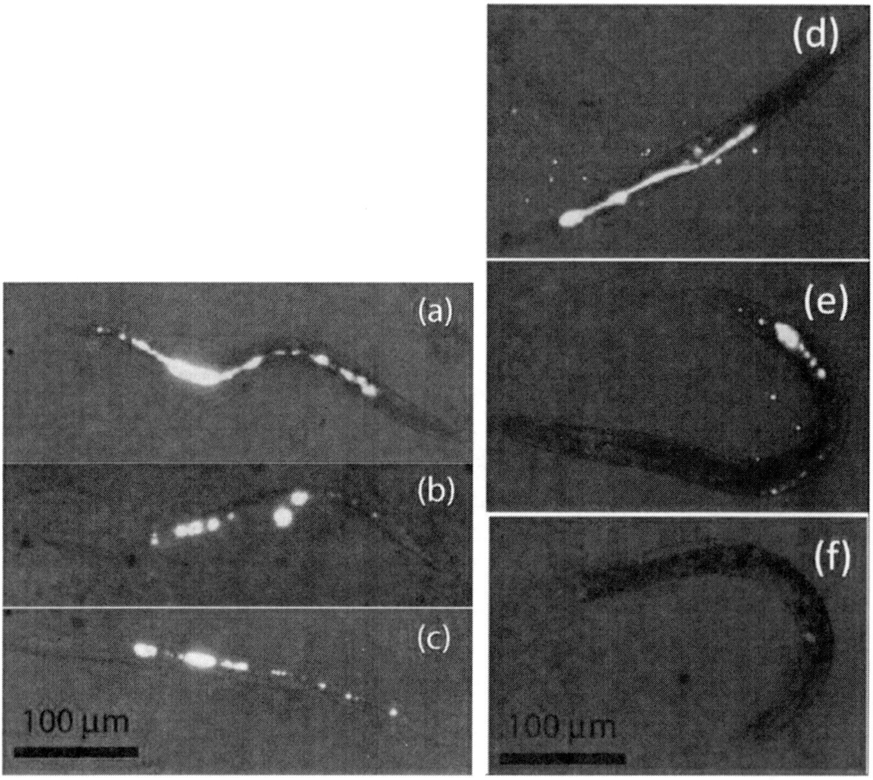

Figure 10. False color two-photon images of *C. elegans* at 980 nm excitation with red representing the bright field and green for the phosphor emission. (**Left**) The worms were deprived of food over a period of 24 h, showing little or no change at (**a**) 0 h, (**b**) 4 h, and (**c**) 24 h. (**Right**) The worms were given food immediately after being fed with phosphors, showing decreasing amounts of phosphors at (**d**) 0 h, (**e**) 1 h, and (**f**) 2 h [22].

signal observed in imaging even at high penetration depth (~600 µm) and the signal-to-noise ratio could be reached ~24 between the tumor and the background, which cannot be obtained in single-photon or two-photon fluorescence imaging (Figure 11). Their study may open up a new perspective for cell recognition and targeted imaging-guided cancer diagnosis.

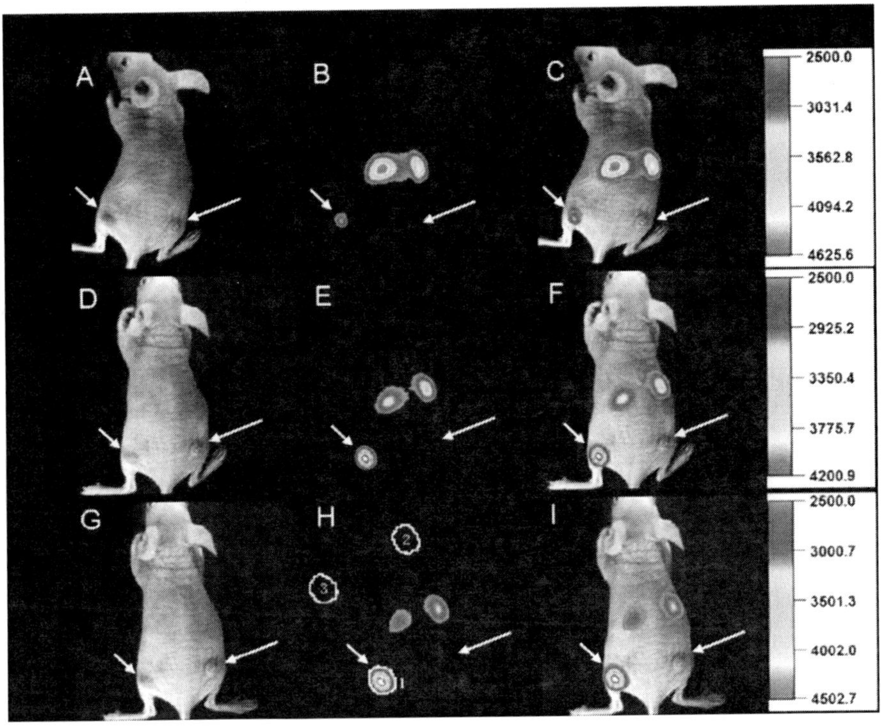

Figure 11. Time-dependent *in vivo* upconversion luminescence imaging of subcutaneous U87MG tumor (left hind leg, indicated by short arrows and MCF-7 tumor (right hind leg, indicated by long arrows) borne by athymic nude mice after intravenous injection of UCN-RGD over a 24 h period. (H) The *in vivo* signal-to-noise ratio (SNR) calculation. Region of interest (ROI)1, specific uptake; ROI 2, nonspecific uptake; ROI 3, background. SNR = (IROI 1 – IROI 3)/(IROI 2 – I ROI 3) [23].

The ability to manipulate color output of nanomaterials can broaden their applications, especially in the case of multiplexed biological labeling and imaging. Various approaches, based on surface plasmon resonance [91–94] or using multicolor-encoded microbeads and nanoparticles [95,96], have been applied to tune multicolor output. However, most of these methods need high energy excitation source in the UV region, which bring limitations to bioimaging studies due to the significant background (autofluorescence) and photo damage to the samples. The UCNs-based method becomes extremely promising to solve these problems, and a few current researchers have already shown the possibility to output multicolor lights. It can be obtained by adjusting the reaction temperature and time, crystal structure and phase, or changing the combinations of Ln^{3+} dopants and dopant concentration (Figure 12) [26,39,83,97,98]. Ehlert *et al.* presented that four different colors of UCNs can be spectrally separated under multiplexing conditions with a single excitation source of 980 nm [99]. Li and co-workers obtained the multicolor output signal by encapsulating organic dyes or QDs into the silica shell and the upconversion fluorescence was generated based on FRET from the UCN-cores to organic dyes or QDs [100].

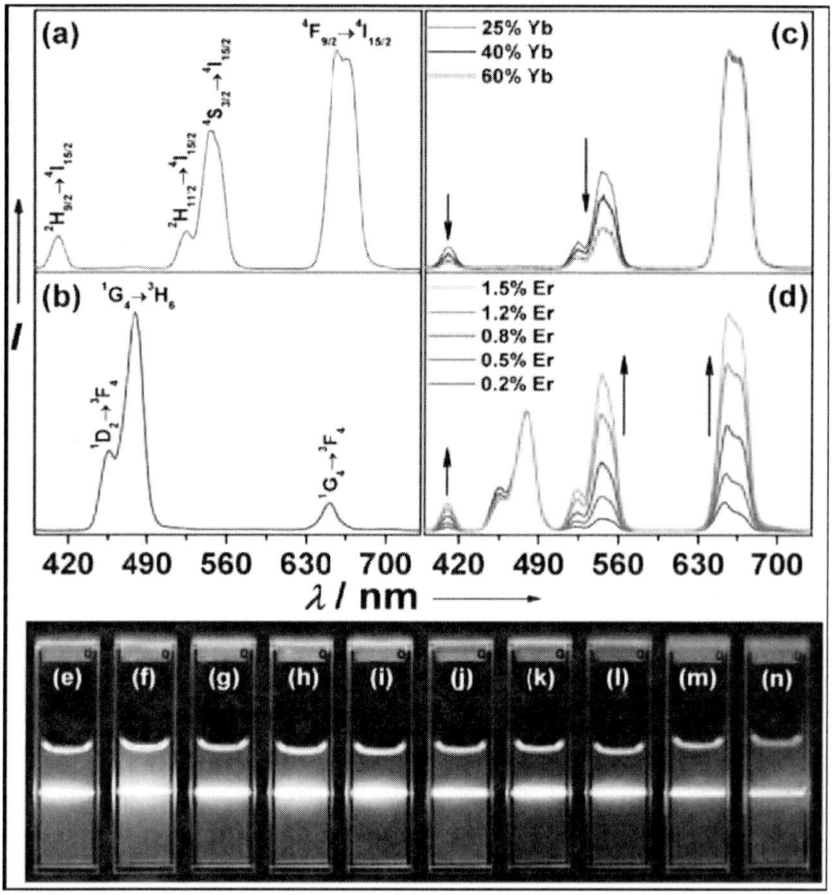

Figure 12. Room temperature upconversion emission spectra of (**a**) $NaYF_4$:Yb/Er (18/2 mol%), (**b**) $NaYF_4$:Yb/Tm (20/0.2 mol%), (**c**) $NaYF_4$:Yb/Er (25–60/2 mol%), and (**d**) $NaYF_4$:Yb/Tm/Er (20/0.2/0.2–1.5 mol%) particles in ethanol solutions. The spectra in (c) and (d) were normalized to Er^{3+} 650 nm and Tm^{3+} 480 emissions, respectively. Compiled luminescent photos showing corresponding colloidal solutions of (**e**) $NaYF_4$:Yb/Tm (20/0.2 mol%), (**f–j**) $NaYF_4$:Yb/Tm/Er (20/0.2/0.2–1.5 mol%), and (**k–n**) $NaYF_4$:Yb/Er (18–60/2 mol%). The samples were excited at 980 nm with a 600 mW diode laser [83].

The emission efficiency can further be enhanced by combining with other metal nanoparticles [12,101], surface modification [102], and multiple Ln^{2+} ions in the UCNs [103]. Chen *et al.* recently found that by simply increasing the content of the doped Yb^{3+} ions from 20 to 100%, their NIR-to-NIR upconversion photoluminescence intensity increased about 8.6 times per Yb^{3+} concentration and 43 times per nanoparticle [104].

4.4. Applications in Sensors

DNA/RNA analysis is of great importance in molecular biology, genetic, and molecular medicine. Great effects have been invested in the precise concentration detection, such as metal nanoparticle-based analysis [105–108]. In recent years, UCNs were also used for the sensitive detection of oligonucleotides. For example, van de Rijke *et al.* [52] and Corstjens *et al.* [109] used UCNs as direct labeling reagents to detect single strand nucleic acids. A new design of a nucleotide sensor by Zhang *et al.* [110] used UCN as energy donor and the other fluorophore as an energy acceptor in a sandwich assay format (Figure 13). In the presence of UCNs and IR irradiation, the fluorophore was brought close to the UCN and energy transfer took place, leading to the light emission from the fluorophore. The target oligonucleotide can be detected by monitoring the fluorophore emission. This sensor displayed high sensitivity (1.3 nM), high specificity, and self-calibration capability. A general aptasensor for detection of various target molecules was reported recently by Liu *et al.* [111], which was based on UCNs-graphene oxide FRET. This proposed design can be further extended for sensing other kinds of molecules as well as causing structure conformations of ssDNA, which have shown its great potential in clinical diagnostic and biosensing techniques.

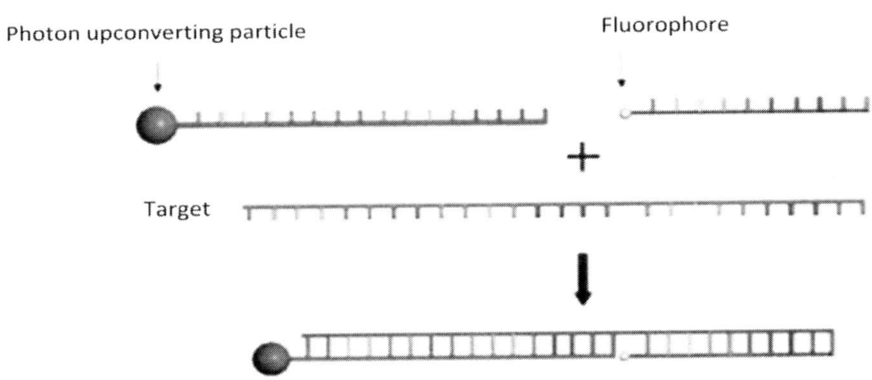

Figure 13. The schematic of the nucleotide sensor design [110].

The UCNs-based sensors are used not only in the biological field, but also in the chemical area. Mader*et al.* [112] developed an ammonia sensor based on the use of UCNs and the pH probe phenol red immobilized in a polystyrene matrix. In the presence of ammonia, a strong increase in the 560 nm absoption of the pH probe caused the green emission of the UCNs to be screened off, while the red emission of the UCNs remained unaffected. The ratio of the intensities of the green emissions and the red emission of the UCNs served as the detection signal. Achatz *et al.* [113] first developed an oxygen sensor by using UCNs as the nanolamps. In the sensing method they presented, the UCNs were used as light source that can cause photoexcitation of an iridium (III) complex. The fluorescence of this complex was dynamically quenched by oxygen. Most recently, Li *et al.* [114] described a highly selective and sensitive CN^- sensor by using the chromophoric iridium (III) complex-coated upconversion nanoparticles. It is the first time that the developed sensor can be used for both bioimaging of an anion in living cells and sensing. This novel CN^- sensor can provide a very low detection limit of 0.18 μM in the aqueous solution. Also, a chromophoric rughenium complex-assembled upconversion nanophosphor was developed and used as a Hg^{2+} sensor by Li's group [115]. The Hg^{2+} nanoprobe showed its capability of not only monitoring the concentration of Hg^{2+} but also monitoring changes in the distribution of Hg^{2+} in living cells by upconversion luminescence bioimaging.

5. CONCLUSIONS

In this review, the mechanism of upconversion phenomenon, the UCNs' unique luminescence properties, and the recent advances in UCNs' chemical synthesis and applications were highlighted. Great progress in size, shape, and phase control of UCNs has been achieved by using various synthesis methods, which also has dramatically broadened their applications in the biological field. In addition to their successful applications in biological labels and fluorescence imaging, they also attracted increasing attention in cancer therapy (mainly focus on PDT). However, there are still some limitations to overcome. First, the synthesized UCNs are normally not water soluble. The one-pot synthesis of water-soluble UCNs always results in broad size distribution and poor shape. Although surface modification can enhance their water solubility and biocompatibility, the procedures are time-consuming and may affect the luminescence efficiency. Secondly, the UCNs-based PDT is still in its infancy. The questions of how to load enough photosensitizers without changing the size significantly, and how to bind the photosensitize-loaded UCNs more specifically to cancer cells still need to be addressed. We believe UCNs-based techniques will be widely exploited and will solve some of today's most challenging problems.

ACKNOWLEDGMENTS

The work was supported by the National Science Foundation Grant CHE-0911472 and CHE0947043, and the U.S. Environmental Protection Agency Grant SU-83508601.

REFERENCES

1. Auzel, F. Upconversion and anti-stokes processes with f and d ions in solids. *Chem. Rev* **2004**, *104*, 139–174.

2. Egger, P.; Rogin, P.; Riedener, T.; Gudel, H.U.; Wickleder, M.; Hulliger, J. Ba_2ErCl_7—A new near IR to UV upconversion material. *Adv. Mater* **1996**, *8*, 668–672.

3. Yang, J.; Deng, Y.; Wu, Q.; Zhou, J.; Bao, H.; Li, Q.; Zhang, F.; Li, F.; Tu, B.; Zhao, D. Mesoporous silica encapsulating upconversion luminescence rare-earth fluoride nanorods for secondary excitation. *Langmuir* **2010**, *26*, 8850–8856.

4. Zhou, J.; Sun, Y.; Du, X.; Xiong, L.; Hu, H.; Li, F. Dual-modality *in vivo* imaging using rare-earth nanocrystals with near-infrared to near-infrared (NIR-to-NIR) upconversion luminescence and magnetic resonance properties. *Biomaterials* **2010**, *31*, 3287–3295.

5. Cao, T.; Yang, Y.; Gao, Y.; Zhou, J.; Li, Z.; Li, F. High-quality water-soluble and surface-functionalized upconversion nanocrystals as luminescent probes for bioimaging. *Biomaterials***2011**, *32*, 2959–2968.

6. Hu, H.; Yu, M.; Li, F.; Chen, Z.; Gao, X.; Xiong, L.; Huang, C. Facile epoxidation strategy for producing amphiphilic up-converting rare-earth nanophosphors as biological labels. *Chem. Mater***2008**, *20*, 7003–7009.

7. Naccache, R.; Vetrone, F.; Mahalingam, V.; Cuccia, L.A.; Capobianco, J.A. Controlled synthesis and water dispersibility of hexagonal phase $NaGdF_4:Ho^{3+}/Yb^{3+}$ nanoparticles. *Chem. Mater* **2009**,*21*, 717–723.

8. Yi, G.; Chow, G.M. Water-soluble $NaYF_4$:Yb,Er(Tm)/$NaYF_4$/polymer core/shell/shell nanoparticles with significant enhancement of upconversion fluorescence. *Chem. Mater* **2007**, *19*, 341–343.

9. Hu, H.; Xiong, L.; Zhou, J.; Li, F.; Cao, T.; Huang, C. Multimodal-luminescence core-shell nanocomposites for targeted imaging of tumor cells. *Chem. Eur. J* **2009**, *15*, 3577–3584.

10. Jiang, S.; Zhang, Y.; Lim, K.M.; Sim, E.K.W.; Ye, L. NIR-to-visible upconversion nanoparticles for fluorescent labeling and targeted delivery of siRNA. *Nanotechnology* **2009**, *20*, 155101–155110.

11. Jiang, S.; Zhang, Y. Upconversion nanoparticle-based FRET system for study of siRNA in live cells. *Langmuir* **2010**, *26*, 6689–6694.

12. Zhang, F.; Braun, G.B.; Shi, Y.; Zhang, Y.; Sun, X.; Reich, N.O.; Zhao, D.; Stucky, G. Fabrication of Ag@SiO_2@Y_2O_3: Er nanostructures for bioimaging: Tuning of the upconversion fluorescence with silver nanoparticles. *J. Am. Chem. Soc* **2010**, *132*, 2850–2851.

13. Li, Z.; Zhang, Y. Monodisperse silica-coated polyvinylpyrrolidone $NaYF_4$ nanocrystals with multicolor upconversion fluorescence emission. *Angew. Chem. Int. Ed* **2006**, *45*, 7732–7735.

14. Liu, Q.; Li, C.; Yang, T.; Yi, T.; Li, F. “Drawing” upconversion nanophosphors into water through host-guest interaction. *Chem. Commun* **2010**, *46*, 5551–5553.

15. Liu, Q.; Chen, M.; Sun, Y.; Chen, G.; Yang, T.; Gao, Y.; Zhang, X.; Li, F. Multifunctional rare-earth self-assembled nanosystem for tri-modal upconversion luminescence/fluorescence/positron emission tomography imaging. *Biomaterials* **2011**, *32*, 8243–8253.

16. Liu, Q.; Sun, Y.; Li, C.; Zhou, J.; Li, C.; Yang, T.; Zhang, X.; Yi, T.; Wu, D.; Li, F. 18F-labeled magnetic-upconversion nanophosphors via rare-earth cation-assisted ligand assembly. *ACS Nano***2011**, *5*, 3146–3157.

17. Wang, X.; Zhuang, J.; Peng, Q.; Li, Y. A general strategy for nanocrystal synthesis. *Nature* **2005**,*437*, 121–124.

18. Jalil, R.A.; Zhang, Y. Biocompatibility of silica coated $NaYF_4$ upconversion fluorescent nanocrystals. *Biomaterials* **2008**, *29*, 4122–4128.

19. Guo, H.; Qian, H.; Idris, N.M.; Zhang, Y. Singlet oxygen induced apoptosis of cancer cells using upconversion fluorescent nanoparticles as a carrier of photosensitizer. *Nanomedicine* **2010**, *6*, 484–495.

20. Li, Z.; Wang, L.; Wang, Z.; Liu, X.; Xiong, Y. Modification of $NaYF_4$:Yb,Er@SiO_2 nanoparticles with gold nanocrystals for tunable green to red upconversion emission. *J. Phys. Chem. C* **2011**, *115*, 3291–3296.

21. Liu, Q.; Sun, Y.; Yang, T.; Feng, W.; Li, C.; Li, F. Sub-10 nm hexagonal lanthanide-doped $NaLuF_4$upconversion nanocrystals for sensitive bioimaging *in vivo*. *J. Am. Chem. Soc* **2011**, *133*, 17122–17125.

22. Lim, S.F.; Riehn, R.; Ryu, W.S.; Khanarian, N.; Tung, C.; Tank, D.; Austin, R.H. *In vivo* and scanning electron microscopy imaging of upconverting nanophosphors in caenorhabditis elegans.*Nano Lett* **2006**, *6*, 169–174.

23. Xiong, L.; Chen, Z.; Tian, Q.; Cao, T.; Xu, C.; Li, F. High contrast upconversion luminescence targeted imaging *in vivo* using peptide-labeled nanophosphors. *Anal. Chem* **2009**, *81*, 8687–8694.

24. Lin, C.; Berry, M.T.; Anderson, R.; Smith, S.; May, P.S. Highly luminescent NIR-to-visible upconversion thin films and monoliths requiring no high-temperature treatment. *Chem. Mater***2009**, *21*, 3406–3413.

25. Yi, G.; Lu, H.; Zhao, S.; Ge, Y.; Yang, W.; Chen, D.; Guo, L. Synthesis, characterization, and biological application of size-controlled nanocrystalline $NaYF_4$:Yb,Er infrared-to-visible up-conversion phosphors. *Nano Lett* **2004**, *4*, 2191–2196.

26. Heer, S.; Kompe, K.; Gvdel, H.U.; Haase, M. Highly efficient multicolour upconversion emission in transparent colloids of lanthanide-doped $NaYF_4$ nanocrystals. *Adv. Mater* **2004**, *16*, 2102–2105.

27. Mai, H.; Zhang, Y.; Sun, L.; Yan, C. Size- and phase-controlled synthesis of monodisperse $NaYF_4$:Yb,Er nanocrystals from a unique delayed nucleation pathway monitored with upconversion spectroscopy. *J. Phys. Chem. C* **2007**, *111*, 13730–13739.

28. Mai, H.X.; Zhang, Y.W.; Si, R.; Yan, Z.G.; Sun, L.D.; You, L.P.; Yan, C.H. High-quality sodium rare-earth fluoride nanocrystals: Controlled synthesis and optical properties. *J. Am. Chem. Soc* **2006**,*128*, 6426–6436.

29. Boyer, J.C.; Vetrone, F.; Cuccia, L.A.; Capobianco, J.A. Synthesis of colloidal upconverting $NaYF_4$nanocrystals doped with Er^{3+}, Yb^{3+} and Tm^{3+}, Yb^{3+} via thermal decomposition of lanthanide trifluoroacetate precursors. *J. Am. Chem. Soc* **2006**, *128*, 7444–7445.

30. Boyer, J.-C.; Cuccia, L.A.; Capobianco, J.A. Synthesis of colloidal upconverting $NaYF_4$: $Er^{3+}/Yb^{3+}$and Tm^{3+}/Yb^{3+} monodisperse nanocrystals. *Nano Lett* **2007**, *7*, 847–852.

31. Yi, G.; Sun, B.; Yang, F.; Chen, D.; Zhou, Y.; Cheng, J. Synthesis and characterization of high-efficiency nanocrystal up-conversion phosphors: Ytterbium and erbium codoped lanthanum molybdate. *Chem. Mater* **2002**, *14*, 2910–2914.

32. Zeng, J.H.; Su, J.; Li, Z.H.; Yan, R.X.; Li, Y.D. Synthesis and upconversion luminescence of hexagonal-phase $NaYF_4$:Yb,Er^{3+} phosphors of controlled size and morphology. *Adv. Mater* **2005**,*17*, 2119–2123.

33. Taylor, J.R.; Fang, M.M.; Nie, S. Probing specific sequences on single DNA molecules with bioconjugated fluorescent manoparticles. *Anal. Chem* **2000**, *72*, 1979–1986.

34. Wei, Y.; Lu, F.; Zhang, X.; Chen, D. Synthesis of oil-dispersible hexagonal-phase and hexagonal-shaped $NaYF_4$:Yb,Er nanoplates. *Chem. Mater* **2006**, *18*, 5733–5737.

35. Fan, X.; Pi, D.; Wang, F.; Qiu, J.; Wang, M. Hydrothermal synthesis and luminescence behavior of lanthanide-doped GdF_3 nanoparticles. *Trans. Nanotechnol* **2006**, *5*, 123–128.

36. Zhang, F.; Wan, Y.; Yu, T.; Zhang, F.; Shi, Y.; Xie, S.; Li, S.; Li, Y.; Xu, L.; Tu, B.; Zhao, D. Uniform nanostructured arrays of sodium rare-earth fluorides for highly efficient multicolor upconversion luminescence. *Angew. Chem. Int. Ed* **2007**, *46*, 7976–7979.

37. Li, C.; Yang, J.; Quan, Z.; Yang, P.; Kong, D.; Lin, J. Different microstrutures of $NaYF_4$ fabricated by hydrothermal process: Effects of pH values and fluoride sources. *Chem. Mater* **2007**, *19*, 4933–4942.

38. Li, C.; Quan, Z.; Yang, J.; Yang, P.; Lin, J. Highly uniform and monodisperse β-$NaYF_4$: Ln^{3+}synthesis and luminescent properties. *Inorg. Chem* **2007**, *46*, 6329–6337.

39. Wang, L.; Li, Y. $NaY_{1.5}Na_{0.5}F_6$ single crystal nanorods as multicolor luminescent material. *Nano Lett* **2006**, *6*, 1645–1649.

40. Liu, X.; Zhao, J.; Sun, Y.; Song, K.; Yu, Y.; Du, C.; Kong, X.; Zhang, H. Ionothermal synthesis of hexagonal-phase $NaYF_4$:Yb^{3+},Er^{3+}/Tm^{3+} upconversion nanophosphors. *Chem. Commun* **2009**, 6628–6630.

41. Sandrock, T.; Scheife, H.; Heumann, E.; Huber, G. High-power continuous-wave upconversion fiber laser at room temperature. *Opt. Lett* **1997**, *22*, 808–810.

42. Shalav, A.; Richards, B.S.; Trupke, T. Application of $NaYF_4:Er^{3+}$ upconversion phosophors for enhanced near-infrared silicon solar cell response. *Appl. Phys. Lett* **2005**, *86*, 13503–13505.

43. Ivanova, S.; Pelle, F. Evaluation upconversion materials developed to improve the efficiency of solar cells: Reply to comment. *J. Opt. Soc. Am. B* **2010**, *27*, 1356–1358.

44. Kik, P.G.; Polman, A. Cooperative upconversion as the gain-limiting factor in Er doped miniature Al_2O_3 optical waveguide amplifiers. *J. Appl. Phys* **2003**, *93*, 5008–5012.

45. Strohhofer, C.; Polman, A. Relationship between gain and Yb^{3+} concentration in Er^{3+}-Yb^{3+} doped waveguide amplifiers. *J. Appl. Phys* **2001**, *90*, 4314–4320.

46. Zhou, Z.; Hu, H.; Yang, H.; Yi, T.; Huang, K.; Yu, M.; Li, F.; Huang, C. Up-conversion luminescent switch based on photochromic diarylethene and rare earth nanophosphors. *Chem. Commun* **2008**, 4786–4788.

47. Zhang, C.; Zhou, H.; Liao, L.; Feng, W.; Sun, W.; Li, Z.; Xu, C.; Fang, C.; Sun, L.; Zhang, Y.; Yan, C. Luminescence modulation of ordered upconversion nanopatterns by a photochromic diarylethene: Rewritable optical storage with nondestructive readout. *Adv. Mater* **2010**, *22*, 633–637.

48. Zijlmans, H.J.M.A.A.; Bonnet, J.; Burton, J.; Kardos, K.; Vail, T.; Niedbala, R.S.; Tanke, H.J. Detection of cell and tissue surface antigens using up-converting phosphors: A new reporter technology. *Anal. Biochem* **1999**, *267*, 30–36.

49. Kuningas, K.; Rantanen, T.; Karhunen, U.; Lövgren, T.; Soukka, T. Simultaneous use of time-resolved fluorescence and anti-stokes photoluminescence in a bioaffinity assay. *Anal. Chem* **2005**,*77*, 2826–2834.

50. Rantanen, T.; Järvenpää, M.-L.; Vuojola, J.; Kuningas, K.; Soukka, T. Fluorescence-quenching-based enzyme-activity assay by using photon upconversion. *Angew. Chem. Int. Ed* **2008**, *47*, 3811–3813.

51. Rantanen, T.; Päkkilä, H.; Jämsen, L.; Kuningas, K.; Ukonaho, T.; Lövgren, T.; Soukka, T. Tandem dye acceptor used to enhance

upconversion fluorescence resonance energy transfer in homogeneous assays. *Anal. Chem* **2007**, *79*, 6312–6318.

52. van de Rijke, F.; Zijlmans, H.; Li, S.; Vail, T.; Raap, A.K.; Niedbala, R.S.; Tanke, H. Up-converting phosphor reporters for nucleic acid microarrays. *Nature* **2001**, *19*, 273–276.

53. Lim, S.F.; Riehn, R.; Tung, C.; Ryu, W.S.; Zhuo, R.; Dalland, J.; Austin, R.H. Upconverting nanophosphors for bioimaging. *Nanotechnology* **2009**, *20*, 405701–405707.

54. Hilderbrand, S.A.; Shao, F.W.; Salthouse, C.; Mahmood, U.; Weissleder, R. Upconverting luminescent nanomaterials: Application to *in vivo* bioimaging. *Chem. Commun* **2009**.

55. Chatterjee, D.K.; Rufaihah, A.J.; Zhang, Y. Upconversion fluorescence imaging of cells and small animals using lanthanide doped nanocrystals. *Biomaterials* **2008**, *29*, 937–943.

56. Xiong, L.; Yang, T.; Yang, Y.; Xu, C.; Li, F. Long-term *in vivo* biodistribution imgaing and toxicity of polyacrylic acid-coated upconversion nanophosphors. *Biomaterials* **2010**, *31*, 7078–7085.

57. Daniell, M.D.; Hill, J.S. A history of photodynamic theray. *Aust. NZ J. Surg* **1991**, *61*, 340–348.

58. Ackroyd, R.; Kelty, C.; Brown, N.; Reed, M. The history of photodetection and photodynamic therapy. *Photochem. Photobiol* **2001**, *74*, 656–669.

59. Raab, O. Uber die wirkung fluoreszierender stoffe auf infusorien. *Zeitung Biol* **1900**, *39*, 524–526.

60. Sharman, W.M.; Allen, C.M.; van Lier, J.E. Role of activated oxygen species in photodynamic theray. *Meth. Enzymol* **2000**, *319*, 376–400.

61. Oleinick, N.L.; Morris, R.L.; Belichenko, T. The role of apoptosis in response to photodynamic therapy: What, where, why, and how. *Photochem. Photobiol. Sci* **2002**, *1*, 1–21.

62. Dougherty, T.J.; Gomer, C.J.; Henderson, B.W.; Jori, G.; Kessel, D.; Korbelik, M.; Moan, J.; Peng, Q. Photodynamic therapy. *J. Natl. Cancer Inst* **1998**, *90*, 889–905.

63. Brown, S.B.; Brown, E.A.; Walker, I. The present and future role of photodynamic therapy in cancer treatment. *Lancet Oncol* **2004**, *5*, 497–508.

64. Dougherty, T.J. An update on photodynamic therapy applications. *J. Clin. Laser. Med. Surg* **2002**,*20*, 3–7.

65. Stummer, W.; Hassan, A.; Kempski, O.; Goetz, C. Photodynamic therapy within edematous brain tissue: considerations on sensitizer dose and time point of laser irradiation. *J. Photochem. Photobiol. B* **1996**, *36*, 179–181.

66. Asadishad, B.; Vossoughi, M.; Alemzadeh, I. Folate-receptor-targeted delivery of doxorubicin using polyethylene glycol-functionalized gold nanoparticles. *Ind. Eng. Chem. Res* **2010**, *49*, 1958–1963.

67. Brown, S.D.; Nativo, P.; Smith, J.A.; Stirling, D.; Edwards, P.R.; Venugopal, B.; Flint, D.J.; Plumb, J.A.; Graham, D.; Wheate, N.J. Gold nanoparticles for the improved anticancer drug delivery of the active component of oxaliplatic. *J. Am. Chem. Soc* **2010**, *132*, 4678–4684.

68. Jang, B.; Park, J.Y.; Tung, C.H.; Kim, I.H.; Choi, Y. Gold nanorod photosensitizer complex for near infrared fluorescence imaging and photodynamic/photothermal therapy *in vivo*. *ACS Nano* **2011**,*5*, 1086–1094.

69. Ratanatawanate, C.; Chyao, A.; Balkus, K.J. S-nitrosocysteine decorated PbS QDs/TiO_2 nanotubes for enhanced production of singlet oxygen. *J. Am. Chem. Soc* **2011**, *133*, 3492–3497.

70. Hsieh, J.M.; Ho, M.L.; Wu, P.W.; Chou, P.T.; Tsaih, T.T.; Chi, Y. Iridium-complex modified CdSe/ZnS quantum dots: A conceptual design for bifunctionality toward imaging and phtoosensitization. *Chem. Commun* **2006**.

71. Tsay, J.M.; Trzoss, M.; Shi, L.X.; Kong, X.X.; Selke, M.; Jung, M.E.; Weiss, S. Singlet oxygen production by peptide-coated quantum dot-photosensitizer conjugates. *J. Am. Chem. Soc* **2007**,*129*, 6865–6871.

72. Samia, A.C.S.; Chen, X.B.; Burda, C. Semiconductor quantum dots for photodynamic therapy. *J. Am. Chem. Soc* **2003**, *125*, 15736–15737.

73. Shi, L.X.; Hernandez, B.; Selke, M. Signlet oxygen generation from water-soluble quantum dot-organic dye nanocomposites. *J. Am. Chem. Soc* **2006**, *128*, 6278–6279.

74. Shen, X.; He, F.; Wu, J.; Xu, G.; Yao, S.Q.; Xu, Q. Enhanced two photon singlet oxygen genergation by photosensitizer doped conjugated polymer nanoparticles. *Langmuir* **2011**, *27*, 1739–1744.

75. Chen, Z.; Chen, H.; Hu, H.; Yu, M.; Li, F.; Zhang, Q.; Zhou, Z.; Yi, T.; Huang, C. Versatile synthesis strategy for carboxylic acid-functionalized upconverting nanophosphors as biological labels. *J. Am. Chem. Soc* **2008**, *130*, 3023–3029.

76. Chatterjee, D.K.; Zhong, Y. Upconversion nanoparticles as nanotransducers for photodynamic therapy in cancer cells. *Nanomedicine* **2008**, *3*, 73–82.

77. Qian, H.S.; Guo, H.C.; Ho, P.C.L.; Mahendran, R.; Zhang, Y. Mesoporous-silica-coated up-conversion fluorescent nanoparticles for photodynamic therapy. *Small* **2009**, *5*, 2285–2290.

78. Zhang, P.; Steelant, W.; Kumar, M.; Scholfield, M. Versatile photosensitizers for photodynamic therapy at infrared excitation. *J. Am. Chem. Soc* **2007**, *129*, 4526–4527.

79. Medintz, I.L.; Uyeda, H.T.; Goldman, E.R.; Mattoussin, H. Quantum dot bioconjugates fro imaging, labeling, and sensing. *Nat. Mater* **2005**, *4*, 435–446.

80. Michalet, X.; Pinaud, F.F.; Bentolila, L.A.; Tsay, J.M.; Doose, S.; Li, J.J.; Sundaresan, G.; Wu, A.M.; Gambhir, S.S.; Weiss, S. Quantum dots for live cells, *in vivo* imaging, and diagnostics. *Science* **2005**,*307*, 538–544.

81. Ow, H.; Larson, D.R.; Srivastava, M.; Baird, B.A.; Webb, W.W.; Wiesner, U. Bright and stable core-shell fluorescent silica nanoparticles. *Nano Lett* **2005**, *5*, 113–117.

82. O'Connell, M.J.; Bachilo, S.M.; Huffman, C.B.; Moore, V.C.; Strano, M.S.; Haroz, E.H.; Rialon, K.L.; Boul, P.J.; Noon, W.H.; Kittrell, C.; Ma, J.; Hauge, R.H.; Weisman, R.B.; Smalley, R.E. Band gap fluorescence from individual single-walled carbon nanotubes. *Science* **2002**, *297*, 593–596.

83. Wang, F.; Liu, X. Upconversion multicolor fine-tuning: Visible to near-infrared emission from lanthanide-doped $NaYF_4$ nanoparticles. *J. Am. Chem. Soc* **2008**, *130*, 5642–5643.

84. Xiong, L.-Q.; Chen, Z.-G.; Yu, M.-X.; Li, F.-Y.; Liu, C.; Huang, C.-H. Synthesis, characterization, and *in vivo* targeted imaging of amine-functionalized rare-earth up-converting nanophosphors.*Biomaterials* **2009**, *30*, 5592–5600.

85. Zhou, J.-C.; Yang, Z.-L.; Dong, W.; Tang, R.-J.; Sun, L.-D.; Yan, C.-H. Bioimaging and toxicity assessments of near-infrared upconversion luminescent $NaYF_4$:Yb,Tm nanocrystals. *Biomaterials***2011**, *32*, 9059–9067.

86. Cheng, L.; Yang, K.; Shao, M.; Lee, S.T.; Liu, Z. Multicolor *in vivo* imaging of upconversion nanoparticles with emissions tuned by luminescence resonance energy transfer. *J. Phys. Chem. C***2011**, *115*, 2686–2692.

87. Ryu, J.; Park, H.Y.; Kim, K.; Kim, H.; Yoo, J.H.; Kang, M.; Im, K.; Grailhe, R.; Song, R. Facile synthesis of ultrasmall and hexagonal $NaGdF_4$:Yb^{3+},Er^{3+} nanoparticles with magnetic and upconversion imaging properties. *J. Phys. Chem. C* **2010**, *114*, 21077–21082.

88. Park, Y.; Kim, J.H.; Taek, L.K.; Jeon, K.S.; Na, H.B.; Yu, J.H.; Kim, H.M.; Lee, N.; Choi, S.H.; Baik, S.; Kim, H.; Park, S.P.; Park, B.J.; Kim, Y.W.; Lee, S.H.; Yoon, S.Y.; Song, I.C.; Moon, K.K.; Suh, Y.D.; Hyeon, T. Nonblinking and nonbleaching upconverting nanoparticles as an optical imaing nanoprobe and T1 magnetic resonance imaging contrast agent. *Adv. Mater* **2009**, *21*, 4467–4471.

89. Yu, M.; Li, F.; Chen, Z.; Hu, H.; Zhan, C.; Yang, H.; Huang, C. Laser scanning up-conversion luminescence microscopy for imaging cells labeled with rare-earth nanophosphors. *Anal. Chem***2009**, *81*, 930–935.

90. Wu, S.; Han, G.; Milliron, D.J.; Aloni, S.; Altoe, V.; Talapin, D.V.; Cohen, B.E.; Schuck, P.J. Non-blinking and photostable upconverted luminescence from single lanthanide doped nanocrystals.*PNAS* **2009**, *106*, 10917–10921.

91. Kim, F.; Song, J.H.; Yang, P. Photochemical synthesis of gold nanorods. *J. Am. Chem. Soc* **2002**, *124*, 14316–14317.

92. Schwartzberg, A.M.; Olson, T.Y.; Talley, C.E.; Zhang, J.Z. Synthesis, characterization, and tunable optical properties of hollow gold nanospheres. *J. Phys. Chem. B* **2006**, *110*, 19935–19944.

93. Wiley, B.J.; Chen, Y.; McLellan, J.M.; Xiong, Y.; Li, Z.Y.; Ginger, D.; Younan, X. Synthesis and optical properties of silver nanobars and nanorice. *Nano Lett* **2007**, *7*, 1032–1036.

94. Tao, A.; Sinsermsuksakul, P.; Yang, P. Tunable plasmonic lattices of silver nanocrystals. *Nat. Nanotechnol* **2007**, *2*, 435–440.

95. Lin, W.; Weihong, T. Multicolor FRET silica nanoparticles by single wavelength excitation. *Nano Lett* **2006**, *6*, 84–88.

96. Han, M.; Gao, X.; Su, J.Z.; Nie, S. Quantum-dot-tagged microbeads for multiplexed optical coding of biomolecules. *Nat. Nanotechnol* **2001**, *19*, 631–635.

97. Niu, W.; Wu, S.; Zhang, S.; Li, J.; Li, L. Multicolor output and shape controlled synthesis of lanthanide ion doped fluorides upconversion nanoparticles. *Dalton Trans* **2011**, *40*, 3305–3314.

98. Mai, H.X.; Zhang, Y.-W.; Sun, L.-D.; Yan, C.-H. Highly efficient multicolor up-conversion emissions and their mechanisms of monodisperse $NaYF_4$:Yb,Er core and core/shell-structured nanocrystals. *J. Phys. Chem. C* **2007**, *111*, 13721–13729.

99. Ehlert, O.; Thomann, R.; Darbandi, M.; Nann, T. A four-color colloidal multiplexing nanoparticle system. *ACS Nano* **2008**, *2*, 120–124.

100. Li, Z.; Zhang, Y.; Jiang, S. Multicolor core/shell-structured upconversion fluorescent nanoparticles. *Adv. Mater* **2008**, *20*, 4765–4769.

101. Zhang, H.; Li, Y.; Ivanov, I.A.; Qu, Y.; Duan, X. Plasmonic modulation of the upconversion fluorescence in $NaYF_4$:Yb/Tm hexaplate nanocrystals using gold nanoparticles or nanoshells.*Angew. Chem. Int. Ed* **2010**, *49*, 2865–2868.

102. Li, D.; Dong, B.; Bai, X.; Wang, Y.; Song, H. Influence of the TGA modification on upconversion luminescence of hexagonal phase $NaYF_4:Yb^{3+},Er^{3+}$ Nanoparticles. *J. Phys. Chem. C* **2010**, *114*, 8219–8226.

103. Huang, Q.; Yu, J.; Ma, E.; Lin, K. Synthesis and characterization of highly efficient near infrared upconversion $Sc^{3+}/Er^{3+}/Yb^{3+}$ tridoped $NaYF_4$. *J. Phys. Chem. C* **2010**, *114*, 4719–4724.

104. Chen, G.; Ohulchanskyy, T.; Kumar, R.; Agren, H.; Prasad, P.N. Ultrasmall monodisperse $NaYF_4:Yb^{3+}/Tm^{3+}$ nanocrystals with enhanced near infrared upconversion photoluminescence. *ACS Nano* **2010**, *4*, 3163–3168.

105. Bi, S.; Zhang, J.; Zhang, S. Ultrasensitive and selective DNA detection based on nicking endonuclease assisted signal amplification and its application in cancer cell detection. *Chem. Commun* **2010**, *46*, 5509–5511.

106. Song, J.; Li, Z.; Cheng, Y.; Liu, C. Self aggregation of oligonucleotide functionalized gold nanoparticles and its applications for highly sensitive detection of DNA. *Chem. Commun* **2010**, *46*, 5548–5550.

107. Li, H.; Rothberg, L.J. Label free colorimetric detection of specific sequences in genomic DNA amplified by the polymerase chain reaction. *J. Am. Chem. Soc* **2004**, *126*, 10958–10961.

108. Rho, S.; Kim, S.J.; Lee, S.C.; Chang, J.H.; Kang, H.-G.; Choi, J. Colorimetric detection of ssDNA in a solution. *Curr. Appl. Phys* **2009**, *9*, 534–537.

109. Corstjens, P.L.A.M.; Zuiderwijk, M.; Nilsson, M.; Feindt, H.; Niedbala, R.S.; Tanke, H.J. Lateral-flow and up-converting phosphor reporters to detect single-stranded nucleic acids in a sandwich-hydridization assay. *Anal. Biochem* **2003**, *312*, 191–200.

110. Zhang, P.; Rogelj, S.; Nguyen, K.; Wheeler, D. Design of a highly sensitive and specific nucleotide sensor based on photon upconverting particles. *J. Am. Chem. Soc* **2006**, *128*, 12410–12411.

111. Liu, C.; Wang, Z.; Jia, H.; Li, Z. Efficient fluorescence resonance energy transfer between upconversion nanophosphors and graphene oxide: A highly sensitive biosensing platform. *Chem. Commun* **2011**, *47*, 4661–4663.

112. Mader, H.S.; Wolfbeis, O.S. Optical ammonia sensor based on upconverting luminescent nanoparticles. *Anal. Chem* **2010**, *82*, 5002–5004.

113. Achatz, D.; Meier, R.J.; Fischer, L.H.; Wolfbeis, O.S. Luminescent sensing of oxygen using a quenchable probe and upconverting nanoparticles. *Angew. Chem. Int. Ed* **2011**, *50*, 260–263.

114. Liu, J.; Liu, Y.; Liu, Q.; Li, C.; Sun, L.; Li, F. Iridium(III) complex-coated nanosystem for ratiometric upconversion luminescence bioimaging of cyanide anions. *J. Am. Chem. Soc* **2011**, *133*, 15276–15279.

115. Liu, Q.; Peng, J.; Sun, L.; Li, F. High-efficiency upconversion luminescent sensing and bioimaging of Hg(II) by chromophoric ruthenium complex-assembled nanophosphors. *ACS Nano* **2011**, *5*, 8040–8048.

Chapter 2

Rare Earth Ion-Doped Upconversion Nanocrystals: Synthesis and Surface Modification

Hongjin Chang [1], Juan Xie [2], Baozhou Zhao [1], Botong Liu [1], Shuilin Xu [1], Na Ren [1], Xiaoji Xie [1], Ling Huang [1,*] and Wei Huang [1,2,*]

[1]*Key Laboratory of Flexible Electronics (KLOFE) and Institute of Advanced Materials (IAM), Jiangsu National Synergistic Innovation Center for Advanced Materials (SICAM), Nanjing Tech University (NanjingTech), Nanjing 211816, China*

[2]*Key Laboratory for Organic Electronics and Information Displays and Institute of Advanced Materials (IAM), National Synergistic Innovation Center for Advanced Materials (SICAM), Nanjing University of Posts and Telecommunications, Nanjing 210023, China*

ABSTRACT

The unique luminescent properties exhibited by rare earth ion-doped upconversion nanocrystals (UCNPs), such as long lifetime, narrow emission line, high color purity, and high resistance to photobleaching, have made them widely used in many areas, including but not limited to high-resolution displays, new-generation information technology, optical communication, bioimaging, and therapy. However, the inherent upconversion luminescent properties of UCNPs are influenced by various parameters, including the size, shape, crystal structure, and chemical composition of the UCNPs, and even the chosen synthesis process and the surfactant molecules used. This review will provide a complete summary on the synthesis methods and the surface modification strategies of UCNPs reported so far. Firstly, we summarize the synthesis methodologies developed in the past decades, such as thermal decomposition, thermal coprecipitation, hydro/solvothermal, sol-gel, combustion, and microwave synthesis. In the second part, five main streams of surface modification strategies for converting hydrophobic

UCNPs into hydrophilic ones are elaborated. Finally, we consider the likely directions of the future development and challenges of the synthesis and surface modification, such as the large-scale production and actual applications, stability, and so on, of the UCNPs.

Keywords: rare earth; nanocrystal; upconversion; synthesis; surface modification

1. INTRODUCTION

On 29 December 1959, Richard Feynman predicted at the annual American Physical Society meeting that: "*if we use a method to control the arrangement of things at a small scale, so that we can get a lot of features beyond imagination, also can see the performance of the material to produce rich change.*" The tiny material is what we call nanomaterial today, which is usually referred to as that in the three dimensional space, where at least one dimension (1D) is at the nanoscale range (1–100 nm) or materials are composed of them as a basic unit. As the size of the material decreases to the nanometer scale, the material will show a lot of new features that the bulk counterpart does not possess, such as the small size effect, surface effect of nanomaterials, quantum confinement effect, and macroscopic quantum tunneling effect [1,2,3]. In the past decades, nanomaterials have gained wide attention from all over the world, and have now been widely applied in many fields.

Rare earth (RE) elements possess unique electronic configuration where the 4f electrons are effectively shielded by the closely lied 5s and 5p subshells. Typically, the electron transitions of RE ions are mainly derived from the inner *4f-4f* or *4f-5d* transitions, and thus the spectroscopic properties of RE ions are barely perturbed by the local chemical microenvironment, which imparts RE compounds including the complexes' and nanomaterials' unique spectroscopic characteristics, such as rich energy levels, long luminescence decay time, narrow emission line, and high color purity in contrast to those of quantum dots and organic dyes [4,5,6]. Due to the preserved electron transitions, the emission wavelengths of each lanthanide ion largely depend on its own electronic configuration, and the combination (at certain ratio) of different lanthanide ions is also widely adopted to realize various luminescent materials with adjustable luminescence for different purposes.

If these RE ions were doped into proper nanocrystals, a series of new luminescent characteristics related with the original features of according RE ions may be observed. In addition, upconversion nanocrystals (UCNPs) have shown high chemical stability, biological compatibility, long luminescence lifetime, and tunable emission wavelength [7,8,9], making them tremendously exploited in biolabeling [10], bio-detection [11,12], bioimaging [13,14], FRET-based sensing [15], drug delivery [16], and volumetric 3D display [17].

Despite the substantially shielded transitions of RE ions, the luminescent properties of UCNPs are also affected by their size, shape, crystal structure, and chemical composition of the materials. For example, reduced particle size will cause increased surface area, which would introduce more defects on the nanoparticle surface and consequently hamper the luminescent efficiencies, though the small nanoparticles are more advantageous to the biological applications. To facile the applications of UCNPs, it is of primary necessity to develop according viable and robust methodologies to synthesize target nanocrystals with desired size, shape, crystal structure, chemical composition, and most importantly, the proper surface functional groups for anticipated applications.

In recent years, various attempts have been reported for synthesizing UCNPs in a controlled manner, including thermal decomposition, thermal coprecipitation, hydro/solvothermal, combustion, microwave, and so on. In the meantime, the nanocrystals, after surface modification, such as SiO_2encapsulation, polymer encapsulation, ligand oxidation, and ligand exchange, can be easily coupled with DNA, protein, and other functional molecules, and facilitate expected applications. However, there are still some challenges in the synthesis of desired nanocrystals, and the surface modification usually involves extra experimental steps and lowers the luminescence efficiency at certain extent depending on the chosen method. Thus, a proper synthesis method and a suitable strategy for designed surface modification are highly desired. Herein, we attempt to provide a comprehensive overview of the state-of-the-art synthetic methods and the surface modification strategies for UCNPs reported in the past decades.

2. SYNTHETIC APPROACHES

2.1. Thermal Decomposition

The thermal decomposition process, comprising of dissolution of organic and/or inorganic precursors in organic solvent with high-boiling point, is a traditional method for preparation of inorganic nanocrystals. The typical experimental procedure of thermal decomposition method is composed of: (1) a given amount of $RE(CF_3COO)_3$ precursors is added into a mixture of oleic acid (OA), 1-octadecene (OD), and sometimes oleylamine (OM) at room temperature; (2) the solution is heated to 165 °C for 30 min with vigorous magnetic stirring to remove water and oxygen under argon protection; (3) the solution is heated to high temperature (usually >300 °C) for a certain period of time under argon protection and the nanocrytals are then collected from the reaction mixture after cooling down to room temperature. Yan [18,19] first prepared high-monodisperse LaF_3 triangular nanoplates (Figure 1a) and hexagonal SmF_3 nanoparticles (Figures 1b,c) via the thermal decomposition process using $La(CF_3COO)_3$and $Sm(CF_3COO)_3$ as precursors. Using this method, Capobianco [20] and Nann [21] synthesized $NaYF_4$nanocrystals with narrow size-distribution. Later, this approach was extended as a common process to synthesize high-quality UCNPs including but not limited to $NaLaF_4$ [22], $NaGdF_4$ [23], $LiYF_4$ [24], KY_3F_{10} [25], $BaYF_5$ [26], REOF [19,27,28], and REOCl [29]. The reaction temperature, time, and the molar ratio of OA, OD, and sometimes OM in the reaction mixture have been demonstrated to exert different effects on the final nanocrystals. It should be noted that sometimes OM shall be introduced as a necessary component to adjust the reaction environment so that the final product with different morphologies and dimensions or a brand new product can be obtained [27,28,29].

The most prominent advantages of this method are that the products are of high quality, with pure crystal phase, and strong upconversion emission. However, this method also suffers from disadvantages including: (1) presynthesis of $RE(CF_3COO)_3$ precursors is typically required; (2) the decomposition of trifluoroacetates simultaneously produces toxic fluorinated and oxyfluorinated carbon species and thus careful handling of the reactions in fully ventilated chemical hood is required; (3) the anaerobic and water-free reaction environment further increases the operation difficulty. In the themolysis method, one of the key factors for achieving size-tunable and monodispersed UCNPs requires a proper selection of the coordinating ligands. Yan and co-worker reported that oleylamine ligand is a delicate buffer for F^-ions, the lighter the rare earth, the more OM it requires, owing to the fact that the basicity of the RE oxide gradually decreases along with the increasing of the atomic number of the RE series [19].

2.2. Thermal Coprecipitation

Due to the limitations of the thermal decomposition method, thermal coprecipitation approach is developed and has now been used as one of the most convenient methods for UCNPs synthesis. The experimental procedure is generally composed of: (1) RE salts were mixed with a solution of OA, OD, and sometimes OM, at certain ratio, which was heated to 165 °C for 30 min and then cooled down to room temperature; (2) a methanol solution of NH_4F and AOH (A = Li, Na, K) was added to the mixture and stirred for 30 min; (3) after removal of the methanol and residual water by evaporation, the reaction mixture was heated to high temperature (usually >300 °C) under argon protection, which produces desired nanocrystals. The benefits of the coprecipitation method include operational simplicity, the lack of toxic by-products, and the wide application across various materials.

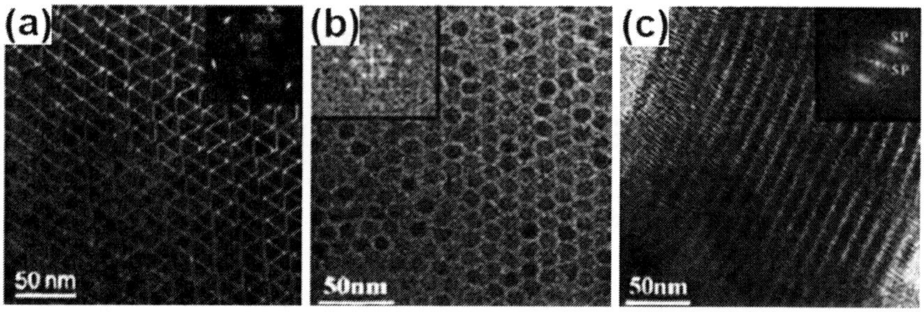

Figure 1. (**a**) TEM image of edge-to-edge super lattices of LaF_3triangular nanoplates. Reproduced from [18]. Copyright 2005, American Chemical Society. (**b,c**) TEM images of edge-to-edge and face-to-face super lattices of SmF_3 hexagonal nanoplates, respectively. Reproduced from [19]. Copyright 2006, John Wiley and Sons.

With continuous efforts, many kinds of UCNPs have been directly synthesized by this method. For example, in 2002, van Veggel and co-workers [30] have synthesized the down conversion LaF_3nanocrystals doped with RE^{3+} (RE = Eu, Er, Nd, and Ho) ions. This approach was later expanded by Yi and Chow [31], who prepared upconversion LaF_3 nanophosphors with smaller particle size (5 nm). In addition to LaF_3, $LuPO_4$:Yb/Tm, $YbPO_4$:Er, $NaYF_4$:Yb/Er(Tm), $NaGdF_4$:Yb/Er, and $Y_3Al_{15}O_{12}$(YAG):Yb/Tm nanophosphors were synthesized using this method [5,32,33,34]. Recently, our group has synthesized the first Scandium-based fluoride nanocrystals Na_xScF_{3+x}:Ln using the coprecipitation method [35] (Figure 2). Interestingly, the crystal structure evolution from pure monoclinic Na_3ScF_6 to pure hexagonal $NaScF_4$ phase was observed by tuning the ratio of OA and OD (Figure 2a). In addition, the hexagonal $NaScF_4$:Yb/Er crystals emit strong red upconversion (665 nm) under 980 nm laser excitation, different from those of the traditional hexagonal $NaYF_4$:Yb/Er nanocrystals, which usually emit strong green (547 nm) upconversion luminescence (Figure 2b).

Despite the general usefulness and obvious improvement of the thermal coprecipitation method for UCNPs synthesis, it suffers from the long-time and continuous operation of the experimental process, which usually takes more than 5 h, including the removal of methanol solvent, water generated during the synthesis, and the controlled crystal growth at a certain high-temperature. Moreover, large scale synthesis of UCNPs using this method is still a great challenge.

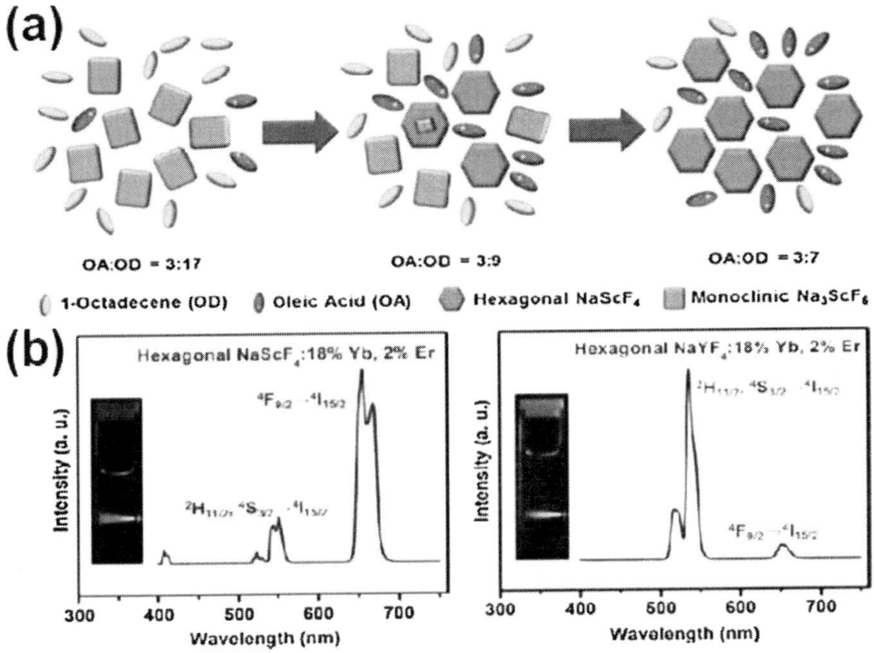

Figure 2. (**a**) Schematic illustration of the crystal structure evolution at varying polarities of the reaction medium. (**b**) Upconversion luminescence spectra of hexagonal-phase $NaScF_4$:Yb/Er and $NaYF_4$:Yb/Er nanocrystals. Reproduced from [35]. Copyright 2012, American Chemical Society.

2.3. Hydro/Solvothermal Synthesis

Hydro/solvothermal is a process of chemical reaction between negative ions and positive ions that usually precipitate from the solvent under high temperature and high pressure, generating nanoscale materials in the solvent after proper processing. This method has now become a widely-employed synthetic approach for UCNPs since it is easy to operate, does not require stringent operation of the experimental process, and moreover, the reaction temperature of hydro/solvothermal method is usually lower than those used for thermal decomposition and thermal co-precipitation synthesis for UCNPs.

The commonly used surfactants for UCNPs preparation include ethylenediamine tetraacetic acid (EDTA) [36,37], cetyltrimethylammonuim bromide (CTAB) [38], trisodium citrate (Na_3Cit) [39], linoleate acid [40], oleic acid [41,42,43,44,45,46]. As early as 2002, Li and coworkers [47,48,49] used RE nitrate and KOH to synthesize $Ln(OH)_3$ nanowires, nanotubes, and nanoparticles (Figure 3a–c). Long and uniform nanobelts were obtained when NaOH/KOH and RE acetates were used (Figures 3d,e). Later on, EDTA was used as complexing agent and CTAB as structure-directing agent to synthesize β-$NaYF_4$:Yb/Er nanotubes, nanorods, and nanospheres [34] (Figure 4a–d). Different from Li's work, Wang and co-workers [50] synthesized upconversion $NaYF_4$ nanocrystals using oleic acid-mediated hydrothermal synthesis (Figure 4e–h).

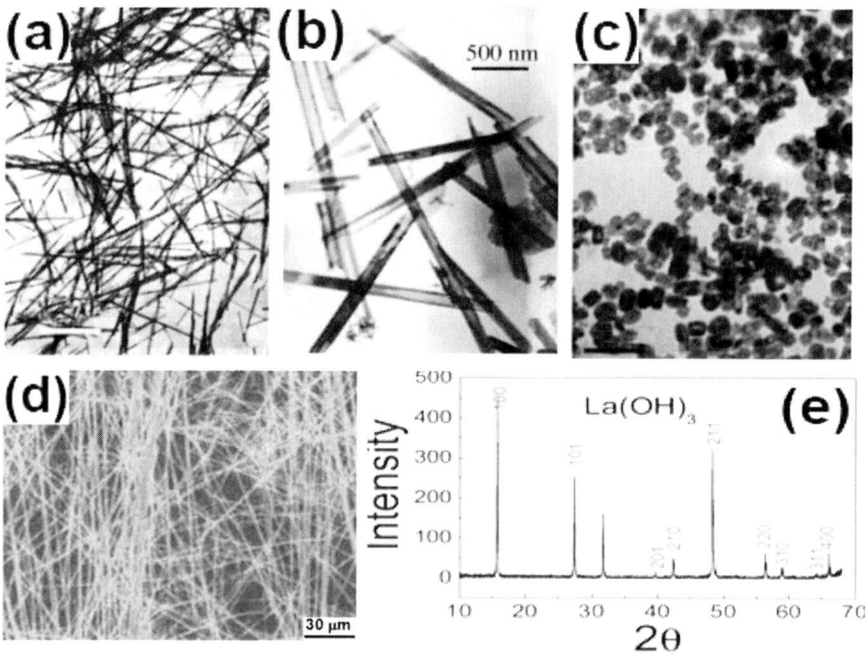

Figure 3. TEM images of (**a**) $La(OH)_3$ nanowires. Reproduced from [47]. Copyright 2002, John Wiley and Sons; (**b**) Y_2O_3 nanotubes. Reproduced from [48]. Copyright 2003, John Wiley and Sons; (**c**) LaF_3 nanoparticles. Reproduced from [49]. Copyright 2003, John Wiley and Sons; (**d**) $La(OH)_3$ nanobelt; (**e**) a typical XRD pattern of the as-synthesized $La(OH)_3$. Reproduced from [50]. Copyright 2007, John Wiley and Sons.

Figure 4. TEM and SEM images of $NaYF_4$:Yb/Er nanocrystals prepared under different hydro/solvothermal conditions. (**a,b**) TEM images of $NaYF_4$:Yb/Er nanocrystals synthesized in acetic acid and ethanol in the presence of CTAB, respectively; (**c,d**) TEM images of $NaYF_4$:Yb/Er nanocrystals using EDTA in acetic acid and ethanol, respectively. Reproduced from [34]. Copyright 2005, John Wiley and Sons; SEM images of (**e,f**) flower-patterned hexagonal disks; (**g**) hexagonal nanotubes; and (**h**) nanorods of β-$NaYF_4$. Reproduced from [51]. Copyright 2007, John Wiley and Sons.

Recently, via a hybrid thermal decomposition/solvothermal method, LaF_3:Yb/Er/Tm/Ho nanoplates with multicolor upconversion luminescence were synthesized [51,52]. The feasibility of the hydro/solvothermal methods was demonstrated by several groups. For example, Zhang has synthesized $BaYF_4$ [53], $CePO_4$ [54], YPO_4 [55], Gd_2O_3 [56], $GdVO_4$ [57], and Lin's group reported the uniform microstructured YPO_4 [58], YVO_4 [59], LnF_3 [60], and $NaYF_4$ [45] via the sodium citrate assisted hydrothermal route. In recent years, Liu and co-workers reported the synthesis of $KMnF_3$ nanocrystals with only single-band UC emission [61]. Zhao and Hao's group [62,63] represented a strategy for the rationale manipulation of green and red upconversion emission, and the pure red emission of $NaYF_4$:Yb/Er nanocrystals has been achieved by Mn^{2+} doping.

However, the disadvantages of this method are substantially difficult to overcome. Firstly of all, there are too many parameters including the reaction temperature, surfactant type and concentration, reactant concentration, solvent and the composition of it, and reaction time, to consider in order to filter out the most optimized experimental conditions for one specific reaction. Furthermore, the obtained nanocrystals usually have large size distribution, and sometimes the by-product residues stay on the surface of the nanocrystals, which are difficult to remove. Therefore, it remains challenging to develop a general and facile hydro/solvothermal method for the synthesis of high-quality UCNPs.

2.4. Sol-Gel Method

Sol-gel process is a typical wet-chemical technique for synthesizing UCNPs, which can be generally divided into three types: (1) sol-gel route based on the hydrolysis and condensation of molecular precursors; (2) gelation route based on condensation of the aqueous solutions containing metal-chelates; and (3) polymerizable complex route [64]. In sol-gel process, a RE nitrate salt or metal alkoxide is generally used as the starting reactants. The reaction is started by mixing the reactants in liquid phase through hydrolysis and condensation reactions, followed by annealing at high temperature for a certain period of time.

In 2002, Prasad *et al.* [65] developed a sol emulsion-gel method that produces Er^{3+} doped ZrO_2nanophosphors. In order to reduce the segregation of particles and ensure compositional homogeneity, Lin and co-workers fabricated an inorganic YVO_4:Eu thin film phosphor by combining the pechini-type sol-gel process with inkjet printing. The mixed solution of metal salt precursors, citric acid, and poly(ethylene glycol) was directly used as ink to deposit patterns on ITO-coated glass substrate. After calcination at 600 °C in air, the YVO_4:Eu patterns at the micrometer-scale were formed on the substrate [66]. In another report, Song and co-works successfully fabricated many kinds of inverse opal photonic crystals (PCs) by the sol-gel method with a PMMA latex sphere template, including YVO_4:Dy [67], TiO_2:Sm [68], YBO_3:Eu [69], and $LaPO_4$:Ce,Tb [70]. In addition, the sol-gel process was also developed for synthesizing various UCNPs using metal oxides as host materials such as TiO_2:Er, $BaTiO_3$:Er (Figure 5), $Lu_3Ga_5O_{12}$:Er, and YVO_4:Yb/Er [71,72,73].

For the sol-gel method, the annealing procedure (temperature and time) is the key step in the preparation process, which can seriously determine the quality of the samples. It should be noted that, although the sol-gel method can be used for large-scale production and the products usually offer high luminescence intensity due to the high crystallinity formed at high annealing temperature, the sol-gel derived nanocrystals generally have broad particle size distribution, irregular morphology, and are insoluble in water, which compose the shortcomings of this method.

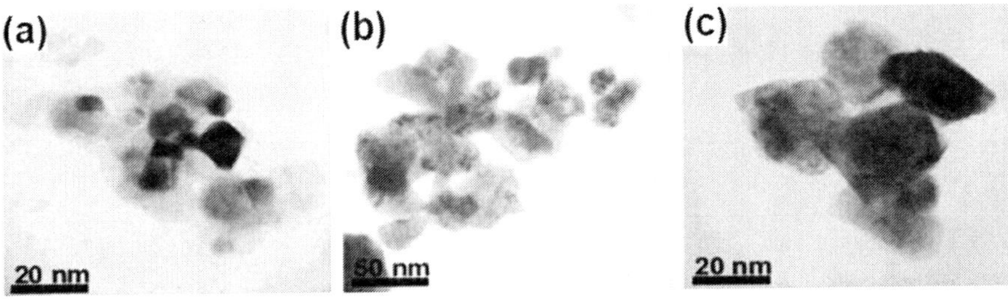

Figure 5. TEM images of the 1.0 mol% Er^{3+}-doped $BaTiO_3$nanoparticles obtained after heating to three different temperatures: (**a**) 700 °C; (**b**) 850 °C; and (**c**) 1000 °C. Reproduced from [71]. Copyright 2003, American Chemical Society.

2.5. Microemulsion

To synthesize nanomaterials via the microemulsion method, it usually requires surfactant, co-surfactant, organic solvent, water, and the initial reagents. In a typical microemulsion solution, amphiphilic surfactants form a monolayer at the oil-water interface, with the hydrophobic tails of the surfactant molecules dissolved in the oil phase and the hydrophilic head groups in the aqueous phase. Preparation of RE fluoride nanomaterials usually needs two separate microemulsion systems and the two microemulsions containing RE ions and fluorine ions, respectively, get mixed to initiate the reaction. Over the past few years, Lemyre and coworkers [74] have reported that upconversion YF_3 nanoparticles can be prepared in a reverse water-in-cyclohexane microemulsion system stabilized by polyoxyethylene isooctylphenyl ether (IGEPAL CO-520 or NP-5was supplied by Aldrich) (Figure 6a). Qin and coworkers [75,76] prepared YF_3 upconversion nanophosphors, using CTAB and 1-pentanol instead of NP-5 (Figure 6b).

The use of microemulsion has many advantages, such as the low-cost for equipment, easy operation, the small size of the UCNPs, and the controlled morphology of products by adjusting the dosage of the surfactant, solvent, as well as the aging time. However, this technique has many challenges including the small amount of products generated, difficulty of sample separation and narrow scope synthesis. More importantly, it is difficult to achieve massive production to meet industry requirements.

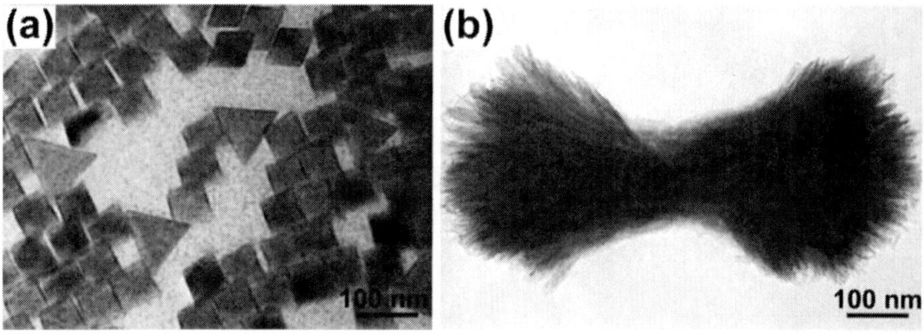

Figure 6. TEM images of (**a**) YF_3 nanophosphors prepared in the NP-5 stabilized microemulsion system. Reproduced from [74]. Copyright 2005, American Chemical Society. (**b**) YF_3 nanobundles synthesized in the CTAB and NP-5 stabilized microemulsion system. Reproduced from [75]. Copyright 2008, American Chemical Society.

2.6. Combustion Synthesis

Compared to sol-gel and hydro/solvothermal synthesis, combustion method for UCNPs synthesis can be finished in a short period of time. The process of combustion synthesis is an oxidation-reduction reaction in essence. Metal nitrates are usually selected as oxidizer and the source for metal ions while the organic compounds as reducing agent and fuel. There are two requirements for organic fuel selection: (1) the reaction occurred between the fuel and nitrate must be relatively mild, producing nontoxic gases; (2) it is better to select the fuels that can complex with the metal ions, enhance the solubility of metal ions, and prevent the separation by crystallization of the metal salts in the precursor solution.

During the combustion synthesis process, combustion wave spreads the reaction materials in a self-sufficient situation without requiring extra heat in the process of the whole reaction. This time- and energy-saving method was used to synthesize oxide and oxysulfide nanomaterials. For example, Capobianco, Luo, and Zhang's groups have respectively synthesized a variety of oxide and oxysulfide nanophosphors (Y_2O_3, $Gd_3Ca_5O_{12}$, La_2O_2S, and Gd_2O_3) [77,78,79,80] via this method.

2.7. Flaming Synthesis

Flaming synthesis is another powerful method for producing RE oxide nanomaterials, which can be divided into four stages: (1) precursor reaction;

(2) nucleation; (3) growth and polymerization; and (4) ion deposition. Notably, the flaming synthesis differs from the typical combustion synthesis that all reactions take place in the gas-phase and form fine powders. The core advantages of the technique are time-saving and low-cost. In 2007, Ju [81] reported the synthesis of Y_2O_3:Yb/Er(Tm, Ho) nanophosphors using this method. However, since the oxidation reaction happens along with the flaming process, this method is mainly limited to oxide nanomaterial synthesis, and it is almost impossible to synthesize other types of UCNPs, such as fluoride, phosphate, vanadate, and so on. It should also be noted that the flaming synthesis offers an opportunity for large-scale synthesis of RE doped oxides.

2.8. Electrospinning

During the electrospinning process, the precursor solution is spun through four stages (cocoons, stretching, refinement, and curing) under the effect of high voltage electrostatic field. Simple operation, good repeatability, and wide application scope are the advantages of this approach. For example, Song [82] have reported $NaYF_4$/PVP composite nanofibers, with diameter in the range of 300–800 nm, prepared through electrospinning. Nevertheless, this method is still in its infantile stage. New recipes for the synthesis of uniform and small-sized (<20 nm) nanoparticles with controllable morphology are still required.

2.9. Microwave Synthesis

Microwave synthesis contains solid microwave and liquid microwave method. The former mixes up RE oxides with ammonium bifluoride and ammonium fluoride, and the nanoparticles were directly synthesized via microwave. The latter dissolves the RE salts and the fluoride source into solvent. Then, these raw materials react with each other when the solvent is heated by microwave. The method was extended to produce PrF_3 hemispheres with diameter of about 31 nm [83].

The advantage of microwave synthesis is generally composed of: (1) it can selectively heat the samples to high temperature while the rest of the microwave device remains at room temperature; (2) microwave can heat the reaction system uniformly, cause less side reaction and the product is relatively simple; (3) fast heat and low energy consumption; (4) improve the structure and properties of synthetic material by adding proper surfactant.

Besides the above synthetic strategies elaborated, combinatorial approaches have also been employed to produce micro- or nano-scale RE fluoride crystals, which after proper lanthanide doping, can upconvert near-infrared light to visible frequencies, enabling the applications of such materials to biological imaging, telecommunication, and solar energy conversion [84]. For example, Liu [85] have synthesized $NaYF_4$:Yb,Er upconversion nanocrystals with characteristic upconversion luminescence spectra in a continuous capillary. Later, the research group improved the condition of the reaction prevented the growth of β-$NaYF_4$ [86]. On the other side, Zhu *et al.* synthesized LaF_3 [87] and $LaPO_4$[88] nanoparticles doped with Ce^{3+} and Tb^{3+} using microcapillary flow reactors heated with microwaves. However, they also suffer from the wide quality variation of the UCNPs synthesized from different vessels and the large-scale production.

3. SURFACE MODIFICATION OF UCNPS

Because of the influence of impurities and lattice defects on the synthesized UCNPs, the quantum yield of UCNPs is lower than the corresponding bulk materials. In addition, the UCNPs are mostly insoluble in water since they are prepared from the organic environment and usually surrounded by hydrophobic surfactant molecules. Thus, it is important to develop appropriate strategies to make them hydrophilic and in the meantime maintain their upconversion efficiency to satisfy various purposes. For example, the ideal luminescent nanocrystals used for biocompatible purposes should meet several requirements, including: (1) high luminescence efficiency and low background noise; (2) good solubility and stability in biological environment; (3) good biological compatibility; and (4) proper size (below 100 nm).

However, a notorious weak point of UCNPs is the inherent low upconversion efficiency. Currently, the urgent task is to improve the upconversion efficiency, and there have been several groups working on this topic via the core-shell strategy.

The introduction of an inert crystalline shell of an undoped material around each doped nanocrystal provides an effective option to improve the luminescence efficiency of UCNPs. The shell usually has the same composition as the core host crystal, which can effectively reduce the surface quenching effect. In such structures, all dopant ions are confined in the interior core of the nanocrystals, effectively suppressing the non-radiative energy transfer from RE ions to the surface quenching sites, which results in improved upconversion luminescence efficiency. A significant demonstration was made by Yi and Chow [89] who reported a luminescence enhancement of nearly 30 times on 8 nm $NaYF_4$:Yb/Tm nanocrystals coated with a 1.5 nm thick $NaYF_4$ shell. Later on, Chen and co-works exploited a strategy to achieve dual-mode luminescence from identical Eu^{3+} ions in monodisperse hexagonal-phase $NaGdF_4$nanocrystals that consist of the $NaGdF_4$:Yb/Tm core and the $NaGdF_4$:Eu shell. Typical red downconversion luminescence of Eu^{3+} has been detected via the sensitization of Gd^{3+} ions. By using Yb^{3+} and Tm^{3+} embedded in the cores as double sensitizers, intense upconversion luminescence of Eu^{3+}in the shells can be achieved in $NaGdF_4$:Yb/Tm@$NaGdF_4$:Eu core-shell nanocrystals upon excitation at 976 nm. The upconversion intensity of Eu^{3+} in core-shell nanocrystals is found about one order of magnitude higher than the counterparts of the triply-doped core only, due to the inhibition of the deleterious cross-relaxations between Tm^{3+} and Eu^{3+} ions in core-shell nanocrystals that are reasonably separated in space [90].

Recently, Capobianco [91] proposed a strategy to significantly enhance the intensity of the upconversion by employing a novel $NaGdF_4$:Yb/Er@$NaGdF_4$:Yb active-core@active-shell architecture (Figure 7). The active-shell serves two purposes: (i) minimize the quenching centers and channels; and (ii) transfer absorbed near infrared energy to the luminescence centers (emitters).

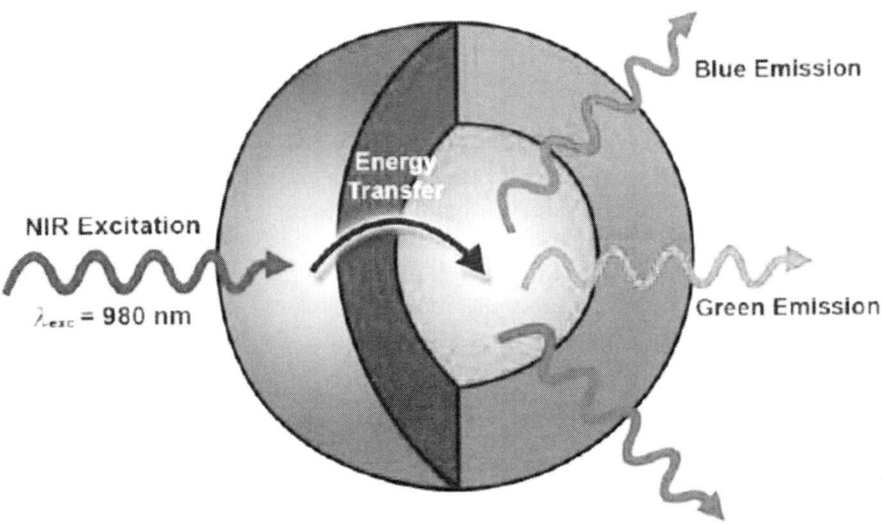

Figure 7. Schematic illustration of the active core-active shell nanoparticle architecture showing the absorption of NIR near infrared light by the Yb^{3+}-rich shell (red) and subsequent energy transfer to the Er^{3+},Yb^{3+}-doped core

(green), which leads to upconverted blue, green, and red emissions. Reproduced from [91]. Copyright 2009, John Wiley and Sons.

Utilizing the core-shell strategy, Zhang and co-workers [92] have broken through the well accepted upper limit of the concentration quenching threshold, e.g., from ~2 mol% to 5 mol% for Er^{3+}, by a designed multi-layer core-shell structure, which contains four parts: the illuminating core ($NaYF_4$:Yb/Er), the first separating shell ($NaYF_4$:Yb), the second illuminating shell ($NaYF_4$:Yb/Er) and the final inactive shell ($NaYF_4$:Yb). The separating layer effectively inhibits the energy transfer process between the Er^{3+} ions in the inside and outside layers and largely reduces the possibility of excitation energy trapping by defects, resulting in an effective upconversion luminescence enhancement [92]. Other core-shell examples include CeF_3:Tb@LaF_3 [93], $NaYF_4$:Yb/Er@$NaYF_4$ [89,94], $NaGdF_4$:Yb/Er@$NaGdF_4$ [95], and KYF_4@KYF_4:Yb/Er [96]. Liu [97] introduced a rationale core-shell strategy that provides precise control over the concentration of dopant in the core and shell layers of nanoparticles. A small amount of Nd^{3+} ions is doped into the core, while a high concentration of Nd^{3+}ions (~20 mol%) is selectively doped in the shell layers for effective harvesting of light at 800 nm. Chen and co-workers developed a facile strategy based on successive layer-by-layer (LBL) injection of shell precursors for the synthesis of $LiLuF_4$:Ln^{3+} core-shell UCNPs [98]. It is reported that the size of nanocrystals can be tuned by adjusting the amount of NH_4F (the more the amount of NH_4F solution used, the bigger the size of crystals) and the shell thickness of core-shell nanocrystals can also be controlled by the volume of the core [99]. Besides, Yan has proved in the structure of $NaYF_4$:Yb,Er@CaF_2that the thickness of the CaF_2 shell were controlled by adjusting the [Ca]/[RE] ratio [100]. Although this approach provides tunable emission intensity in nanocrystals, the luminescence quantum yields of the nanocrystals are limited due to the weak ligand fields and high energy oscillations. Further improvement of the luminescence efficiency of UCNPs can be expected through the controlled growth of a rationally designed crystalline inner shell. Generally speaking, such core-shell structure improves the optical properties of the nanocrystals, but usually does not change the chemical functional group binding on the surface. So, proper surface modification is needed in order to make practical applications of UCNPs, especially for bio-related purposes.

3.1. SiO_2 Encapsulation

SiO_2 encapsulation involves the growth of an amorphous silica shell on the UCNPs core. Due to the rich –OH groups on the SiO_2 coated UCNPs, it becomes possible to further functionalize the UCNPs with expected chemical groups such as $-NH_2$, $-COOH$, polymers, and more importantly a wide variety of biomolecules. Nann and Capobianco [101] obtained surface-functionalized YF_3 nanocrystals using this technique. One year later, Zhang and coworkers [102] succeeded in growing a silica shell with adjustable thickness in the range of 2–10 nm on the surface of PVP stabilized cubic $NaYF_4$:Yb/Er/Tm nanocrystals. Later on, they used the microemulsion method to coat $NaYF_4$ nanocrystals with SiO_2, which resulted in monodispersed SiO_2-coated UCNPs (Figure 8) [103].

Alternatively, Song and co-workers [104] showed the conversion of hydrophobic $NaYF_4$:Yb/Er upconversion nanophosphors into hydrophilic ones by amphiphilic silane modification with ultrathin thickness (~1 nm) at room temperature. In this strategy, the coating layers can also provide the possibility for loading $Eu(TTA)_3(TPPO)_2$ complex with downconversion luminescence where they realized the dual mode temperature sensing and dual mode cell imaging within the physiological environment [104]. There are other successful examples to modify the surface of the UCNPs [105,106,107,108] via the SiO_2 encapsulation method, such as LaF_3:Ln@SiO_2, $NaYF_4$:Yb/Er/Fe_3O_4@SiO_2, $NaYF_4$:Er,Tm,Ho@SiO_2, NdF_3@SiO_2, and $NaYF_4$@SiO_2@ quantum dots [103].

The SiO_2 coating provides a versatile platform for multi-functionalization of UCNPs. Nevertheless, the method is time consuming and difficult for large scale synthesis. On the other hand, the coated SiO_2layer may also affect the luminescence intensity of the UCNPs through light scattering, which prevents it from becoming the ideal surface modification recipe because of the already low upconversion efficiency of the UNCPs. Thus, there is still much work to do to improve the general applicability of this technique.

3.2. Polymer Encapsulation

Polymer encapsulation typically involves the absorption of an additional amphiphilic polymer onto the nanophosphor surface through the hydrophobic-hydrophobic attraction between the original surfactant molecules and the hydrocarbon chains of the polymer. Using this strategy, Chow and co-workers successfully coated a polyacrylicacid (PAA) layer on $NaYF_4$:Yb/Er@$NaYF_4$ core-shell nanoparticles [89]. The hydrophobic core-shell nanoparticles were rendered hydrophilic by amphiphilic PAA coating.

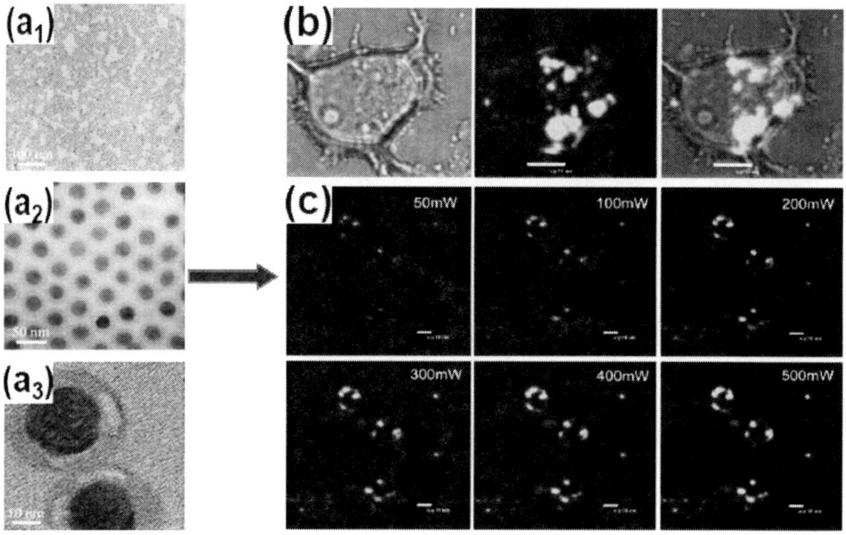

Figure 8. Silica-coated $NaYF_4$:Yb/Er nanocrystals and their application for cell imaging. (**a**1–C**a**3) TEM images of silica-coated $NaYF_4$:Yb/Er UCNPs upconversion nanoparticles at different magnifications; (**b**) Confocal fluorescence image of MCF-7 cells using silica-coated $NaYF_4$:Yb/Er nanospheres (Left: bright-field, middle: upconversion image under 980 nm excitation, and Right: superimposed images of MCF-7 cells incubated with the nanoparticles for 24 h); (**c**) Confocal fluorescence images of MCF-7 cells with the nanospheres, excited by a 980 nm laser with different power intensities. Reproduced from [103]. Copyright 2008, John Wiley and Sons.

Through amphiphilic coating, the multi-functionalization of upconversion nanoparticles could also be realized. For instance, Feng and co-workers fabricated upconversion detection nanocomposites, which were formed by coating the amphiphilic polymer (C18PMH-PEG) on the $NaYF_4$:Yb/Er,Tm nanophosphors based on the hydrophobic–hydrophobic interaction. The polymer modified nanoparticles were then assembled for the selective luminescence detection of mercury ions in water. Using the ratiometric upconversion luminescence emission as a detection signal, the detection limit of Hg^{2+} for this nanoprobe in aqueous solution is 8.2 ppb, which is much lower than that (329 ppb) determined by the UV/Vis technology [109]. Liu modified UCNPs with a polyethylene glycol (PEG) grafted amphiphilic polymer, via a hydrophobic interaction-based supramolecular chemistry strategy, for targeted intracellular drug delivery and UCL imaging. This work reveals the great potential of UCNPs for multifunctional drug delivery and biomedical imaging applications (Figure 9) [110]. In contrast to the conventional methods, K. Prud'homme group [111] reported the successful preparation of colloidal UCNPs stable in buffers and serum media (Leibovitz L-15 media with added fetal bovine serum) using FNP and PEG surface coatings. These polymer-modified UCNPs provide promising new materials for applications in bioimaging and photodynamic therapy.

The polymer-coating method is easier when compared to SiO_2 encapsulation. The *in-situ* coating of polymers on the surface of UCNPs sheds light on the development of multi-functional upconversion nanoparticles. However, the coating through hydrophobic interaction is not stable, and better strategies for a robust polymer coating are needed to satisfy various application requirements.

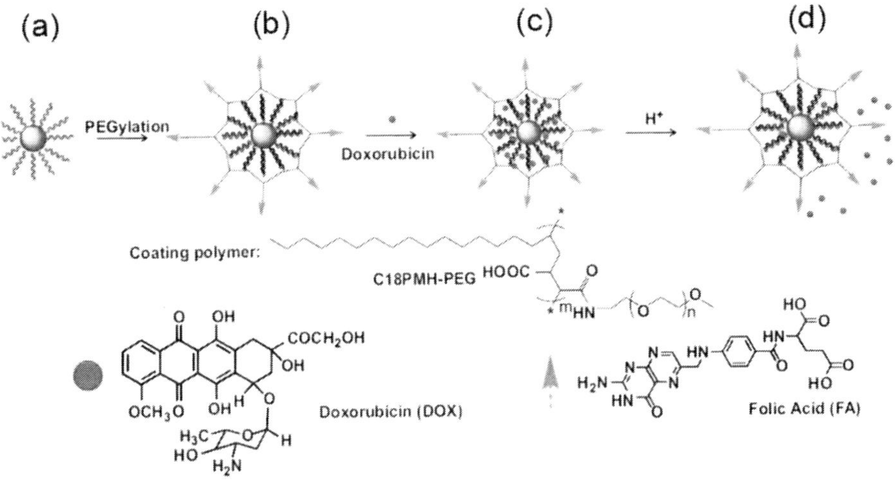

Figure 9. Schematic illustration of the UCNP-based drug delivery system. (**a**) As-synthesized oleic acid capped UCNPs; (**b**) C18PMH-PEG-FA functionalized UCNPs; (**c**) DOX loading on UCNPs. DOX molecules are physically adsorbed into the oleic acid layer on the nanoparticle surface by hydrophobic interactions; (**d**) Release of DOX from UCNPs triggered by decreasing pH. Reproduced from [110]. Copyright 2011, Elsevier Ltd.

3.3. Ligand Oxidation

The ligand oxidation technique involves oxidation of the native ligands that contain unsaturated carbon-carbon bonds to generate a pendant hydrophilic functional group. It is reported that a large number of carboxylic acid groups can be generated on the UCNPs surface after oxidation, which not only renders the nanocrystals good solubility in water, but also provides tailorability to biological molecules via direct chemical coupling (Figure 10). It is also reported that the oxidation process has no obvious negative effect on the shape, crystal structure, chemical composition, and luminescence properties of the upconversion nanomaterials [112]. However, this method is limited only to those ligands that contain unsaturated carbon-carbon bonds. The oxidation process may also cause ligand removal due to the harsh experimental conditions. It would be ideal if the modulation of ligand could be achieved under mild experimental conditions, which cause no or less damage to the surfaces of the UCNPs. Biomedical research faces a major challenge in that the lengthy period of oxidation may lead to the excessive formation of the brown MnO_2 side product, which is not easy to separate and further weakens the upconversion fluorescence [23]. So, the surface functionalization of UCNPs to render these nanoparticles dispersible in aqueous media is very important. Capobianco and co-workers show that the Ln-OA (Ln = lanthanide; OA = oleate) surface of the Ln-UCNPs is replaced by Ln-OH whose state of charge can be tuned by pH and the efficiency of upconversion luminescence can be enhanced by replacing OH with OD. Furthermore, they have studied the effect of different acids, *i.e.*, HCl, H_3PO_4, and HF on the surface properties of the oleate-free Ln-UCNPs [113].

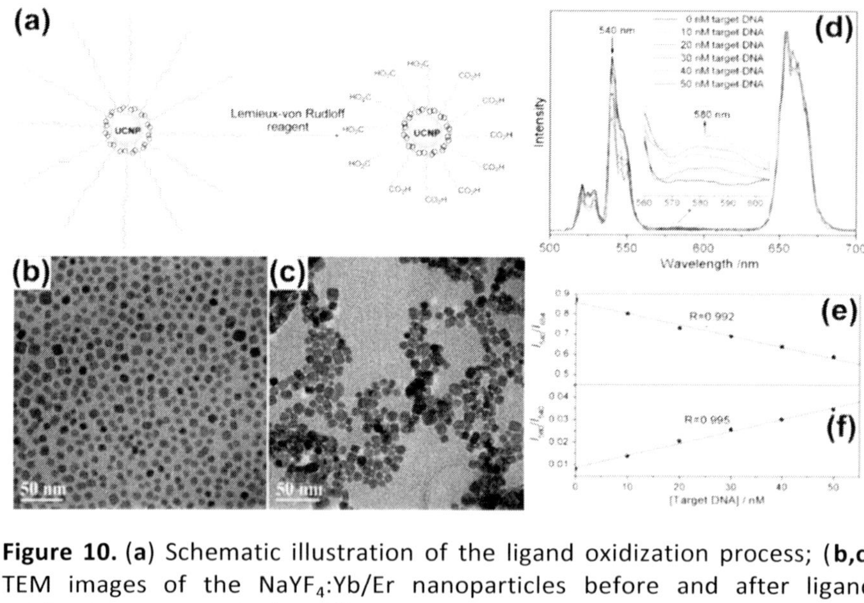

Figure 10. (**a**) Schematic illustration of the ligand oxidization process; (**b,c**) TEM images of the $NaYF_4$:Yb/Er nanoparticles before and after ligand oxidization, respectively; (**d**) Luminescence spectra of a mixture of streptavidin-functionalized $NaYF_4$:Yb/Er nanoparticles, capture-DNA, and reporter-DNA in the presence of different concentrations of target-DNA under continuous-wave excitation at 980 nm. The linear relationships between target-DNA concentration and the intensity ratios of (**e**) I540/I654 and (**f**) I580/I540. Reproduced from [112]. Copyright 2008, American Chemical Society.

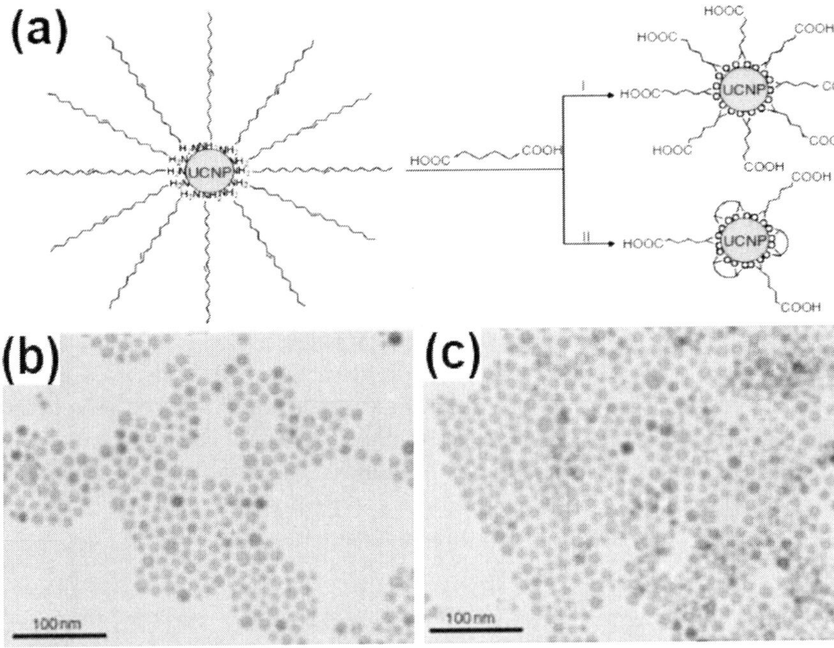

Figure 11. (**a**) Schematic illustration showing the ligand-exchange reactions on OM-stabilized upconversion nanophosphors. (**b,c**) TEM images of $NaYF_4$:Yb/Er nanophosphors prior to and after ligand exchange reactions, respectively. Reproduced from [114]. Copyright 2006, John Wiley and Sons.

3.4. Ligand Exchange

Ligand exchange involves the displacement of original hydrophobic ligands that have weak coordination with the RE ions on the surface of UCNPs, by ligands that have stronger coordination capability (with the RE ions) and hydrophilic functional groups. As an early example, Chow and co-workers [114] have demonstrated the preparation of water-soluble $NaYF_4$:Yb/Er nanoparticles with oleylamine ligands via the ligand exchange method (Figure 11). A large variety of ligands have been reported, including poly(acrylicacid) (PAA) [115], poly(ethyleneglycol) (PEG)-phosphate [116], mercaptopropionic acid (MPA) [117], citrate [118]. Li's research group [119] demonstrated a new generation of ^{18}F-labeled lanthanide nanoparticles of $NaYF_4$ co-doped with $Gd^{3+}/Yb^{3+}/Er^{3+}$ as multimodality nanoprobes for PET, MR and UCL imaging. The presence of Yb^{3+} and Er^{3+} co-doped in the $NaYF_4$ nanoparticles gives rise to intense UCL emission in the visible region for luminescent imaging, and 60% Gd^{3+} doping provides the paramagnetic relaxation for MRI. Successful labeling of the lanthanide nanoparticles with ^{18}F gave a product suitable for PET imaging [119]. It is noted that after the ligand exchange, most of these commonly used ligand molecules on the UCNP surface carry additional functional groups to facilitate further biofunctionalization and bioconjugation. However, the exchange efficiency of the technique is difficult to evaluate since the surfactant molecules cannot be completely replaced by the oleylamine molecules, and the exchange process is also affected by the pH value of the solution, the concentrations of both the surfactant and the oleylamine molecules, and even the ionic strength.

Besides the above surface modification strategies, Prabhas V. Moghe and coworkers [120] have developed a new approach to rendering $NaYF_4$:Yb,Er nanoparticles stable in coacervated HSA nanoshells functionalized with cyclic arginine-glycine-aspartic acid (cRGD) tripeptide. They observed that the composite particles were highly biocompatible *in vitro*, capable of selectively targeting cancerous cell lines exhibiting higher expression of cancer-specific integrin markers, and amenable to fluorescence imaging with high fidelity.

4. CONCLUSIONS

In summary, this review has summarized the recent development of UCNPs with emphasis on the synthetic methods and the surface modification strategies. With careful control of the reaction conditions, we can now obtain high quality, monodispersed UCNPs with various chemical components. However, each individual synthetic strategy has their unique advantages associated with substantial shortcomings. Our subsequent goal is to develop a general synthetic approach which could meet the requirements of large scale synthesis and multifunctionalities for UCNPs.

In parallel, the controlled surface modification method now can produce UCNPs with high colloidal stability, biocompatibility, and tailorable chemical functionalities. The above discussion has also demonstrated that the rapid development of surface modification method facilitates the applications of UCNPs in detection, bioimaging, therapy, and solar cells. Despite the rapid development, these exiting methods still can only be used in a narrow scope. Strategies with universal comparability for various UCNPs may be the next goal of this field. Alternatively, methods that provide easy surface modulation capability for certain types of nanocrystals could be another option for multifunctionalization.

ACKNOWLEDGES

We thank the financial support from the National Natural Science Foundation of China (No.: 21301090) and the Natural Science Foundation of Jiangsu Province (No.: BK20130923).

AUTHOR CONTRIBUTIONS

Hongjin Chang, Ling Huang, and Wei Huang proposed the structure of the manuscript and wrote the paper. Juan Xie, Baozhou Zhao, Botong Liu, Shuilin Xu, and Na Ren helped to search for the reference and copyright collection. Xiaoji Xie performed the manuscript editing and language coloring.

CONFLICTS OF INTEREST

The authors declare no conflict of interest.

REFERENCES

1. Law, M.; Goldberger, J.; Yang, P.D. Semiconductor nanowires and nanotubes. *Annu. Rev. Mater. Sci.* **2004**, *34*, 83–122.

2. Yoffe, A.D. Low-dimensional systems: Quantum size effects and electronic properties of semiconductor microcrystallites (zero-dimensional systems) and some quasi-two-dimensional systems. *Adv. Phys.* **2002**, *42*, 173–266.

3. Su, V.C.; Chen, P.H.; Lin, R.M.; Lee, M.L.; You, Y.H.; Ho, C.I.; Chen, Y.C.; Chen, W.F.; Kuan, C.H. Suppressed quantum-confined stark effect in InGaN-based LEDs with nano-sized patterned sapphire substrates. *Opt. Exp.* **2013**, *21*, 30066–30071.

4. Wang, F.; Liu, X.G. Recent advances in the chemistry of lanthanide-doped upconversion nanocrystals. *Chem. Soc. Rev.* **2009**, *38*, 976–989.

5. Yi, G.S.; Lu, H.C.; Zhao, S.Y.; Yue, G.; Yang, W.J.; Chen, D.P. Synthesis, characterization, and biological application of size-controlled nanocrystalline $NaYF_4$:Yb,Er infrared-to-visible up-conversion phosphors. *Nano Lett.* **2004**, *4*, 2191–2196.

6. Carlos, L.D.; Ferreira, R.A.; Ribeiro, S.J. Lanthanide-containing light-emitting organic-inorganic hybrids: A bet on the future. *Adv. Mater.* **2009**, *21*, 509–534.

7. Zhou, J.; Liu, Z.; Li, F.Y. Upconversion nanophosphors for small-animal imaging. *Chem. Soc. Rev.***2012**, *41*, 1323–1349.

8. Wang, F.; Han, C.M.; Li, F.Y. Upconversion-nanophosphor-based functional nanocomposites. *Adv. Mater.* **2013**, *25*, 5287–5303.

9. Chen, J.; Guo, C.R.; Wang, M.; Wang, L.P.; Mao, C.B.; Xu, S.K. Controllable synthesis of $NaYF_4$:Yb,Er upconversion nanophosphors and their application to *in vivo* imaging of*Caenorhabditis elegans*. *J. Mater. Chem.* **2011**, *21*, 2632–2638.

10. Wang, F.; Han, Y.; Liu, X.G. Simultaneous phase and size control of upconversion nanocrystals through lanthanide doping. *Nature* **2010**, *463*, 1061–1065.

11. Hampl, J.; Hall, M.; Mufti, N.A.; Yao, Y.M.; Mac Queen, D.B.; Wright, W.H.; Cooper, D.E. Upconverting phosphor reporters in immunochromatographic assays. *Anal. Biochem.* **2001**, *288*, 176–187.

12. VandeRijke, F.; Zijlmans, H.; Li, S.; Vail, T.; Raap, A.K.; Niedala, R.S.; Tanke, H.J. Up-converting phosphor reporters for nucleic acid microarrays. *Nat. Biotechnol.* **2001**, *19*, 273–276.

13. Lim, S.F.; Riehn, R.; Ryu, W.S.; Khanarian, N.; Tung, C.-K.; Tank, D.; Austin, R.H. *In vivo* and scanning electron microscopy imaging of upconverting nanophosphors in *Caenorhabditis elegans*.*Nano Lett.* **2006**, *6*, 169–174.

14. Chatterjee, D.K.; Rufaihah, A.J.; Zhang, Y. Upconversion fluorescence imaging of cells and small animals using lanthanide doped nanocrystals. *Biomaterials* **2008**, *29*, 937–943.

15. Zhang, P.; Rogelj, S.; Nguyen, K.; Wheeler, D. Design of a highly sensitive and specific nucleotide sensor based on photon upconverting particles. *J. Am. Chem. Soc.* **2006**, *128*, 12410–12411.

16. Gai, S.L.; Yang, P.P.; Li, C.X.; Wang, W.X.; Dai, Y.L.; Niu, N.; Lin, J. Nanomorphology and charge generation in bulk heterojunctions based on low-bandgap dithiophene polymers with different bridging atoms. *Adv. Funct. Mater.* **2010**, *20*, 1166–1172.

17. Hinklin, T.R.; Rand, S.C.; Laine, R.M. Transparent, polycrystalline upconverting nanoceramics: Towards 3-D displays. *Adv. Mater.* **2008**, *20*, 1270–1273.

18. Zhang, Y.W.; Sun, X.; Si, R.; You, L.P.; Yan, C.H. Single-crystalline and monodisperse LaF_3triangular nanoplates from a single-source precursor. *J. Am. Chem. Soc.* **2005**, *127*, 3260–3261.

19. Sun, X.; Zhang, Y.W.; Du, Y.P.; Yan, Z.G.; Si, R.; You, L.P.; Yan, C.H. From trifluoroacetate complex precursors to monodisperse rare-earth fluoride and oxyfluoride nanocrystals with diverse shapes through controlled fluorination in solution phase. *Chem. Eur. J.* **2007**, *13*, 2320–2332.

20. Boyer, J.C.; Cuccia, L.A.; Capobianco, J.A. Synthesis of colloidal upconverting $NaYF_4:Er^{3+}/Yb^{3+}$and Tm^{3+}/Yb^{3+} monodisperse nanocrystals. *Nano Lett.* **2007**, *7*, 847–852.

21. Wang, H.Q.; Thomas, N. Monodisperse upconverting nanocrystals by microwave-assisted synthesis. *ACS Nano* **2009**, *3*, 3804–3808.

22. Yi, G.S.; Lee, W.B.; Chow, G.M. Synthesis of $LiYF_4$, $BaYF_5$, and $NaLaF_4$ optical nanocrystals. *J. Nanosci. Nanotechnol.* **2007**, *7*, 2790–2794.

23. Naccache, R.; Vetrone, F.; Mahalingam, V.; Capobianco, J.A. Controlled synthesis and water dispersibility of hexagonal phase $NaGdF_4$:Ho/Yb nanoparticles. *Chem. Mater.* **2009**, *21*, 717–723.

24. Mahalingam, V.; Vetrone, F.; Naccache, R.; Speghini, A.; Capobianco, J.A. Colloidal Tm^{3+}/Yb^{3+}-doped $LiYF_4$ nanocrystals: Multiple luminescence spanning the UV to NIR regions via low-energy excitation. *Adv. Mater.* **2009**, *21*, 4025–4028.

25. Mahalingam, V.; Vetrone, F.; Naccache, R.; Speghini, A.; Capobianco, J.A. Structural and optical investigation of colloidal Ln^{3+}/Yb^{3+} co-doped KY_3F_{10} nanocrystals. *J. Mater. Chem.* **2009**, *19*, 3149–3152.

26. Vetrone, F.; Mahalingam, V.; Capobianco, J.A. Near-infrared-to-blue upconversion in colloidal $BaYF_5:Tm^{3+},Yb^{3+}$ nanocrystals. *Chem. Mater.* **2009**, *21*, 1847–1851.

27. Zhang, P.D.; Zhang, Y.W.; Zheng, Z.G.; Sun, L.D.; Yan, C.H. Highly luminescent self-organized sub-2-nm EuOF nanowires. *J. Am. Chem. Soc.* **2009**, *131*, 16364–16365.

28. Du, Y.P.; Zhang, Y.W.; Sun, L.D.; Yan, C.H. Luminescent monodisperse nanocrystals of lanthanide oxyfluorides synthesized from trifluoroacetate precursors in high-boiling solvents. *J. Phys. Chem. C* **2008**, *112*, 405–415.

29. Du, Y.P.; Zhang, Y.W.; Sun, L.D.; Yan, C.H. Atomically efficient synthesis of self-assembled monodisperse and ultrathin lanthanide oxychloride nanoplates. *J. Am. Chem. Soc.* **2009**, *131*, 3162–3163.

30. Stouwdam, J.W.; van Veggel, F.C.J.M. Near-infrared emission of redispersible Er^{3+}, Nd^{3+}, and Ho^{3+} doped LaF_3 nanoparticles. *Nano Lett.* **2002**, *2*, 733–737.

31. Yi, G.S.; Chow, G.M. Colloidal LaF_3:Yb,Er, LaF_3:Yb,Ho and LaF_3:Yb,Tm nanocrystals with multicolor upconversion fluorescence. *J. Mater. Chem.* **2005**, *15*, 4460–4464.

32. Heer, S.; Kompe, K.; Gudel, H.U.; Haase, M. Highly efficient multicolour upconversion emission in transparent colloids of lanthanide-doped $NaYF_4$ nanocrystals. *Adv. Mater.* **2004**, *16*, 2102–2105.

33. Heer, S.; Lehmann, O.; Haase, M.; Gudel, H.U. Blue, green, and red upconversion emission from lanthanide-doped $LuPO_4$ and $YbPO_4$ nanocrystals in a transparent colloidal solution. *Angew. Chem. Int. Ed.* **2003**, *42*, 3179–3182.

34. Zeng, J.H.; Su, J.; Li, Z.H.; Yan, R.X.; Li, Y.D. Synthesis and upconversion luminescence of hexagonal-phase $NaYF_4$:Yb^{3+},Er^{3+} phosphors of controlled size and morphology. *Adv. Mater.***2005**, *17*, 2119–2123.

35. Teng, X.; Zhu, Y.H.; Wei, W.; Wang, S.C.; Huang, J.F.; Naccache, R.; Hu, W.B.; Han, Y.; Zhang, Q.C.; Fan, Q.L.; *et al.* Lanthanide-doped Na_xScF_{3+x} nanocrystals: Crystal structure evolution and multicolor tuning. *J. Am. Chem. Soc.* **2012**, *134*, 8340–8343.

36. Liang, L.F.; Xu, H.F.; Su, Q.; Konishi, H.; Jiang, Y.B.; Wu, M.M.; Wang, Y.F.; Xia, D.Y. Hydrothermal synthesis of prismatic $NaHoF_4$ microtubes and $NaSmF_4$ nanotubes. *Inorg. Chem.***2004**, *43*, 1594–1596.

37. Zhang, M.F.; Fan, H.; Xi, B.J.; Wang, X.Y.; Dong, C.; Qian, Y.T. Synthesis, characterization, and luminescence properties of uniform Ln^{3+}-doped YF_3 nanospindles. *J. Phys. Chem. C* **2007**, *111*, 6652–6657.

38. Liang, X.; Wang, X.; Zhuang, J.; Peng, Q.; Li, Y.D. Branched $NaYF_4$ nanocrystals with luminescent properties. *Inorg. Chem.* **2007**, *46*, 6050–6055.

39. Zhao, J.W.; Suan, Y.J.; Kong, X.G.; Tian, L.J.; Wang, Y.; Tu, L.P.; Zhao, J.L.; Zhang, H. Controlled synthesis, formation mechanism, and great enhancement of red upconversion luminescence of $NaYF_4:Yb^{3+},Er^{3+}$ nanocrystals/submicroplates at low doping level. *J. Phys. Chem. B* **2008**, *112*, 15666–15672.

40. Wang, X.; Zhuang, J.; Peng, Q.; Li, Y.D. Hydrothermal synthesis of rare-earth fluoride nanocrystals. *Inorg. Chem.* **2006**, *45*, 6661–6665.

41. Wang, X.; Li, Y.D. A general strategy for nanocrystal synthesis. *Nature* **2005**, *437*, 121–124.

42. Huo, Z.Y.; Chen, C.; Li, Y.D. Systematic synthesis of lanthanide phosphate nanocrystals. *Chem. Eur. J.* **2007**, *13*, 7708–7714.

43. Wang, L.Y.; Li, Y.D. Systematic synthesis of lanthanide phosphate nanocrystals. $Na(Y_{1.5}Na_{0.5})F_6$single-crystal nanorods as multicolor luminescent materials. *Nano Lett.* **2006**, *6*, 1645–1649.

44. Wang, L.Y.; Li, Y.D. Controlled synthesis and luminescence of lanthanide doped $NaYF_4$nanocrystals. *Chem. Mater.* **2007**, *19*, 727–734.

45. Wang, G.F.; Peng, Q.; Li, Y.D. Luminescence tuning of upconversion nanocrystals. *Chem. Eur. J.***2010**, *16*, 4923–4931.

46. Wang, L.Y.; Li, P.; Li, Y.D. Down and upconversion luminescent nanorods. *Adv. Mater.* **2007**, *19*, 3304–3307.

47. Wang, X.; Li, Y.D. Synthesis and characterization of lanthanide hydroxide single-crystal nanowires. *Angew. Chem. Int. Ed.* **2002**, *41*, 4790–4793.

48. Wang, X.; Sun, X.M.; Yu, D.P.; Zou, B.S.; Li, Y.D. Rare earth compound nanotubes. *Adv. Mater.***2003**, *15*, 1442–1445.

49. Wang, X.; Li, Y.D. Fullerene-like rare-earth nanoparticles. *Angew. Chem. Int. Ed.* **2003**, *42*, 3497–3500.

50. Hu, C.G.; Liu, H.; Wang, Z.L. $La(OH)_3$ and La_2O_3 nanobelts-synthesis and physical properties.*Adv. Mater.* **2007**, *19*, 470–474.

51. Zhang, F.; Wan, Y.; Yu, T.; Zhang, F.; Shi, Y.; Xie, S.; Li, Y.; Xu, L.; Tu, B.; Zhao, D. Uniform nanostructured arrays of sodium rare-earth

fluorides for highly efficient multicolor upconversion luminescence. *Angew. Chem. Int. Ed.* **2007**, *46*, 7976–7979.

52. Liu, C.; Chen, D. Controlled synthesis of hexagon shaped lanthanide-doped LaF_3 nanoplates with multicolor upconversion fluorescence. *J. Mater. Chem.* **2007**, *17*, 3875–3880.

53. Pan, S.H.; Deng, R.P.; Feng, J.; Wang, S.Y.; Wang, S.; Zhu, M.; Zhang, H.J. Microwave-assisted synthesis and down and upconversion luminescent properties of $BaYF_5$:Ln (Ln = Yb/Er, Ce/Tb) nanocrystals. *CrystEngComm* **2013**, *15*, 7640–7643.

54. Yang, M.; You, H.P.; Zheng, Y.H.; Liu, K.; Jia, G.; Song, Y.H.; Huang, Y.J.; Zhang, L.H.; Zhang, H. Hydrothermal synthesis and luminescent properties of novel ordered sphere $CePO_4$ hierarchical architectures. *J. Inorg. Chem.* **2009**, *48*, 11559–11565.

55. Yang, M.; You, H.P.; Huang, Y.H.; Zhang, H.J. Synthesis and luminescent properties of orderly $YPO_4:Eu^{3+}$ olivary architectures self-assembled by nanoflakes. *CrystEngComm* **2010**, *12*, 4141–4145.

56. Zheng, Y.H.; You, H.P.; Jia, K.; Liu, K.; Song, Y.H.; Yang, M.; Zhang, H.J. Highly uniform Gd_2O_3hollow microspheres: Template-directed synthesis and luminescence properties. *Langmuir* **2010**,*26*, 5122–5128.

57. Jia, G.; You, H.P.; Liu, K.; Zheng, Y.H.; Guo, N.; Zhang, H.J. Facile hydrothermal synthesis and luminescent properties of large-scale $GdVO_4:Eu^{3+}$ nanowires. *Cryst. Growth Des.* **2009**, *9*, 5101–5107.

58. Li, C.X.; Hou, Z.Y.; Zhang, C.M.; Yang, P.P.; Li, G.G.; Xu, Z.H.; Fan, Y.; Lin, J. Controlled synthesis of Ln^{3+} (Ln = Tb, Eu, Dy) and V^{5+} ion-doped YPO_4 nano-/microstructures with tunable luminescent colors. *Chem. Mater.* **2009**, *21*, 4598–4607.

59. Li, C.X.; Yang, J.; Yang, P.P.; Lian, H.Z.; Lin, J. Hydrothermal synthesis of lanthanide fluorides LnF_3 (Ln = La to Lu) nano/microcrystals with multiform structures and morphologies. *Chem. Mater.* **2008**, *20*, 4317–4326.

60. Li, C.X.; Yang, J.; Quan, Z.W.; Yang, P.P.; Kong, D.Y.; Lin, J. Different microstructures of ss-$NaYF_4$fabricated by hydrothermal process: Effects of pH values and fluoride sources. *Chem. Mater.* **2007**,*19*, 4933–4942.

61. Wang, J.; Wang, F.; Wang, C.; Liu, Z.; Liu, X.G. Single-band upconversion emission in lanthanide-doped $KMnF_3$ nanocrystals. *Angew. Chem. Int. Ed.* **2011**, *50*, 10369–10372.

62. Tian, G.; Gu, Z.J.; Zhou, L.J.; Yin, W.Y.; Liu, X.X.; Yan, L.; Jin, S.; Ren, W.L.; G, M.; Li, S.G.; *et al.* Mn^{2+} dopant-controlled synthesis of $NaYF_4$:Yb/Er upconversion nanoparticles for *in vivo* imaging and drug delivery. *Adv. Mater.* **2012**, *24*, 1226–1231.

63. Zeng, S.J.; Yi, Z.G.; Lu, W.; Qian, C.; Wang, H.B.; Rao, L.; Zeng, T.M.; Liu, H.R.; Liu, H.G.; Fei, B.;*et al.* Simultaneous realization of phase/size manipulation, upconversion luminescence enhancement, and blood vessel imaging in multifunctional nanoprobes through transition metal Mn^{2+} doping. *Adv. Funct. Mater.* **2014**, *24*, 4051–4059.

64. Lin, J.; Yu, M.; Lin, C.K.; Liu, X.M. Multiform oxide optical materials via the versatile Pechini-type sol-gel process: Synthesis and characteristics. *J. Phys. Chem. C* **2007**, *111*, 5835–5845.

65. Patra, A.; Friend, C.S.; Kapoor, R.; Prasad, P.N. Upconversion in Er^{3+}:ZrO_2 nanocrystals. *J. Phys. Chem. B* **2002**, *106*, 1909–1912.

66. Cheng, Z.Y.; Xing, R.B.; Hou, Z.Y.; Huang, S.S.; Lin, J. Patterning of light-emitting YVO_4:Eu^{3+} thin films via inkjet printing. *J. Phys. Chem. C* **2010**, *114*, 9883–9888.

67. Zhu, Y.S.; Xu, W.; Zhang, H.Z.; Wang, W.; Xu, S.; Song, H.W. Inhibited long-scale energy transfer in dysprosium doped yttrium vanadate inverse opal. *J. Phys. Chem. C* **2012**, *116*, 2297–2302.

68. Wang, W.; Song, H.W.; Liu, Q.; Bai, X.; Wang, Y.; Dong, B. Modified optical properties in a samarium doped titania inverse opal. *Opt. Lett.* **2010**, *35*, 1449–1451.

69. Liu, Q.; Song, H.W.; Wang, W.; Bai, X.; Wang, Y.; Dong, B.; Xu, L.; Han, W. Observation of lamb shift and modified spontaneous emission dynamics in the YBO_3:Eu^{3+} inverse opal. *Opt. Lett.* **2010**,*35*, 2898–2900.

70. Zhu, Y.S.; Sun, Z.P.; Yin, Z.; Song, H.W.; Xu, W.; Wang, Y.F.; Zhang, L.G.; Zhang, H.Z. Self-assembly, highly modified spontaneous emission and energy transfer properties of $LaPO_4$:Ce^{3+},Tb^{3+} inverse opals. *Dalton Trans.* **2013**, *42*, 8049–8057.

71. Patra, A.; Friend, C.S.; Kapoor, R.; Prasad, P.N. Fluorescence upconversion properties of Er^{3+}-doped TiO_2 and $BaTiO_3$ nanocrystallites. *Chem. Mater.* **2003**, *15*, 3650–3655.

72. Venkatramu, V.; Falcomer, D.; Speghini, A.; Bettinelli, M.; Jayasankar, C.K. Synthesis and luminescence properties of Er^{3+}-doped $Lu_3Ga_5O_{12}$ nanocrystals. *J. Lumin.* **2008**, *128*, 811–813.

73. Yang, K.S.; Zheng, F.; Wu, R.; Li, H.; Zhang, X. Upconversion luminescent properties of YVO_4:Yb^{3+},Er^{3+} nano-powder by sol-gel method. *J. Rare Earth* **2006**, *24*, 162–166.

74. Lemyre, J.L.; Ritcey, A.M. Synthesis of lanthanide fluoride nanoparticles of varying shape and size. *Chem. Mater.* **2005**, *17*, 3040–3043.

75. Wang, G.F.; Qin, W.P.; Zhang, J.S. Synthesis, growth mechanism, and tunable upconversion luminescence of Yb^{3+}/Tm^{3+}-codoped YF_3 nanobundles. *J. Phys. Chem C* **2008**, *112*, 12161–12167.

76. Wang, G.F.; Qin, W.P.; Wei, G.D. Synthesis and upconversion luminescence properties of Yb^{3+}/Tm^{3+}-codoped $BaSiF_6$ nanorods. *J. Fluorine Chem.* **2009**, *130*, 158–161.

77. Vetrone, F.; Boyer, J.C.; Capobianco, J.A.; Speghini, A.; Bettinelli, M. Significance of Yb^{3+}concentration on the upconversion mechanisms in codoped Y_2O_3:Er^{3+},Yb^{3+} nanocrystals. *J. Appl. Phys.* **2004**, *96*, 661–667.

78. Pandozzi, F.; Vetrone, F.; Boyer, J.C.; Naccache, R.; Capobianco, J.A.; Speghini, A.; Bettinellin, M. A spectroscopic analysis of blue and ultraviolet upconverted emissions from $Gd_3Ga_5O_{12}$:Tm^{3+},Yb^{3+} nanocrystals. *J. Phys. Chem. B* **2005**, *109*, 17400–17405.

79. Luo, X.X.; Cao, W.H. Ethanol-assistant solution combustion method to prepare La_2O_3:Yb,Pr nanometer phosphor. *J. Alloys. Compds.* **2008**, *46*, 529–534.

80. Xu, L.L.; Yu, Y.N.; Li, X.G. Synthesis and upconversion properties of monoclinic Gd_2O_3:Er^{3+}nanocrystals. *Opt. Mater.* **2008**, *30*, 1284–1288.

81. Qin, X.; Yokomori, T.; Ju, Y.G. Flame synthesis and characterization of rare-earth (Er^{3+}, Ho^{3+}, and Tm^{3+}) doped upconversion nanocryphosphors. *Appl. Phys. Lett.* **2007**, *90*, 073104.

82. Dong, B.; Song, H.W.; Yu, H.Q.; Zhang, H.; Qin, R.F.; Bai, X.; Pan, G.H.; Lu, S.Z.; Wang, F.; Fan, L.B.; *et al*. Upconversion properties of Ln^{3+} doped $NaYF_4$/polymer composite fibers prepared by electrospinning. *J. Phys. Chem. C* **2008**, *112*, 1435–1440.

83. Ma, L.; Chen, W.X.; Zheng, Y.Z.; Zhao, J.; Xu, Z.D. Microwave-assisted hydrothermal synthesis and characterizations of PrF_3 hollow nanoparticles. *Mater. Lett.* **2007**, *61*, 2765–2768.

84. Chan, E.M. Combinatorial approaches for developing upconverting nanomaterials: High-throughput screening, modeling, and applications. *Chem. Soc. Rev.* 2014.

85. Liu, H.; Jakobsson, O.; Xu, C.T.; Xie, H.; Laurell, T.; Andersson-Engels, S. Synthesis of $NaYF_4$:Yb^{3+}/Er^{3+} Upconverting Nanoparticles in a Capillary-Based Continuous Microfluidic Reaction System. In *Colloidal Quantum Dots/Nanocrystals for Biomedical Applications*; SPIE: Bellingham, WA, USA, 2011; Volume 7909, p. 790917.

86. Xu, C.T.; Zhan, Q.; Liu, H.; Somesfalean, G.; Qian, J.; He, S.; Andersson-Engels, S. Upconverting nanoparticles for pre-clinical diffuse optical imaging, microscopy and sensing: Current trends and future challenges. *Laser Photon. Rev.* **2013**, *7*, 663–697.

87. Zhu, X.; Zhang, Q.; Li, Y.; Wang, H. Redispersible and water-soluble LaF_3:Ce,Tb nanocrystalsvia a microfluidic reactor with temperature steps. *J. Mater. Chem.* **2008**, *18*, 5060–5062.

88. Zhu, X.; Zhang, Q.; Li, Y.; Wang, H.J. Facile crystallization control of LaF_3/$LaPO_4$:Ce,Tb nanocrystals in a microfluidic reactor using microwave irradiation. *J. Mater. Chem.* **2010**, *20*, 1766–1771.

89. Yi, G.S.; Chow, G.M. Water-soluble $NaYF_4$:Yb,Er(Tm)/$NaYF_4$/polymer core/shell/shell nanoparticles with significant enhancement of upconversion fluorescence. *Chem. Mater.* **2007**, *19*, 341–343.

90. Liu, Y.S.; Tu, D.T.; Zhu, H.M.; Li, R.F.; Luo, W.Q.; Chen, X.Y. A strategy to achieve efficient dual-mode luminescence of Eu^{3+} in lanthanides doped multifunctional $NaGdF_4$ nanocrystals. *Adv. Mater.* **2010**, *22*, 3266–3271

91. Vetrone, F.; Naccache, R.; Mahalingam, V.; Morgan, C.G.; Capobianco, J.A. The active-core/active-shell approach: A strategy to enhance the upconversion luminescence in lanthanide-doped nanoparticles. *Adv. Funct. Mater.* **2009**, *19*, 2924–2929.

92. Liu, X.M.; Kong, X.G.; Zhang, Y.L.; Tu, L.P.; Wang, Y.; Zeng, Q.H.; Li, C.G.; Shi, Z.; Zhang, H. Breakthrough in concentration quenching threshold of upconversion luminescence via spatial separation of the emitter doping area for bio-applications. *Chem. Commun.* **2011**, *47*, 11957–11959.

93. Wang, Z.L.; Quan, Z.W.; Jia, P.Y.; Lin, C.K.; Luo, Y.; Chen, Y.; Fang, J.; Zhou, W.; O'Connor, C.J.; Lin, J. A facile synthesis and photoluminescent properties of redispersible CeF_3,CeF_3:Tb^{3+}, and CeF_3:Tb^{3+}/LaF_3 (core/shell) nanoparticles. *Chem. Mater.* **2006**, *18*, 2030–2037.

94. Mai, H.X.; Zhang, Y.W.; Yan, C.H. Highly efficient multicolor up-conversion emissions and their mechanisms of monodisperse $NaYF_4$:Yb,Er core and core/shell-structured nanocrystals. *J. Phys. Chem. C* **2007**, *111*, 13721–13729.

95. Boyer, J.C.; Gagnon, J.; Capobianco, J.A. Synthesis, characterization, and spectroscopy of $NaGdF_4$:Ce^{3+},Tb^{3+}/$NaYF_4$ core/shell nanoparticles. *Chem. Mater.* **2007**, *19*, 3358–3360.

96. Helmut, S.; Pavel, P.; Otmanef, Z.; Markus, H. Synthesis and optical properties of KYF_4/Yb,Er nanocrystals, and their surface modification with undoped KYF_4. *Adv. Funct. Mater.* **2008**, *18*, 2913–2918.

97. Xie, X.J.; Gao, N.Y.; Deng, R.R.; Sun, Q.; Xu, Q.H.; Liu, X.G. Mechanistic investigation of photon upconversion in Nd^{3+}-sensitized core-shell nanoparticles. *J. Am. Chem. Soc.* **2013**, *135*, 12608–12611.

98. Huang, P.; Zheng, W.; Zhou, S.Y.; Tu, D.T.; Chen, Z.; Zhu, H.M.; Li, R.F.; Ma, E.; Huang, M.D.; Chen, X.Y. Lanthanide-doped $LiLuF_4$ upconversion nanoprobes for the detection of disease biomarkers. *Angew. Chem. Int. Ed.* **2014**, *53*, 1252–1257.

99. Wang, F.; Deng, R.R.; Liu, X.G. Preparation of core-shell $NaGdF_4$ nanoparticles doped with luminescent lanthanide ions to be used as upconversion-based probes. *Nat. Protoc.* **2014**, *9*, 1634–1644.

100. Wang, Y.F.; Sun, L.D.; Xia, J.W.; Feng, W.; Zhou, J.C.; Shen, J.; Yan, C.H. Rare-earth nanoparticles with enhanced upconversion emission and suppressed rare-earth-ion leakage. *Chem. Eur. J.* **2012**,*18*, 5558–5564.

101. Darbandi, M.; Nann, T. One-pot synthesis of YF_3@silica core/shell nanoparticles. *Chem. Commun.***2006**, *7*, 776–778.

102. Li, Z.Q.; Zhang, Y. Monodisperse silica-coated polyvinylpyrrolidone/$NaYF_4$ nanocrystals with multicolor upconversion fluorescence emission. *Angew. Chem. Int. Ed.* **2006**, *45*, 7732–7735.

103. Li, Z.Q.; Zhang, Y.; Jiang, S. Multicolor core/shell-structured upconvers fluorescent nanoparticles.*Adv. Mater.* **2008**, *20*, 4765–4769.

104. Chen, B.T.; Dong, B.; Wang, J.; Zhang, S.; Xu, L.; Yu, W.; Song, H.W. Amphiphilic silane modified $NaYF_4$:Yb,Er loaded with $Eu(TTA)_3(TPPO)_2$ nanoparticles and their multi-functions: Dual mode temperature sensing and cell imaging. *Nanoscale* **2013**, *5*, 8541–8549.

105. Sivakumar, S.; Diamenteand, P.R.; van Veggel, F.C.J.M. Silica-coated Ln^{3+}-doped LaF_3nanoparticles as robust down and upconverting biolabels. *Chem. Eur. J.* **2006**, *12*, 5878–5884.

106. Liu, Z.Y.; Yi, G.S.; Zhang, H.T.; Ding, J.; Zhang, Y.W.; Xue, J.M. Monodisperse silica nanoparticles encapsulating upconversion fluorescent and superparamagnetic nanocrystals. *Chem. Commun.***2008**, *6*, 694–696.

107. Wang, M.; Mi, C.C.; Zhang, Y.Z.; Liu, J.L.; Li, F.; Mao, C.B.; Xu, S.K. NIR-responsive silica-coated $NaYbF_4$:Er/Tm/Ho upconversion fluorescent nanoparticles with tunable emission colors and their applications in immunolabeling and fluorescent imaging of cancer cells. *J. Phys. Chem. C* **2009**,*113*, 19021–19027.

108. Yu, X.F.; Chen, L.D.; Li, M.; Xie, M.Y.; Zhou, L.; Li, Y.; Wang, Q.Q. Highly efficient fluorescence of NdF_3/SiO_2 core/shell nanoparticles and the applications for *in vivo* NIR detection. *Adv. Mater.***2008**, *20*, 4118–4123.

109. Li, X.H.; Wu, Y.Q.; Liu, Y.; Zou, X.M.; Yao, L.M.; Lia, F.Y.; Feng, W. Cyclometallated ruthenium complex-modified upconversion nanophosphors for selective detection of Hg^{2+} ions in water.*Nanoscale* **2014**, *6*, 1020–1028.

110. Wang, C.; Cheng, L.; Liu, Z. Drug delivery with upconversion nanoparticles for multi-functional targeted cancer cell imaging and therapy. *Biomaterials* **2011**, *32*, 1110–1120.

111. Budijono, S.J.; Shan, J.N.; Yao, N.; Miura, Y.; Hoye, T.; Austin, R.H.; Ju, Y.G.; Prud'homme, R.K. Synthesis of stable block-copolymer-protected $NaYF_4:Yb^{3+},Er^{3+}$ up-converting phosphor nanoparticles. *Chem. Mater.* **2010**, *22*, 311–318.

112. Chen, Z.G.; Chen, H.L.; Hu, H.; Yu, M.X.; Li, F.Y.; Zhang, Q.; Zhou, Z.G.; Yi, T.; Huang, C.H. Versatile synthesis strategy for carboxylic acid-functionalized upconverting nanophosphors as biological labels. *J. Am. Chem. Soc.* **2008**, *130*, 3023–3029.

113. Bogdan, N.; Vetrone, F.; Ozin, G.A.; Capobianco, J.A. Synthesis of ligand-free colloidally stable water dispersible brightly luminescent lanthanide-doped upconverting nanoparticles. *Nano Lett.***2011**, *11*, 835–840.

114. Yi, G.S.; Chow, G.M. Synthesis of hexagonal-phase $NaYF_4$:Yb,Er and $NaYF_4$:Yb,Tm nanocrystals with efficient up-conversion fluorescence. *Adv. Funct. Mater.* **2006**, *16*, 2324–2329.

115. Chen, G.Y.; Ohulchanskyy, T.Y.; Law, W.C.; Agren, H.; Prasad, P.N. Monodisperse $NaYbF_4$: $Tm^{3+}/NaGdF_4$ core/shell nanocrystals with near-infrared to near-infrared upconversion photoluminescence and magnetic resonance properties. *Nanoscale* **2011**, *3*, 2003–2008.

116. Boyer, J.C.; Manseau, M.P.; Murray, J.I.; van Veggel, F.C.J.M. Surface modification of upconverting $NaYF_4$ nanoparticles with PEG-phosphate ligands for NIR (800 nm) biolabeling within the biological window. *Langmuir* **2010**, *26*, 1157–1164.

117. Nyk, M.; Kumar, R.; Ohulchanskyy, T.Y.; Bergey, E.J.; Prasad, P.N. High contrast *in vitro* and *in vivo* photoluminescence bioimaging using near Infrared to near infrared up-conversion in $Tm^{3+}$and Yb^{3+} doped fluoride nanophosphors. *Nano Lett.* **2008**, *8*, 3834–3838.

118. Cao, T.Y.; Yang, T.S.; Gao, Y.; Yang, Y.; Hu, H.; Li, F.Y. Water-soluble $NaYF_4$:Yb/Er upconversion nanophosphors: Synthesis, characteristics and application in bioimaging. *Inorg. Chem. Commun.***2010**, *13*, 392–394.

119. Zhou, J.; Yu, M.X.; Sun, Y.; Zhang, X.Z.; Zhu, X.J.; Wu, Z.H.; Wu, D.M.; Li, F.Y. Fluorine-18-labeled $Gd^{3+}/Yb^{3+}/Er^{3+}$ co-doped $NaYF_4$ nanophosphors formultimodality PET/MR/UCL imaging.*Biomaterials* **2011**, *32*, 1148–1156.

120. Naczynski, D.J.; Andelman, T.; Pal, D.; Chen, S.; Richard, E.; Riman, R.E.; Roth, C.M.; Moghe, P.V. Albumin nanoshell encapsulation of near-infrared-excitable rare-earth nanoparticles enhances biocompatibility and enables targeted cell imaging. *Small* **2010**, *6*, 1631–1640.

Chapter 3

Fabrication of Upconverting Hybrid Nanoparticles for Near-Infrared Light Triggered Drug Release

Ranran Zhang,[1] Risheng Yao,[1] Binbin Ding,[1] Yuxin Shen,[1] Shengwen Shui,[1] Lei Wang,[1] Yu Li,[2] Xianzhu Yang,[1] and Wei Tao[1]

[1]School of Medical Engineering, Hefei University of Engineering, Hefei, Anhui, China
[2]School of Material Science and Engineering, Wuhan University of Technology, Wuhan, Hubei, China

ABSTRACT

Low tissue penetration and harmful effects of (ultraviolet) UV or visible light on normal tissue limit exploiting nanocarriers for the application of light-controlled drug release. Two strategies may solve the problem: one is to improve the sensitivity of the nanocarriers to light to decrease the radiation time; the other one is using more friendly light as the trigger. In this work, we fabricated a core-shell hybrid nanoparticle with an upconverting nanoparticle (UCNP) as the core and thermo- and light-responsive block copolymers as the shell to combine the two strategies together. The results indicated that the sensitivity of the block copolymer to light could be enhanced by decreasing the photolabile moieties in the polymer, and the UCNP could transfer near-infrared (NIR) light, which is more friendly to tissue and cell, to UV light to trigger the phase conversion of the block polymers in situ. Using Nile Red (NR) as the model drug, the hybrid nanoparticles were further proved to be able to act as carriers with the character of NIR triggered drug release.

1. INTRODUCTION

Stimuli responsive nanoparticles are attracting more and more attentions because of their potential for controlled drug delivery applications [1–5]. Among the various stimuli, light is a typical external stimulus and has many superior characters. For example, light can be easily controlled and regulated by external apparatuses, which will offer the temporal and spatial selectivity [6–10]; light can induce various photochemical reactions, which makes it easy to exploit diversiform photosensitive materials [11–14]. However, the tissue penetration of light is low, and providing enough photons to the stimuli responsive nanoparticles in deep tissue is difficult. Furthermore, most of photochemical reactions are induced by UV or visible light, whose harmful effects on normal tissue and health cells limit their applications in light-controlled drug release.

Comparing to UV and visible light, NIR light is more ideal because of its deeper penetration into tissue and less detrimental to healthy cells [15–17]. However, as mentioned above, most of photochemical reactions require high-energy UV or visible light. Recently, UCNPs have emerged as a powerful tool to convert NIR light to higher-energy light in the UV, visible, or NIR regions [18–20]. When UCNPs are incorporated in UV sensitive materials, NIR light may be able to induce the photoreaction of the UV sensitive materials in deeper tissues and decrease the damage in the radiation path from radiation source to the target tissue [21–25].

Another way to overcome the drawbacks of light is to increase the sensitivity of material to light, because less requirement for photons means less radiation time of light or more efficient photochemical reaction in deeper regions. Internal stimuli are biologic properties of disease areas, such as temperature [26, 27], pH [28, 29], and redox potential [30, 31], which generally are stable in the body. Therefore, when internal stimuli sensitive materials combine with a small amount of photosensitive moieties, the effect of photoreaction may be amplified by the internal stimuli [14]. For example, poly(N-isopropylacrylamide) (PNIPAM) is a typical thermosensitive polymer, which is soluble in water below its lower critical temperature (LCST) at about 32°C but insoluble above it. When NIPAMs are copolymerized with hydrophobic or hydrophilic monomers, the LCST of the copolymers will accordingly decrease or increase [32]. Based on the property of PNIPAM, Ionov and Diez synthesized photoresist material by copolymerizing NIPAMs and nitrobenzyl acrylates (NBAs) [33]. They found that only 5 mol% of NBA would decrease the LCST of the copolymer below room temperature (~10°C) because of the hydrophobic nature of the NBA. Upon irradiation of UV light at 365 nm, the NBA moieties left and the residual hydrophilic carboxyl groups increased the LCST of the copolymer above 40°C. This character of PNIPAM also was used for photocontrolled drug release. Recently, Li et al. prepared thermo- and light-responsive micelles as drug carriers by self-assembling PEO-b-P(NIPAM-co-NBA) in water [34]. Upon the irradiation of UV light, the micelles were disintegrated because of the LCST change of the P(NIPAM-co-NBA) block and realized UV-triggered drug release.

To combine the advantages of the less detriment of NIR light to body and enhanced sensitivity of thermo- and light-responsive polymers to light, we fabricated upconverting hybrid nanoparticles by self-assembling PEO-b-P(NIPAM-co-NBA) with the presence of UCNPs in water. The results indicated that the hybrid nanoparticles had a core-shell structure with UCNP as the core and PEO-b-P(NIPAM-co-NBA) as the shell. Furthermore, we used Nile Reds as the model drugs to investigate their light induced release. The result indicated that, as shown in Scheme 1, when the upconverting hybrid nanoparticles were irradiated by NIR light, the UCNPs converted NIR to UV light to induce the photocleavage reaction of NBA moieties in situ; then, as a result of the LCST change of the block copolymers, nanoparticles disintegrated and released the loaded NR molecules.

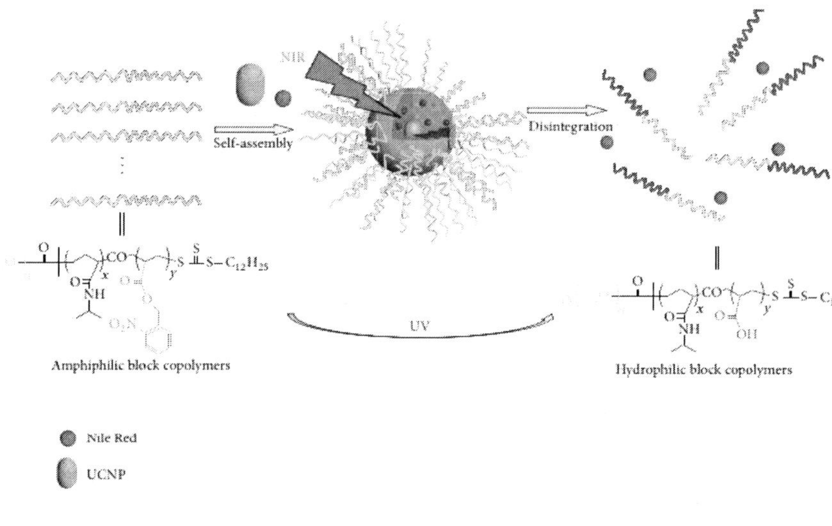

Scheme 1: Schematic illustration for fabrication of the core-shell hybrid nanoparticle and NIR triggered release of the loaded model drugs (Nile Red).

2. MATERIALS AND METHODS

2.1. Materials

Poly(ethylene oxide) monomethyl ether (PEO, Mn: 1000, M_w/M_n: 1.05) was purchased from Alfa Aesar and used as received. N-Isopropylacrylamide

(NIPAM, TCI) was recrystallized twice from hexanes. Azobisisobutyronitrile (AIBN, Sinopharm Chemical Reagent, 95%) was recrystallized from 95% ethanol. Triethylamine (TEA, Aldrich) and dichloromethane (DCM, Sinopharm Chemical Reagent) were dried over CaH_2. Acryloyl chloride, 2-nitrobenzyl alcohol, N,-dicyclohexylcarbodiimide (DCC, 99%), and N,-dimethylaminopyridine (DMAP, 98%) were purchased from Aldrich. Acetone, carbon disulfide, 1-dodecanethiol, isopropanol, tetrabutyl ammonium bromide, chloroform, hexane, and diethyl ether were purchased from Sinopharm Chemical Reagent (AR) and used without further purification. 2-Nitrobenzyl acrylate (NBA) [33], S-1-dodecyl-(,-dimethyl--acetic acid) trithiocarbonate (TTCA) [35], PEO-TTCA, and -$NaYF_4:Tm^{3+}$ 0.5 mol%, Yb^{3+} 30 mol% upconverting nanoparticles (UCNPs) [18] were synthesized according to previously reported literature procedures. Water was deionized with a Millipore system to a resistivity of 18.2 MΩ·cm. All reagents were used as received unless otherwise stated.

2.2. Synthesis of the Amphiphilic PEO_{1000}-b-P(NIPAM-co-NBA) Block Copolymers

In this work, two PEO_{1000}-b-P(NIPAM-co-NBA) block copolymers with different NBA contents were synthesized in 1,4-dioxane with the PEO_{1000}-TTCA as the macroRAFT agent and AIBN as the initiator. The [PEO_{1000}-TTCA] : [NIPAM + NBA] : [AIBN] ratio was maintained at 1 : 180 : 0.2 (mole basis) while the [NIPAM] : [NBA] ratio was 11 : 1 and 20 : 1, respectively. The polymerization procedure was as follows. NIPAM, NBA, AIBN, PEO_{1000}-TTCA, and 1,4-dioxane were charged into a round-bottom flask equipped with a magnetic stirring bar. After being sealed with a rubber septum and then deoxygenated with a stream of bubbled nitrogen for 30 min, the flask was immersed in an oil bath at 70°C and stirred for 24 h. Then, the mixture was cooled to room temperature and precipitated into an excess of diethyl ether. The precipitate was collected and dried under vacuum for 24 h at room temperature to afford the product.

2.3. Preparation of Polymer Micelles and Encapsulation of Nile Red

The procedure for preparation of polymer micelles was as follows. 10 mg of the block copolymer (BCP) was dissolved in 1 mL of THF, which is a good solvent for the two blocks. Under vigorous stirring, deionized water was then added slowly until the solution exhibited a bluish tinge, which indicated the formation of the micelles. After 2 h of stirring, a 4-fold volume of water was added for quenching of micelles. THF and most of water were then removed by evaporation under vacuum at 30°C until the solution regained the initial concentration. BCP micelles with encapsulated NR were prepared by the procedure analogue to the one for BCP micelles. In brief, the BCP and NR were first dissolved in THF at a concentration of 0.8 mg/mL and 0.125 mg/mL. Water was then added to induce the formation of micelles and simultaneous encapsulation of NR by the hydrophobic core of the micelles. After 2 h of stirring, a 4-fold volume of water was added for quenching of micelles and precipitation of unloaded NR. Precipitated NR was removed by filtration through 0.45 μm membrane, and THF and most of water were evaporated by evaporation under vacuum at 30°C until the solution regained the initial concentration.

2.4. Preparation of the NIR Responsive Hybrid Nanoparticles

The NIR responsive hybrid nanoparticles were prepared as follows. BCP (8 mg) and UCNPs (8 mg) were dissolved in THF (5 mL). Under vigorous stirring at room temperature, 40 mL of deionized water was then added dropwise to the solution over 2 h. The obtained mixture was centrifuged at 4800 rpm for 10 min. The supernatant was discarded and the precipitate was redispersed in deionized water (10 mL) under mild stirring to afford the solution of the hybrid nanoparticles. Nile Red loaded hybrid nanoparticles were prepared by the analogue method mentioned above. In brief, the THF solution of Nile Red (10 mg/mL, 25 μL) was added into 5 mL of THF solution containing 8 mg of BCP and 8 mg of UCNPs. Then, 40 mL of deionized water was added dropwise over 2 h. After the solution was centrifuged at 4800 rpm for 10 min, the supernatant was discarded. Finally, the precipitate was redispersed in deionized water (10 mL) under mild stirring to afford the solution of the NR loaded hybrid nanoparticles.

2.5. Determination of Critical Micellization Concentration (CMC)

The fluorescence probe method was employed to determine the CMC of the micelles as follows [36]. A predetermined amount of pyrene stock solution in acetone was added into an empty vial, and then the acetone was evaporated completely. A calculated amount of copolymer solutions at varying concentrations was added to the vial, and the final concentration of pyrene in each flask was fixed at 6.0×10^{-7} mol L^{-1}. The excitation spectra were recorded at 25°C on a spectrofluorophotometer with at 390 nm and a slit width of 5 nm.

3. CHARACTERIZATION

^1H NMR spectra were characterized in $CDCl_3$ by VNMRS600 spectrometer (600 MHz). The molecular weight and molecular weight distribution of the polymer were determined by gel permeation chromatography (GPC), THF was used as eluent (elution rate, 1.0 mL min^{-1}), and polystyrene standards were employed for calibration. Dynamic light scattering (DLS) measurements were carried out using a Zetasizer (Nano-ZS 90, Malvern Instruments), and these samples were averaged over three measurements. Transmission electron microscopy (TEM) observations were conducted on a JEM-2100F electron microscope. The sample was prepared on copper grids coated with thin films of formvar and carbon successively. Fourier transform infrared (FT-IR) spectra were recorded on a Nicolet 67 spectrometer. UV-vis spectra were acquired on Hitachi U-5100 spectrophotometer. Fluorescence emission spectra were recorded using Hitachi F-2700 spectrofluorometer. The slit widths were both kept at 5 nm for excitation and emission, and scanning speed was set at 300 nm/min. UV-vis spot curing system (Futansi) was used to study the photoinduced dissociation of polymer micelles and the release of encapsulated NR. The micelle solutions were exposed to UV light under stirring. All samples were filtered through 0.45 μm filter before the measurement to remove dust.

4. RESULT AND DISCUSSION

4.1. Synthesis of PEO-b-P(NIPAM-co-NBA) Diblock Copolymers

Two PEO-b-P(NIPAM-co-NBA) block copolymers (denoted as p1 and p2) with different NBA moieties were synthesized by RAFT polymerizations using PEO_{1000}-TTCA as the macroRAFT agent. GPC analysis showed that the molecular weights of p1 and p2 were 23.7 and 23.4 kDa, respectively, and the corresponding polydispersities indexes (PDI) were 1.62 and 1.59. The monomodal and symmetric elution peaks of the products clearly shifted to the higher molecular weight region in comparison with that of PEO macroRAFT agent (Figure 1(a)), indicating that block copolymers were successfully synthesized. The chemical structures of the polymers were also characterized by ^1H NMR. As shown in Figure 1(b), the peaks at 3.37 ppm, 3.98 ppm, and 5.42 ppm are characteristic proton signals of PEO ($-CH_2-CH_2-O-$), NBA (benzyl CH_2 protons), and NIPAM ($-CH(CH_3)_2$) moieties, respectively. The NBA molar contents in P(NIPAM-co-NBA) block of p1 and p2 were calculated according to the equation of $0.5 \times A_{5.42} : A_{3.98}$, where the $A_{5.42}$ and $A_{3.98}$ are areas of the peaks at 5.42 and 3.98 ppm, respectively, and the results were 5.06 and 9.41 mol%. The molecular weights of p1 and p2 were also calculated according to ^1H NMR; the results were 23.5 kDa and 23.4 kDa, respectively, which coincided with the results obtained by GPC.

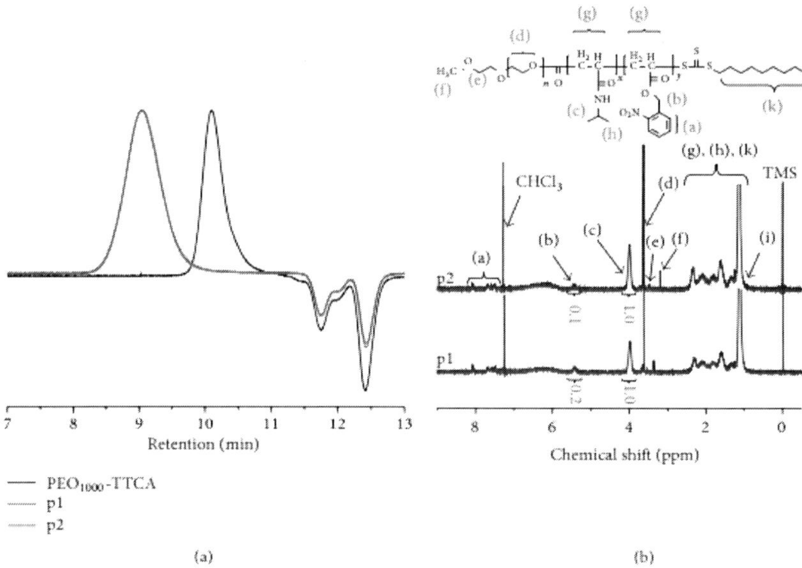

Figure 1: (a) GPC curves of PEO macroinitiator and the two block copolymers with different NBA moieties; (b) ^{1}H NMR (in $CDCl_3$) spectra of p1 and p2.

4.2. Micellization of PEO-b-P(NIPAM-co-NBA) and UV-Triggered Micellar Disintegration in Aqueous Media

NBA moieties in P(NIPAM-co-NBA) block are in the form of o-nitrobenzyl (o-NB) ester. Only 5 mol% of o-NB moieties will decrease the LCST of the block copolymer below 10°C, and the more o-NB groups there are the lower LCST the block copolymer exhibits [33]. Therefore, the block copolymers in current work would be amphiphilic and able to self-assemble to micelles at room temperature. In this work, we used pyrene as the probe to investigate self-assembly behaviors of the block copolymers at room temperature. Figure 2 shows plots of the ratio of fluorescent intensities of pyrene at 336 nm and 332.5 nm (*I*) versus concentrations of the block copolymers. The abrupt rises of intensity ratios indicate that the block copolymers self-assemble in water at room temperature, and the corresponding concentrations show that the critical micelle concentrations (CMCs) of p1 and p2 are about 0.24 g/L and 0.056 g/L, respectively. The CMC of p1 is about 4 times the one of p2, indicating that the P(NIPAM-co-NBA) block of p1 is more hydrophobic because of its more NBA moieties comparing to the one of p2. The self-assemblies were further observed by TEM and DLS. As shown in Figures3(a) and 3(b), the self-assemblies of p1 and p2 are spherical micelles with the diameters about

115.7 nm and 140.5 nm, respectively, which coincide with the results from DLS (insets in Figures 3(a) and 3(b)).

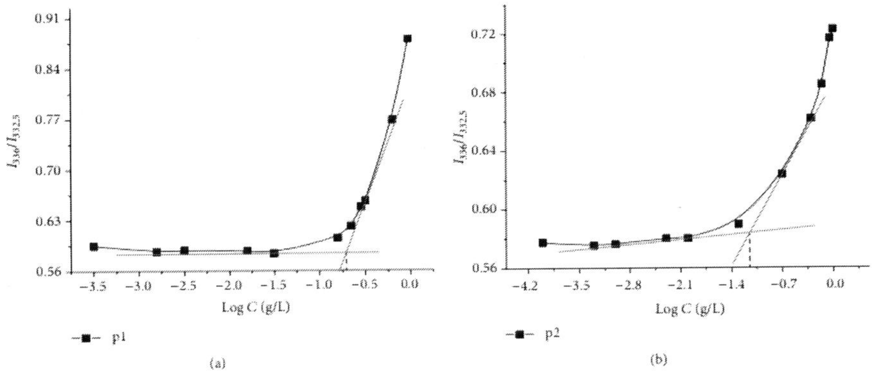

Figure 2: Plots of intensity ratio (*I*) from the pyrene excitation spectra as a function of concentrations of p1 (a) and p2 (b), respectively. Pyrene concentration was fixed to be 6.0×10^{-7} M, and the temperature is at 25°C.

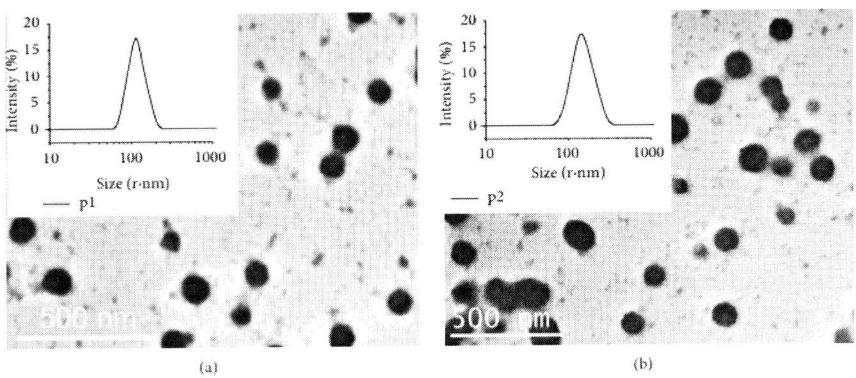

Figure 3: High-resolution TEM micrographs of the self-assemblies of (a) p1 and (b) p2. Insets are corresponding hydrodynamic diameter distributions recorded by DLS. Note. Staining was performed with phosphotungstic acid and the tiny particles may be the crystal of excess phosphotungstic acid.

o-NB esters are a typical photolabile group, which usually yields carboxylic acid under irradiation of UV light with o-nitrosobenzaldehyde as a byproduct [11]. When P(NIPAM-co-NBA) are treated with UV light, they will convert to P(NIPAM-co-AA) (Scheme 1), where the AA denotes acrylic acid. The LCST of P(NIPAM-co-AA) is higher than 40°C even when AA content is just 5 mol% [33].

That is, the UV light will convert the amphiphilic block copolymers to hydrophilic ones at room temperature and cause the disintegration of the corresponding micelles. To verify the UV light induced photocleavage reactions of the o-NB esters in this work, the UV-vis absorption measurements were performed. Figures 4(a) and 4(b) show the UV-vis absorption spectra of the micellar solutions of p1 and p2 irradiated by UV light (365 nm, 2000 mW/cm^2) for various time. The decrease in absorption at 265 nm and the increase at 313 nm prove cleavage of photolabile o-nitrobenzyl ester moieties and the generation of o-nitrosobenzaldehyde [21, 34]. Disintegration of the micelles was able to be observed directly. For example, the original aqueous solution of p1 was turbid and exhibited strong Tyndall effect (Figure4(c)); however, when the p1 solution was treated by UV light for 7 min, it became transparent and its Tyndall effect nearly disappeared, indicating the disintegration of the micelles.

4.3. Photocontrolled Release upon UV Irradiation

The micelles with the character of UV-triggered disintegration may be used as drug carriers with the character of photocontrolled drug release. In this work, we investigated the UV-triggered drug release from the micelles using Nile Red (NR) as the model compound. The principle of using NR as a model compound is based on the fact that the fluorescence emission of NR is low in water due to its very low solubility but becomes much more intense when NR is solubilized in hydrophobic micelle core [37, 38]. Figure 5(a) shows the spectral change of the micellar solution of p1 exposed to UV light for various time. It can be seen that the fluorescent intensity of NR gradually decreases with the increase of the UV irradiation time, indicating that the NR were released from the hydrophobic micelle core to the aqueous solution. Furthermore, we investigated the influence of o-NB ester content on the kinetics of NR release. Figure 5(b) shows the plots of normalized fluorescent intensities versus irradiation time for the NR loaded micelles of the two block copolymers. It can be seen that the less the NBA moieties the polymer has, the faster the fluorescence decreases. Therefore, it can be deduced that the micelles self-assembled by the block copolymer with less NBA content requiring less irradiation time of light to release the cargoes.

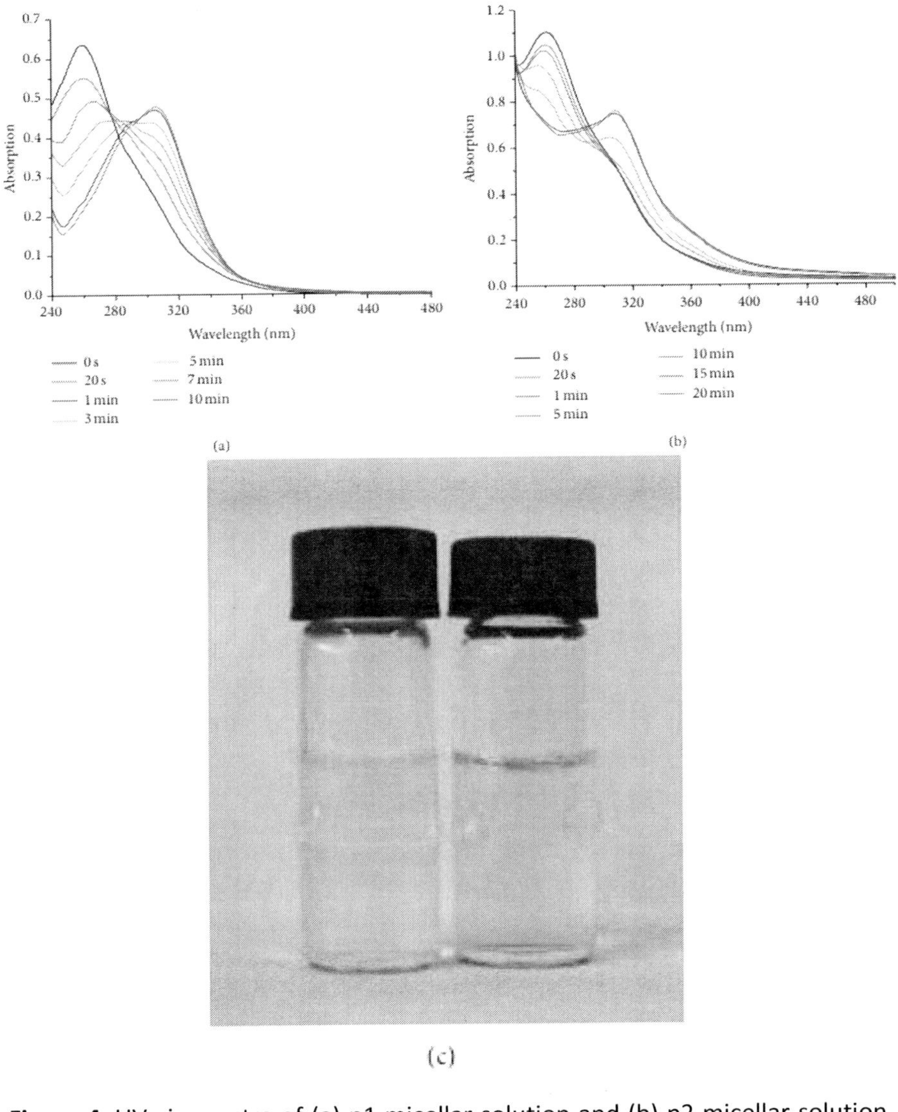

Figure 4: UV-vis spectra of (a) p1 micellar solution and (b) p2 micellar solution in water after various UV irradiation time; (c) photograph of the Tyndall effect of the p1 solution (0.8 mg/mL) before (left) and after (right) irradiation of UV light at 365 nm (2000 mW/cm^2) for 10 min.

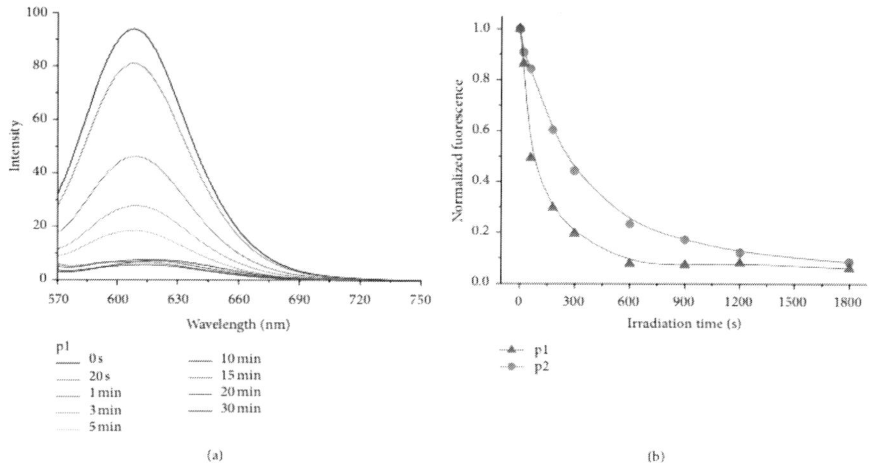

Figure 5: (a) Evolution of fluorescence emission spectra (= 550 nm) of NR loaded PEO-b-P(NIPAM-co-NBA) diblock copolymer micelles under various UV irradiation time (2000 mW/cm^2; 0.8 mg/mL, 2 mL); (b) plots of normalized fluorescence intensity versus UV light irradiation time for the NR loaded micelles of the two block copolymers in aqueous solution.

4.4. Preparation of the NIR Responsive Hybrid Nanoparticles and the NIR-Induced Release

To convert NIR to UV light, $NaYF_4\,Tm^{3+}Yb^{3+}$ UCNPs were prepared. The morphologies observed by TEM revealed that the particles were nearly ellipsoid with major axis and minor axis about 36 nm and 24 nm, respectively (Figure 6(a)). The fluorescent spectrum of the UCNPs excited by NIR light at 980 nm showed output of UV light at about 365 nm (Figure 6(b)), indicating that the UCNPs might be used to trigger the photocleavage reaction of the o-NB groups. In this work, we prepared hybrid nanoparticles using p1 as the organic part. TEM observation indicated that the hybrid nanoparticles were about 120 nm and exhibited a core-shell structure with UCNP as the core and polymers as the corona (Figure 7). To investigate the potential of the hybrid nanoparticles as carriers with the character of NIR-induced drug release, we loaded NR in the hybrid nanoparticles and investigated the NR release under irradiation of NIR light. As shown in Figure 8(a), the fluorescence of the NR loaded hybrid nanoparticles gradually decreases with the varying irradiation time of NIR. However, when NR were loaded in micelles of p1, there was nearly no change in the fluorescence intensity of NR after irradiation of NIR light (Figure 8(b)). The results proved that the hybrid nanoparticles released NR by the stimulus of NIR.

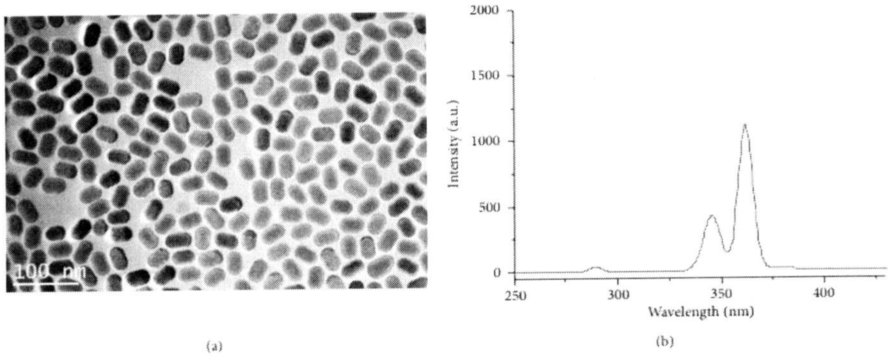

Figure 6: (a) TEM micrographs of the $NaYF_4$:Tm^{3+} 0.5 mol%, Yb^{3+} 30 mol%@ $NaYF_4$core/shell UCNPs; (b) emission spectra of UCNPs under 980 nm NIR excitation (8 W).

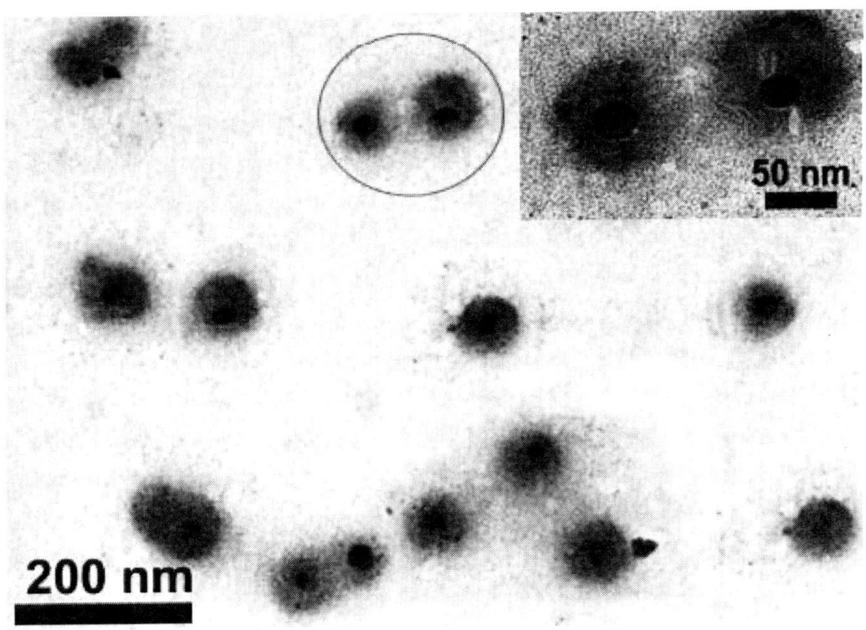

Figure 7: High-resolution TEM micrograph of the hybrid nanoparticles. Inset is the magnified image of the region in red circle. Note. Staining was performed with phosphotungstic acid.

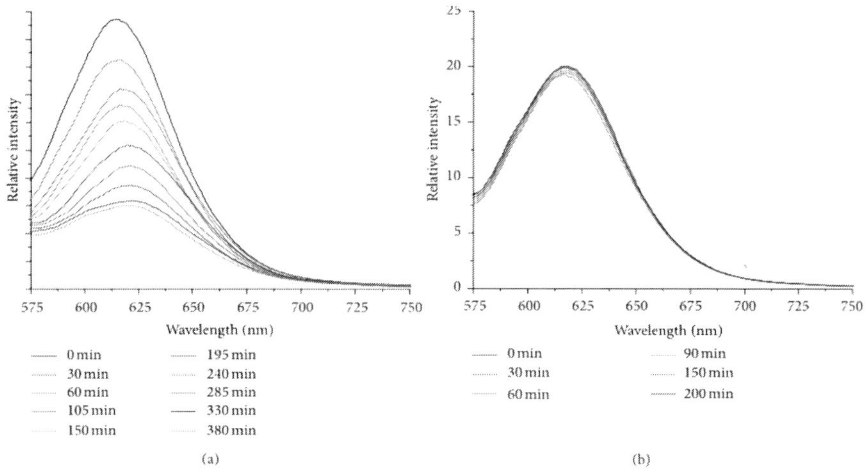

Figure 8: Fluorescence emission spectra (= 550 nm) of the p1 micellar solution (a) loaded with both UCNPs and Nile Red; (b) with only encapsulated Nile Red under 980 nm NIR light (8 W).

5. CONCLUSION

In conclusion, we synthesized UCNPs and thermo- and UV light-responsive block copolymers. The UCNPs were able to convert NIR light at 980 nm to UV light at 365 nm, and the sensitivity of the polymers to UV light was able to be controlled by the content of the photocleavable moieties. By self-assembling the polymers with the presence of the UCNPs in water, core-shell hybrid nanoparticles with the UCNP as the core and the polymers as the shell were obtained. Using NR as the model compound, the hybrid nanoparticles were proved to be potential drug carriers with the character of the NIR-triggered drug release.

CONFLICT OF INTERESTS

The authors declare that there is no conflict of interests regarding the publication of this paper.

ACKNOWLEDGMENTS

This work is supported by the National Natural Scientific Foundation of China Project (21304028 and 51203145), the Fundamental Research Funds for the

Central Universities (2013HGCH0001), and the Student Innovation Training Program of Hefei University of Technology (2013CXSY722).

REFERENCES

1. R. Lehner, X. Y. Wang, M. Wolf, and P. Hunziker, "Designing switchable nanosystems for medical application," Journal of Controlled Release, vol. 161, no. 2, pp. 307–316, 2012.

2. P. Schattling, F. D. Jochum, and P. Theato, "Multi-stimuli responsive polymers—the all-in-one talents,"Polymer Chemistry, vol. 5, no. 1, pp. 25–36, 2014.

3. S. Mura, J. Nicolas, and P. Couvreur, "Stimuli-responsive nanocarriers for drug delivery," Nature Materials, vol. 12, no. 11, pp. 991–1003, 2013.

4. A. Chan, R. P. Orme, R. A. Fricker, and P. Roach, "Remote and local control of stimuli responsive materials for therapeutic applications," Advanced Drug Delivery Reviews, vol. 65, no. 4, pp. 497–514, 2013.

5. R. Cheng, F. H. Meng, C. Deng, H. A. Klok, and Z. Y. Zhong, "Dual and multi-stimuli responsive polymeric nanoparticles for programmed site-specific drug delivery," Biomaterials, vol. 34, no. 14, pp. 3647–3657, 2013.

6. J. Gohy and Y. Zhao, "Photo-responsive block copolymer micelles: design and behavior," Chemical Society Reviews, vol. 42, no. 17, pp. 7117–7129, 2013.

7. N. Fomina, J. Sankaranarayanan, and A. Almutairi, "Photochemical mechanisms of light-triggered release from nanocarriers," Advanced Drug Delivery Reviews, vol. 64, no. 11, pp. 1005–1020, 2012.

8. W. Cao, Y.-W. Gu, M. Meineck, and H.-P. Xu, "The combination of chemotherapy and radiotherapy towards more efficient drug delivery," Chemistry, vol. 9, no. 1, pp. 48–57, 2014.

9. B. Wang, K.-F. Chen, R.-D. Yang, and F. Yang, "Stimulus-responsive polymeric micelles for the light-triggered release of drugs," Carbohydrate Polymers, vol. 103, pp. 510–519, 2014.

10. Q.-J. Xing, N.-J. Li, and D.-Y. Chen, "Light-responsive amphiphilic copolymer coated nanoparticles as nanocarriers and real-time monitors for controlled drug release," Journal of Materials Chemistry B, vol. 2, no. 9, pp. 1182–1189, 2014.

11. H. Zhao, E. S. Sterner, E. B. Coughlin, and P. Theato, "O-Nitrobenzyl alcohol derivatives: opportunities in polymer and materials science," Macromolecules, vol. 45, no. 4, pp. 1723–1736, 2012.

12. L.-L. Meng, W. Huang, D.-L. Wang, X.-H. Huang, X.-Y. Zhu, and D.-Y. Yan, "Chitosan-based nanocarriers with pH and light dual response for anticancer drug delivery," Biomacromolecules, vol. 14, no. 8, pp. 2601–2610, 2013.

13. Y. Liu, C. Y. Yu, H. B. Jin et al., "A supramolecular Janus hyperbranched polymer and its photoresponsive self-assembly of vesicles with narrow size distribution," Journal of the American Chemical Society, vol. 135, no. 12, pp. 4765–4770, 2013.

14. Y. Zhao, "Light-responsive block copolymer micelles," Macromolecules, vol. 45, no. 9, pp. 3647–3657, 2012.

15. T. Furuta, S. S.-. Wang, J. L. Dantzker et al., "Brominated 7-hydroxycoumarin-4-ylmethyls: photolabile protecting groups with biologically useful cross-sections for two photon photolysis," Proceedings of the National Academy of Sciences of the United States of America, vol. 96, no. 4, pp. 1193–1200, 1999.

16. Z.-J. Zhang, J. Wang, and C.-Y. Chen, "Near-infrared light-mediated nanoplatforms for cancer thermo-chemotherapy and optical imaging," Advanced Materials, vol. 25, no. 28, pp. 3869–3880, 2013.

17. J. You, P.-Z. Zhang, F.-Q. Hu et al., "Near-infrared light-sensitive liposomes for the enhanced photothermal tumor treatment by the combination with chemotherapy," Pharmaceutical Research, vol. 31, no. 3, pp. 554–565, 2014.

18. H.-S. Qian and Y. Zhang, "Synthesis of hexagonal-phase core-shell $NaYF_4$ nanocrystals with tunable upconversion fluorescence," Langmuir, vol. 24, no. 21, pp. 12123–12125, 2008.

19. H. S. Qian, H. C. Guo, P. C. Ho, R. Mahendran, and Y. Zhang, "Mesoporous-silica-coated up-conversion fluorescent nanoparticles for photodynamic therapy," Small, vol. 5, no. 20, pp. 2285–2290, 2009.

20. X. Liu, H. Qian, Y. Ji et al., "Mesoporous silica-coated $NaYF_4$ nanocrystals: facile synthesis, in vitro bioimaging and photodynamic therapy of cancer cells," RSC Advances, vol. 2, no. 32, pp. 12263–12268, 2012.

21. B. Yan, J. C. Boyer, N. R. Branda, and Y. Zhao, "Near-infrared light-triggered dissociation of block copolymer micelles using upconverting nanoparticles," Journal of the American Chemical Society, vol. 133, no. 49, pp. 19714–19717, 2011.

22. T.-Q. Wu, M. Barker, K. M. Arafeh, J. C. Boyer, C. J. Carling, and P. N. Branda, "A UV-blocking polymer shell prevents one-photon photoreactions while allowing multi-photon processes in encapsulated upconverting nanoparticles," Angewandte Chemie—International Edition, vol. 125, no. 42, pp. 11312–11315, 2013.

23. Y.-Z. Min, J.-M. Li, F. Liu, E. L. Yeow, and B.-G. Xing, "Near-infrared light-mediated photoactivation of a platinum antitumor prodrug and simultaneous cellular apoptosis imaging by upconversion-luminescent nanoparticles," Angewandte Chemie, vol. 126, no. 4, pp. 1030–1034, 2014.

24. T.-Q. Wu, M. Barker, K. M. Arafeh, J. C. Boyer, C. J. Carling, and P. N. Branda, "A UV-blocking polymer shell prevents one-photon photoreactions while allowing multi-photon processes in encapsulated upconverting nanoparticles," Angewandte Chemie International Edition, vol. 125, no. 42, pp. 11312–11315, 2013.

25. Y.-M. Yang, B. Velmurugan, X.-G. Liu, and B.-G. Xing, "NIR photoresponsive crosslinked upconverting nanocarriers toward selective intracellular drug release," Small, vol. 9, no. 17, pp. 2937–2944, 2013.

26. M. Bikram and J. L. West, "Thermo-responsive systems for controlled drug delivery," Expert Opinion on Drug Delivery, vol. 5, no. 10, pp. 1077–1091, 2008.

27. S. H. Choi, S. H. Lee, and T. G. Park, "Temperature-sensitive pluronic/poly(ethylenimine) nanocapsules for thermally triggered disruption of intracellular endosomal compartment," Biomacromolecules, vol. 7, no. 6, pp. 1864–1870, 2006.

28. E. S. Lee, Z. Gao, and Y. H. Bae, "Recent progress in tumor pH targeting nanotechnology," Journal of Controlled Release, vol. 132, no. 3, pp. 164–170, 2008.

29. D.-Q. Chen and H.-B. Wang, "Novel pH-sensitive biodegradable polymeric drug delivery systems based on ketal polymers," Journal of Nanoscience and Nanotechnology, vol. 14, no. 1, pp. 983–989, 2014.

30. T. Xing, B. Lai, and L.-F. Yan, "Disulfide cross-linked polypeptide nanogel conjugated with a fluorescent probe as a potential image-guided drug-delivery agent," Macromolecular Chemistry and Physics, vol. 214, no. 5, pp. 578–588, 2013.

31. J. Liu, Y. Pang, W. Huang et al., "Redox-responsive polyphosphate nanosized assemblies: a smart drug delivery platform for cancer therapy," Biomacromolecules, vol. 12, no. 6, pp. 2407–2415, 2011.

32. H. Wei, S. X. Cheng, X. Z. Zhang, and R. X. Zhuo, "Thermo-sensitive polymeric micelles based on poly(N-isopropylacrylamide) as drug carriers," Progress in Polymer Science, vol. 34, no. 9, pp. 893–910, 2009.

33. L. Ionov and S. Diez, "Environment-friendly photolithography using poly(N-isopropylacrylamide)-based thermoresponsive photoresists," Journal of the American Chemical Society, vol. 131, no. 37, pp. 13315–13319, 2009.

34. Y.-M. Li, Y.-F. Qian, T. Liu, G.-Y. Zhang, and S.-Y. Liu, "Light-triggered concomitant enhancement of magnetic resonance imaging contrast performance and drug release rate of functionalized amphiphilic diblock copolymer micelles," Biomacromolecules, vol. 13, no. 11, pp. 3877–3886, 2012.

35. L.-F. Yan and W. Tao, "One-step synthesis of pegylated cationic nanogels of poly(N,N′-dimethylaminoethyl methacrylate) in aqueous solution via self-stabilizing micelles using an amphiphilic macroRAFT agent," Polymer, vol. 51, no. 10, pp. 2161–2167, 2010.

36. J. Du, L. Tang, W. Song, Y. Shi, and J. Wang, "Evaluation of polymeric micelles from brush polymer with poly(ε-caprolactone)-b-poly(ethylene glycol) side chains as drug carrier," Biomacromolecules, vol. 10, no. 8, pp. 2169–2174, 2009.

37. A. P. Goodwin, J. L. Mynar, Y. Ma, G. R. Fleming, and J. M. J. Fréchet, "Synthetic micelle sensitive to IR light via a two-photon process," Journal of the American Chemical Society, vol. 127, no. 28, pp. 9952–9953, 2005.

38. J. Q. Jiang, X. Tong, D. Morris, and Y. Zhao, “Toward photocontrolled release using light-dissociable block copolymer micelles,” Macromolecules, vol. 39, no. 13, pp. 4633–4640, 2006.

Chapter 4

Synthesis of Upconversion β-$NaYF_4$:Nd^{3+}/Yb^{3+}/Er^{3+} Particles with Enhanced Luminescent Intensity through Control of Morphology and Phase

Yunfei Shang [1], Shuwei Hao [1,*], Jing Liu [1], Meiling Tan [1], Ning Wang [1], Chunhui Yang [1,2] and Guanying Chen [1,3,*]

[1]*School of Chemical Engineering and Technology, Harbin Institute of Technology, Harbin 150001, China*

[2]*Harbin Huigong Technology Co. Ltd, Harbin 150001, China*

[3]*Institute for Lasers, Photonics and Biophotonics, University at Buffalo, The State University of New York, Buffalo, NY 14260, USA*

ABSTRACT

Hexagonal $NaYF_4$:Nd^{3+}/Yb^{3+}/Er^{3+} microcrystals and nanocrystals with well-defined morphologies and sizes have been synthesized via a hydrothermal route. The rational control of initial reaction conditions can not only result in upconversion (UC) micro and nanocrystals with varying morphologies, but also can produce enhanced and tailored upconversion emissions from the Yb^{3+}/Er^{3+} ion pairs sensitized by the Nd^{3+} ions. The increase of reaction time converts the phase of $NaYF_4$:Nd^{3+}/Yb^{3+}/Er^{3+} particles from the cubic to the hexagonal structure. The added amount of oleic acid plays a critical role in the shape evolution of the final products due to their preferential attachment to some crystal planes. The adjustment of the molar ratio of F^-/Ln^{3+} can range the morphologies of the β-$NaYF_4$:Nd^{3+}/Yb^{3+}/Er^{3+} microcrystals from spheres to nanorods. When excited by 808 nm infrared laser, β-$NaYF_4$:Nd^{3+}/Yb^{3+}/Er^{3+} microplates exhibit a much stronger UC emission intensity than particles with other morphologies. This phase- and morphology-dependent UC emission holds promise for applications in photonic devices and biological studies.

Keywords: $NaYF_4$ microcrystals; Nd^{3+} sensitizer; morphology control; upconversion (UC)

1. INTRODUCTION

Lanthanide doped upconversion (UC) materials have attracted increasing interest due to their ability to convert low-energy excitations into high-energy emissions at shorter wavelength [1,2,3]. This unique property makes UC materials suitable for a wide range of applications, such as biological sensing [4,5],*in vivo* imaging [6,7], drug delivery [8,9], and photodynamic therapy [10,11]. Particularly, lanthanide doped fluoride materials revive these applications owing to their higher UC efficiency and excellent physicochemical stabilities [12]. Among them, β-$NaYF_4$ is considered to be one of the most efficient host lattices with low phonon energies (<350 cm^{-1}) to minimize energy losses at the intermediate states of lanthanide ions [7]. Moreover, Yb^{3+} ions were used as an efficient sensitizer to enhance UC emissions by taking advantage of its high absorption cross-section and efficient energy transferring to the activators of Er^{3+}, Ho^{3+}, or Tm^{3+}. However, the absorption ability of Yb^{3+} sensitizers limited the excitation of most current UC materials to be performed at ~980 nm [13,14]. This elicits heating problems, which cause possible damage to cells and tissues due to strong water absorption at the excitation wavelength.

Recently, extensive efforts have been devoted to development of UC materials that can be excited at better excitation wavelengths but cause minimized adverse effects on the biological cells and tissues. Doping of Nd^{3+} ion as a sensitizer into the current UC systems would be appealing towards this purpose, as Nd^{3+} can be excited by the well-established commercial 808 nm excitation source, the wavelength of which is negligibly absorbed by water molecules. Indeed, the absorption coefficient of water at ~808 nm is around 0.02 cm^{-1}, more than 24 times lower than the value of 0.482 cm^{-1} at ~980 nm. The Nd^{3+} sensitizer displays high absorption cross-section (~1.2×10^{-19} cm^2) at 808 nm, which is one order of magnitude higher than that of Yb^{3+} sensitizer (~1.2×10^{-20} cm^2) [15,16]. Moreover, efficient energy transfers have been established from Nd^{3+} to Yb^{3+} and then from Yb^{3+} to the acceptor (Er^{3+}, Ho^{3+} or Tm^{3+}) [17]. This enables the production of UC materials that can be excited at ~808 nm, and thus draws attention to the need for developing Nd^{3+}-sensitized UC materials for biological applications. Yan*et al.* introduced the Nd^{3+} sensitizer into $NaGdF_4$ nanoparticles and showed that it minimized the overheating effect in *in vivo* bioimaging compared with Yb^{3+}-sensitized nanoparticles [18]. Liu and colleagues presented a new type of core-shell UC nanoparticles based on doping high concentrations of Nd^{3+} ions in the shell structure to enhance the UC emission of activators [13]. Utilizing Nd^{3+}-sensitized UC nanoparticles to remove the constraints in conjunction with conventional UC nanoparticles is an important technology for biological studies; this conclusion has also been independently verified in core-shell UC nanoparticles by Han *et al.* and Wang's group [19,20]. Strategies to prepare Nd^{3+}-sensitized particles and to improve and tailor upconverted luminescence from them are in demand in order to produce impetus for their biomedical applications [21,22]. Thus far, very few studies reported on size- and morphology-controlled synthesis of UC Nd^{3+}-sensitized particles with tailored luminescence when excited at ~808 nm.

The particle morphologies have been found to play a significant role on the luminescent properties of lanthanide-doped UC materials [23,24,25]. The β-$NaLuF_4$:Yb^{3+}, Er^{3+} nanomaterials with tube shapes display higher luminescent intensity than those of other morphologies [26]. Lin *et al.* [26] reveal that the luminescent intensity was increased to 6 times in microtubes' structure, compared with that of limb-like shape. In our previous studies, the microrods exhibit a better photoluminescence signal than other types that resulted in β-$NaYF_4$ micro- and nanocrystals [27]. Until now, no attempts have been made to prepare a range of Nd^{3+}-sensitized β-$NaYF_4$ particles with tuned morphologies and enhanced UC luminescence. Herein, we report on the controlled synthesis of monodisperse Nd^{3+}-sensitized β-$NaYF_4$microcrystals and nanocrystals with uniform size and tunable shapes under a hydrothermal condition. We show that variations of the oleic acid ligand and the concentration ratio of F^-/Ln^{3+} play a synergistic role in defining the morphologies of resulting particles. The crystal growth mechanisms of β-$NaYF_4$:10% Nd^{3+}, 10% Yb^{3+}, 2% Er^{3+} particles with sphere-like shape, octadecahedral shape, and hexagonal shape with protruding centers, were disclosed by collecting samples at pre-set time intervals. Moreover, spectroscopic investigations revealed that upconverted luminescence intensity is dependent on the resulting particle morphology, size, and crystallinity, manifesting the highest intensity from $NaYF_4$:10% Nd^{3+}, 10% Yb^{3+}, 2% Er^{3+} microplates.

2. RESULTS AND DISCUSSION

2.1. Effects of Reaction Time

Because the crystal structure of $NaYF_4$ exhibits cubic (α-) and hexagonal (β-) polymorphic forms, reaction times were changed to probe the phase transformation process. Figure 1 and Figure 2 show the X-ray diffraction (XRD) patterns and field emission scanning electron microscopy (FESEM) images of the resulting $NaYF_4$:10% Nd^{3+}, 10% Yb^{3+}, 2% Er^{3+} samples, respectively. Figure 1 reveals that the phase transformation process (α→β) occurred gradually with an extended reaction time. The sample obtained with a shorter reaction time (t = 3 h) shows a mixture of the α-phase (JCPDS No. 06-0342) and the β-phase (JCPDS No. 16-0334). Judging from the full width of the XRD peaks in Figure 1a, the size of α-phase $NaYF_4$ is very small, which agrees well with the SEM result of Figure 2a. As one can see fromFigure 2a, the sample includes the aggregates formed by a large quantity of spherical nanoparticles. As the reaction time was extended to 6 h and 12 h, the peak intensity of α-$NaYF_4$ became weak (Figure 1b, 1c). In contrast, the peak intensities corresponding to the β-phase $NaYF_4$ became stronger. In addition, from the images shown in Figure 2b, 2c, it is clearly seen that hexagonal $NaYF_4$ microprisms are obtained, which become more regular and smoother with a longer reaction time. This indicated that the samples transformed gradually from α to β phase as the reaction time was prolonged. When the time was extended to 24 h, only pure β-phase $NaYF_4$ existed. Moreover, the intensities of the diffraction peaks in Figure 1d dramatically increased with respect to those samples of short times, implying that the crystallinity of the sample increases with the increase of reaction time. There is a large quantity of microprisms with uniform, smooth, and flat surfaces shown in Figure 2d, which agrees well with the corresponding XRD result of Figure 1d.

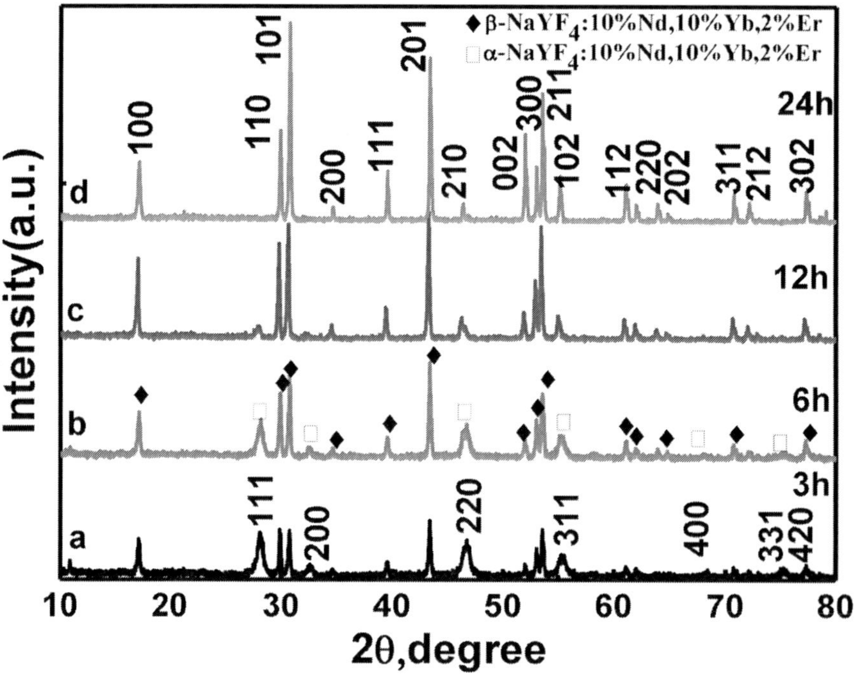

Figure 1. X-ray diffraction (XRD) patterns of $NaYF_4$: 10% Nd^{3+},10% Yb^{3+}, 2% Er^{3+} microcrystals synthesized at different hydrothermal times: (**a**) 3 h; (**b**) 6 h; (**c**) 12 h; and (**d**) 24 h (180 °C F^-/Ln^{3+} = 5:1 OA/Ln^{3+} = 40:1).

The above analysis indicates that the hydrothermal reaction time has significant effects on the size, shape, and phase structure of final products. A long hydrothermal time benefits crystallization and the growth of bigger size microcrystals. Moreover, this result suggests that the β-phase $NaYF_4$ crystal particles possess higher thermal energy than the α-phase $NaYF_4$; a long hydrothermal time is needed to accumulate adequate energy to complete the phase transformation process.

Figure 2. Typical field emission scanning electron microscopy (FESEM) images for the resulting $NaYF_4$:10% Nd^{3+}, 10% Yb^{3+}, 2% Er^{3+} microcrystals prepared with different hydrothermal times: (**a**) 3 h; (**b**) 6 h; (**c**) 12 h; and (**d**) 24 h (Other synthetic parameters were kept identical, 180 °C, F^-/Ln^{3+} = 5:1, OA/Ln^{3+} = 40:1).

2.2. Effect of the Ratio of OA/Ln^{3+}

Because OA can rationally control and modify the shape of $NaYF_4$ crystals, we probed the role of variation of the molar ratio of OA/Ln^{3+} to impact the resulted morphology of $NaYF_4$:10% Nd^{3+}, 10% Yb^{3+}, 2% Er^{3+} microcrystals in our synthesis system. Figure 3a shows the SEM images of the sample synthesized with 6:1 of OA/Ln^{3+}. The image indicates that the resulting products are uniform nanolines. The nanoline can be identified as pure $Y(OH)_3$:10% Nd^{3+}, 10% Yb^{3+}, 2% Er^{3+} by XRD (Figure 4a). It might be explained by the fact that the reaction system was alkaline when it contained little OA. The rare earth ions can thus easily combine with hydroxyl to generate $Y(OH)_3$:10% Nd^{3+}, 10% Yb^{3+}, 2% Er^{3+}precipitation. With the ratio of

OA/Ln^{3+} exceeding 20:1, pure β-phase $NaYF_4$ was successfully produced, as shown in Figure 4b. Figure 3b shows the SEM images of the sample synthesized with 20:1 of OA/Ln^{3+}. The image reveals that the sample contains uniform nanorods with a length of 1.2 μm and an average lateral diameter of 0.245 μm. As the ratio of OA/Ln^{3+} was increased to 30:1, hexagonal microprisms with a length of 0.25 μm and an average lateral diameter of 0.25 μm were obtained. When the ratio of OA/Ln^{3+} exceeded 40:1, uniform hexagonal microprisms with a length of 0.2 μm and the lateral diameter of 1 μm were produced. Hence the aspect ratio L/D (length/diameter) increases gradually with the molar ratio of OA/Ln^{3+}. This indicates that the OA plays an important role in the morphology evolution from nanorods to hexagonal microplates. The involved underlined growing mechanism will be discussed in Section 2.4.

Figure 3. Typical FESEM images for the as-prepared $NaYF_4$: 10% Nd^{3+},10% Yb^{3+}, 2% Er^{3+} microcrystals at different molar ratios of OA/Ln^{3+}: (**a**) 6:1; (**b**) 20:1; (**c**) 30:1; and (**d**) 40:1 (Other synthetic parameters were kept identical, 180 °C, 24 h, F^-/Ln^{3+} = 5:1).

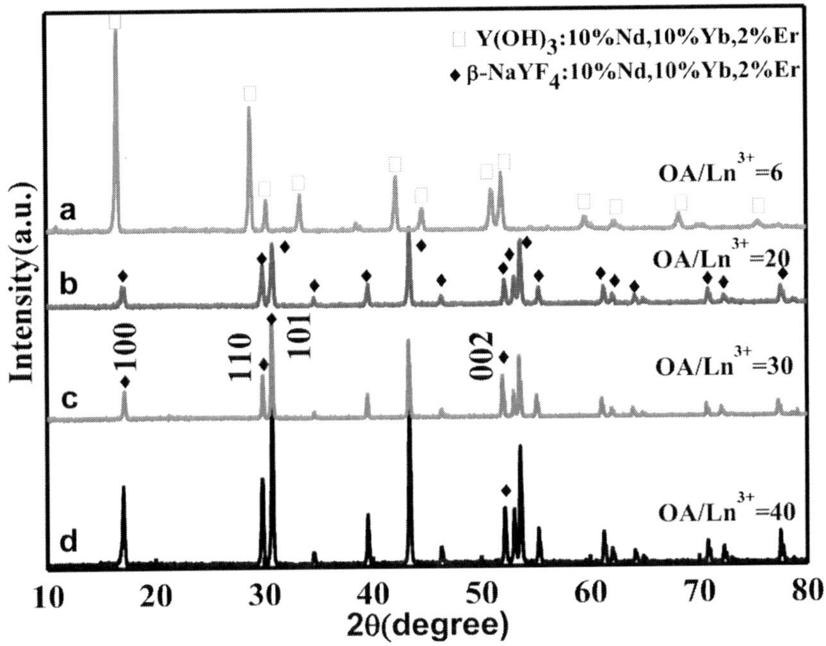

Figure 4. XRD patterns of $NaYF_4$:10% Nd^{3+}, 10% Yb^{3+}, 2% Er^{3+}microcrystals synthesized at different molar ratios of OA/Ln^{3+}: (**a**) 6:1; (**b**) 20:1; (**c**) 3:1; and (**d**) 40:1 (Other synthetic parameters were kept identical, 180 °C, 24 h, F^-/Ln^{3+} = 5:1).

2.3. Effect of the Ratio of F^-/Ln^{3+}

To investigate the effect of the F^-/Ln^{3+} ($F^-/Ln^{3+} \leq 5$) ratio on the morphology, size, and structure of the as-prepared $NaYF_4$:10% Nd^{3+}, 10% Yb^{3+}, 2% Er^{3+} particles, the ratio was taken as 1:1, 2:1, 3:1, and 5:1, respectively. As shown in Figure 5, with a gradual increase of the molar ratio of F^-/Ln^{3+}, a significant change takes place in the morphology and size of $NaYF_4$:10% Nd^{3+}, 10% Yb^{3+}, 2% Er^{3+} microcrystals. When the F^-/Ln^{3+} ratio is fixed at 1:1, the as-prepared microcrystals are spherical with coarse surfaces, implying low crystallization (Figure 5a). When the molar ratio of F^-/Ln^{3+} increased to 2:1 and 3:1, hexagonal microprisms with a protruding center and distortional tubular structure—with the end face of the central convex and concave shape between the center and the edge—were obtained. In addition, with the increase of F^- concentration, the crystallization is also improved. As the molar ratio of F^-/Ln^{3+}exceeded 5:1, uniform nanorods with high crystallization were generated. It can be seen that, with the increase of the molar ratio of F^-/Ln^{3+}, each surface gradually became smooth as along with the higher crystallization, which has been proved by XRD patterns (Figure 6). All the peaks of each sample are characteristic of a pure β-phase of $NaYF_4$; however, the relative intensities are different from each other, implying a different preferential orientation growth.

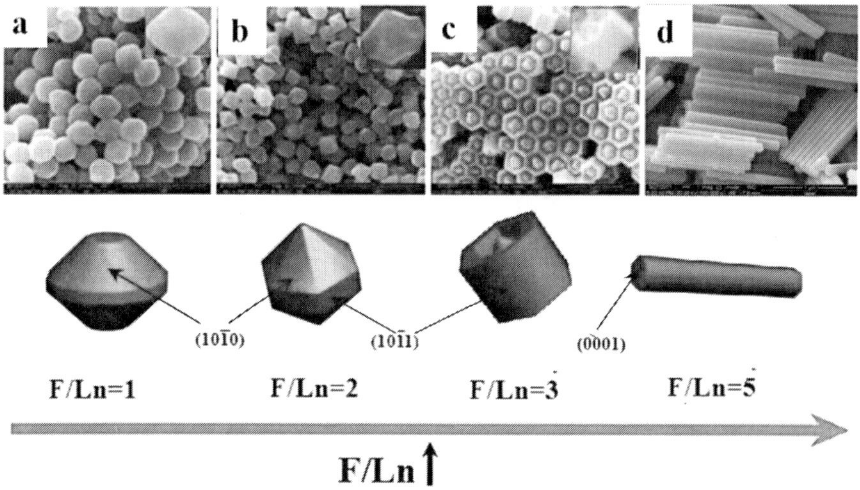

Figure 5. Typical FESEM images for the as-prepared $NaYF_4$: 10% Nd^{3+},10% Yb^{3+}, 2% Er^{3+} microcrystals at different molar ratios of F^-/Ln^{3+}: (**a**) 1:1; (**b**) 2:1; (**c**) 3:1; and (**d**) 5:1 (Other synthetic parameters were kept identical, 180 °C, 24 h, OA/Ln^{3+} = 20:1).

2.4. The Mechanism of $NaYF_4$ Crystal Growth

The hydrothermal growth process of $NaYF_4$ microcrystals includes the nucleation and the growth steps. In the process of nucleation, the OA anions could serve as chelating agent in reaction solution and formed complexes with Y^{3+} ions due to the strong coordination interaction. The chelating ability of the Y^{3+}-OA complex would be weakened under high hydrothermal temperature and pressure, and thus gradually release the Y^{3+} ions to react with Na^+ and F^- ions to generate the $NaYF_4$ nuclei. During the process of growth, OA as stabilizing agents can attach to the surfaces of $NaYF_4$ microcrystal with the alkyl chains left outside, which controls the growth of microcrystals and provides steric forces to prevent their aggregations.

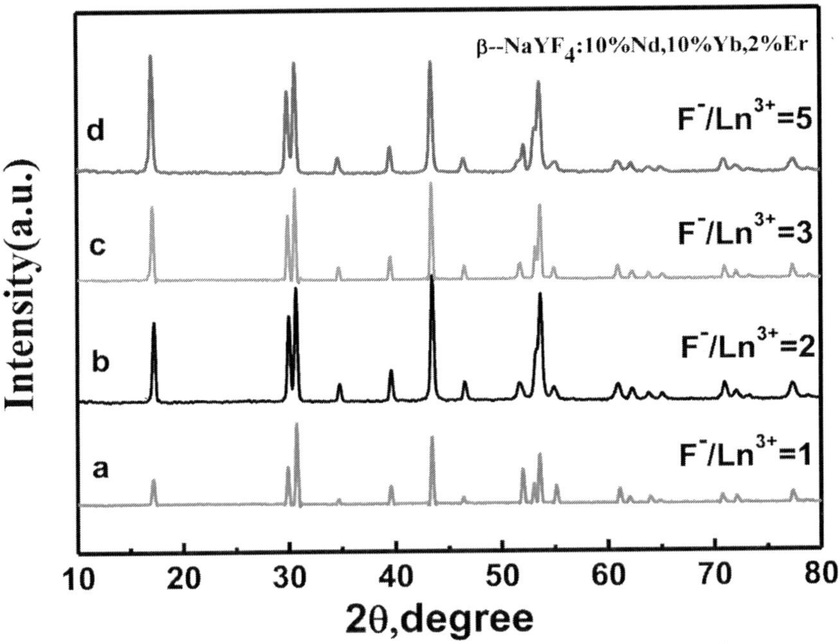

Figure 6. XRD patterns of $NaYF_4$:10% Nd^{3+},10% Yb^{3+}, 2% Er^{3+}microcrystals synthesized at different molar ratios of F^-/Ln^{3+}: (**a**) 1:1; (**b**) 2:1; (**c**) 3:1; and (**d**) 5:1 (Other synthetic parameters were kept identical, 180 °C, 24 h, OA/Ln^{3+} = 20:1).

To test our supposition, we have carried out two parts of the synthetic experiment.

Part 1: the molar ratio of F^-/Ln^{3+} was fixed at 5, and then the molar ratio of OA/Ln^{3+} was changed from 20:1 to 40:1. The hexagonal $NaYF_4$ seed generally includes (0001) and (1010) facets, as shown inScheme 1. It is noted that the facets with smaller surface area will have a higher surface energy and a faster growth rate, thus impacting the shapes of the final products. As for the hexagonal $NaYF_4$, the area of (1010) exceeds that of (0001). This indicates that (0001) facets have higher surface energy, and thus a higher growth velocity, *i.e.*, $v(0001) >> v(1010)$. When the amount of oleic acid (OA) is relatively low (OA/Ln^{3+} < 20:1), OA molecules preferentially adsorb onto the (0001) facets, thus producing uniform microrods (Figure 3b). The short microrods with an average diameter of 0.25 μm and a height of 0.245 μm can be achieved when the OA $/Ln^{3+}$ is increased to 30:1 (Figure 3c). When more OA molecules are introduced, the ratio of $v(1010)/v(0001)$ will be increased, as the number of OA molecules on the (0001) facet will saturate but will get larger on the (1010) facet with larger surface area. When $v(0001) = v(1010)$, the resulting particles will simultaneously grow along the (0001) and (1010) directions to shorter microrods. When this ratio was decreased below 1 as more OA was introduced, particles with microplates formed.

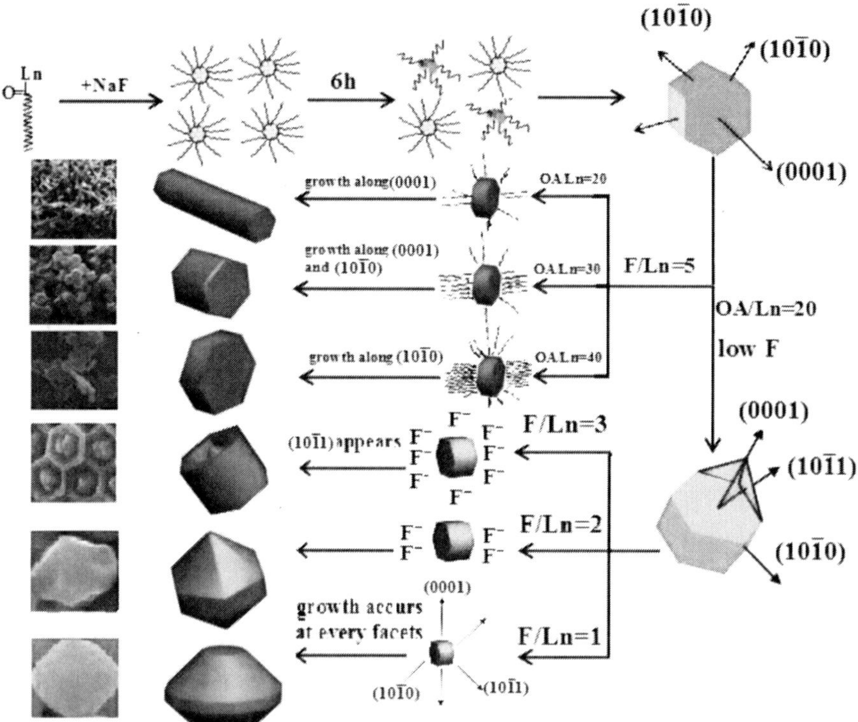

Scheme 1. Schematic illustration of the possible formation mechanism of $NaYF_4$ with various morphologies under different OA/Ln^{3+} and F^-/Ln^{3+} molar ratio conditions.

Part 2: the molar ratio of OA/Ln^{3+} was fixed at 20, while the molar ratio of F^-/Ln^{3+} varied from 1:1 to 5:1. We found that this condition provided an ideal system to induce OA and F^- preferential adsorption onto (0001) facets to control the growth rates of different facets of $NaYF_4$ microcrystals. As can be seen in Figure 5, the hexagonal $NaYF_4$ seed crystals have another type of surface (1010) in the case of $F^-/Ln^{3+} < 5$. The nucleation and growth rate was thus controlled through the competition between F^- and OA towards (0001), (1010), and (1011) facets of $NaYF_4$ microcrystals. The XRD patterns of the products obtained at various ratios of F^-/Ln^{3+} (1:1, 2:1, 3:1, and 5:1) are shown in Figure 6a–d, respectively. As can be seen, all the peaks of each sample are characteristic of a pure β-phase of $NaYF_4$, but the relative intensities are different from each other, suggesting different preferential orientation growth. When the amount of OA(OA/Ln^{3+} = 20) is not enough to hinder the growth of (0001) facets, the differences in the molar ratios of F^-/Ln^{3+} have a significant effect on the growth of $NaYF_4$ microcrystals. With a low molar ratio of F^-/Ln^{3+}(1:1), all of the F^- ions are used in nucleation. Accordingly, the growth of every facet is similar, thus resulting in sphere-like microcrystals (Figure 5a). When the molar ratio of F^-/Ln^{3+} is increased to 2:1, the remainder of the F^- ions attacks the high-energy (0001) facet. Simultaneously, owing to the low free energies of the circumferential edges of seeds, the redundant F^- ions preferentially bound to the circumferential edges of seeds, which would result in growth along the circumferential edge direction [27,28]. The concave octadecahedral β-$NaYF_4$ microcrystals can thus be produced. At F^-/Ln^{3+} = 3:1, the different crystallographic planes can be recognized in Figure 5c. The presence of conical ends with a raised ridge demonstrates that the (1011) facet appears. Moreover, in comparison with octadecahedral β-$NaYF_4$ microcrystals (F^-/Ln^{3+} = 2:1), the length enhancement further confirms that the growth rate is in the (0001) facet. However, in the condition F^-/Ln^{3+} = 5:1, uniform β-$NaYF_4$ nanorods are produced. This is possibly because of the large amount of F^- ions binding strongly to (0001) surfaces to speed up the growth ofβ-$NaYF_4$ crystallites.

2.5. UC Luminescence Properties

β-$NaYF_4$ is regarded as an efficient host for the UC process due to low phonon energies. Nd^{3+} ions were selected as the sensitizer to investigate the spectral and UC luminescent properties of products with various morphologies. Figure 7 shows the UC PL emission spectra of the resulting β-$NaYF_4$:10% Nd^{3+}, 10% Yb^{3+}, 2% Er^{3+} phosphors prepared at 180 °C and the F^-/Ln^{3+} ratio = 5:1 when changing the OA/Ln^{3+} ratio from 6:1 to 40:1 upon irradiation of 808 nm wavelength. UC emission peaks of purple, green, and red centered at 410 nm, 520/540 nm, and 655 nm were observed, which can be assigned to the $^2H_{9/2} \rightarrow {}^4I_{15/2}$, $^2H_{11/2}/^4S_{3/2} \rightarrow {}^4I_{15/2}$, and $^2H_{9/2} \rightarrow {}^4I_{15/2}$ transitions of Er^{3+} ions, respectively. It is interesting to note that the shapes of the emission spectra are similar in all four samples, with the only difference being in the relative intensities of the bands. Additionally, the microplate crystals show highest emission intensity in the four samples. This difference in UC luminescence intensity may be attributed to the changes in morphology, crystallinity, and particle size [29,30]. Excluding the $Y(OH)_3$nanoline, the results show that the β-$NaYF_4$: 10% Nd^{3+},10% Yb^{3+}, 2% Er^{3+} microcrystals with high crystallinity and large size emit stronger UC light owing to the low surface-to-volume ratio. The smaller the surface-to-volume ratio of a microcrystal, the fewer active ions are located on the surface of particles. Therefore, microplates with large size and high crystallinity show a small surface defect, which is efficient to restrain the photon quenching. Moreover, in our case, similar results have also been observed by tuning the molar ratio of F^-/Ln^{3+} from 1:1 to 5:1 (shown in Figure 8). The strongest emission has been observed for nanorods with relatively large size and high crystallinity.

In order to investigate the UC luminescence mechanism, pump power dependences of the luminescence bands centered at 520, 540, and 655 nm were measured and displayed in a double logarithmic plot in Figure 9. For the unsaturated case, the number of photons that are required to populate the upper emitting state can be obtained by the relation $I_f \propto P^n$, where I_f stands for the fluorescent intensity, P is the pump laser power, and n is the number of laser photons required. Fitting the data points yielded slope values of 1.98, 1.93, and 1.96 for

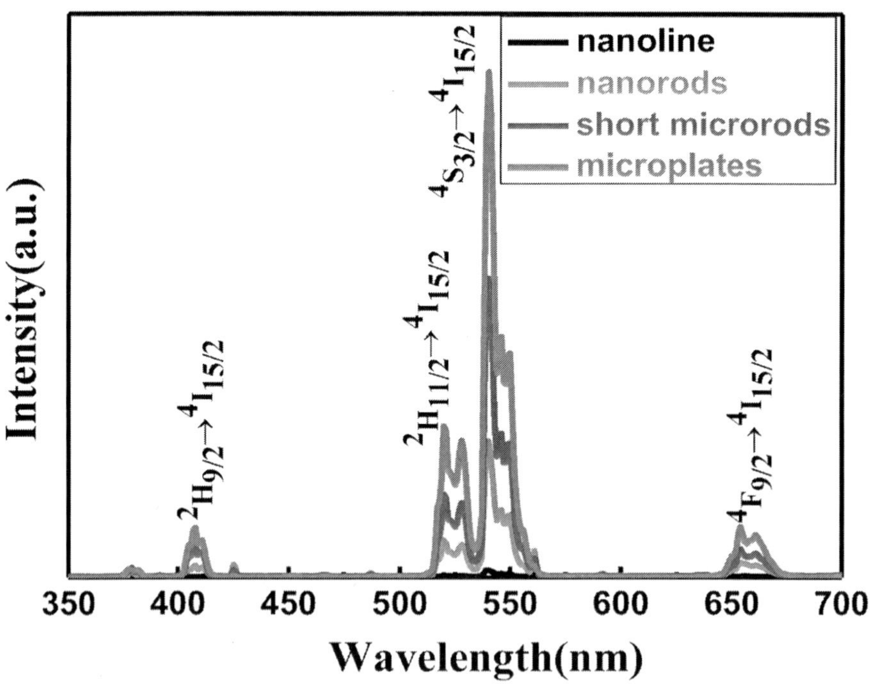

Figure 7. UC Photoluminescence (PL) spectra of $NaYF_4$:10% Nd^{3+},10% Yb^{3+}, 2% Er^{3+} microcrystals with different molar ratios of OA/Ln^{3+} under diode laser excitation at 808 nm. Excitation power density, ~390 W/cm^2.

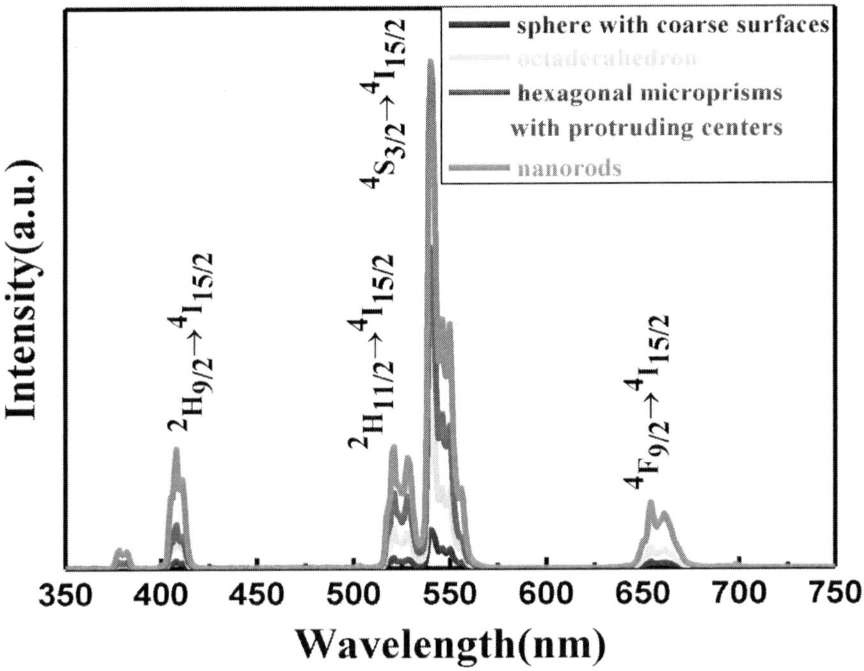

Figure 8. UC Photoluminescence (PL) spectra of $NaYF_4$:10% Nd^{3+},10% Yb^{3+}, 2% Er^{3+} microcrystals with different molar ratios of F^-/Ln^{3+} (180 °C, 24 h, OA/Ln^{3+} = 40) under diode laser excitation at 808 nm. Excitation power density, ~390 W/cm^2.

the $^2H_{11/2} \rightarrow {}^4I_{15/2}$, $^4S_{3/2} \rightarrow {}^4I_{15/2}$, and $^4F_{9/2} \rightarrow {}^4I_{15/2}$ transitions, respectively. This demonstrates that the generation of UC luminescence through the Nd^{3+}-sensitized system involves a two-photon process. In the Nd^{3+}, Yb^{3+}, and Er^{3+} tri-doped system, the energy transfer from Nd^{3+} to Yb^{3+} and then to Er^{3+} should be taken into account under 808 nm diode laser excitation.

The energy transfer mechanism was displayed in Figure 10. The sensitizer Nd^{3+} ions were firstly excited from the $^4I_{9/2}$ to the $^4F_{5/2}$ state after absorbing the excitation energy at 808 nm, and then relaxed to the $^4F_{3/2}$ level through multiphonon processes. The Nd^{3+} ions transfer their absorbed energy to neighboring Yb^{3+} ions and excite them from the $^2F_{7/2}$ (Yb^{3+}) to the $^2F_{5/2}$ (Yb^{3+}) state. Subsequently, the Yb^{3+} ion at $^2F_{5/2}$ state transfers the received energy to its neighboring Er^{3+} ion, exciting it from $^4I_{15/2}$(Er^{3+}) to $^4I_{11/2}$ (Er^{3+}) stare. Receiving transferred energy from the Yb^{3+} ion can further excite the Er^{3+} ion to the $^4F_{7/2}$ (Er^{3+}) level. Multiphonon-assisted relaxations from the $^4F_{7/2}$ state can then populate the $^{2}H_{11/2}$ (Er^{3+}) and $^{4}S_{3/2}$ (Er^{3+}) levels, which generate the 520 and 540 nm emissions by radiative decay to the ground state. In addition, the red emission around 655 nm can be acquired through transition from $^{4}F_{9/2}$ (Er^{3+}) to $^{4}I_{15/2}$ (Er^{3+}). At the same time, a portion of the Er^{3+} ions at $^{4}F_{9/2}$ state can receive energy transfers from Yb^{3+} ions and be promoted to the $^{4}G_{11/2}$ state, from which the $^{2}H_{9/2}$ state can be populated through nonradiative relaxation process. Decay of the excited Er^{3+} ions in $^{2}H_{9/2}$ state to $^{4}I_{15/2}$state results in the purple emission centered at 410 nm.

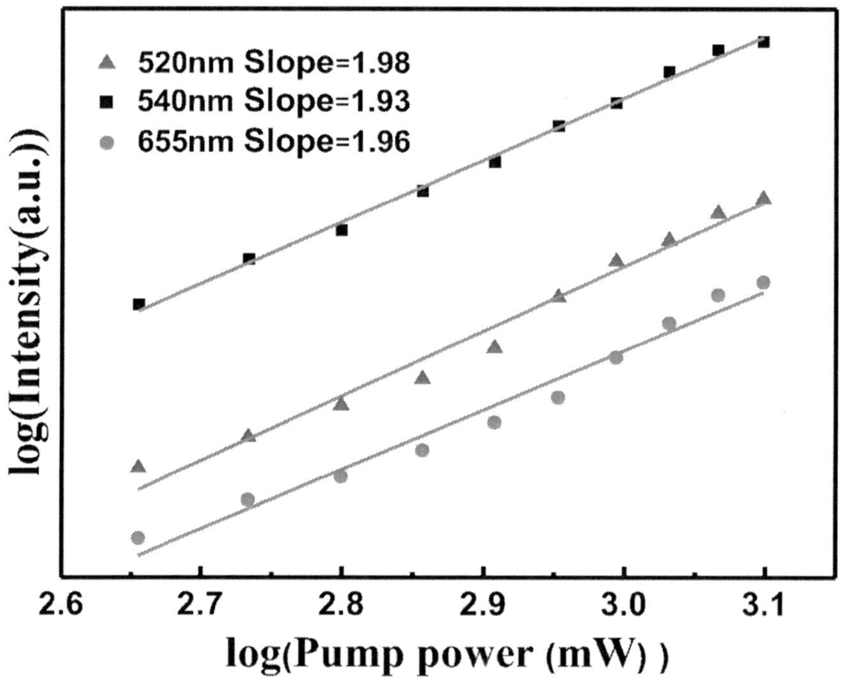

Figure 9. Pump power dependence of the fluorescent bands centered at 520, 540, and 655 nm from $NaYF_4$:10% Nd^{3+}, 10% Yb^{3+}, 2% Er^{3+} on pumping power.

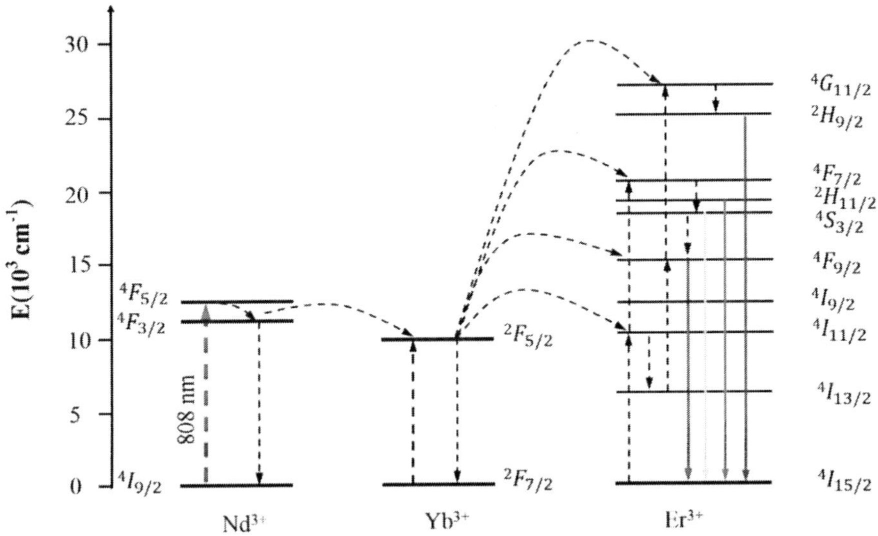

Figure 10. Proposed energy transfer mechanism of Nd^{3+}, Yb^{3+}, Er^{3+}ions following diode laser excitation of 808 nm. The dashed-dotted, dashed, dotted, and full arrows represent the photon excitation, energy transfer, and emission processes.

3. EXPERIMENTAL SECTION

3.1. Preparation

The $NaYF_4$ microcrystals were hydrothermally prepared using oleic acid as shape modifier, and NaF, $Y(NO_3)_3$ as precursors at 180 °C under a basic condition. In a typical synthesis of $NaYF_4$ nanorods, 1.08 g (27 mmol) of NaOH, 9 mL (24 mmol) of oleic acid (OA) (90 wt.%), and 10 mL (120 mmol) of ethanol were well mixed at room temperature to get a white viscous solution. Then, 4 mL (3.6 mmol) of 0.252g NaF (F^-/Ln^{3+} = 5:1) solution was added with vigorous stirring until a translucent solution was obtained. Then 6.41 mL (1.2 mmol) of 0.1872 M $Y(NO_3)_3$ was poured into the above solution keeping vigorous stirring. After aging for 1 h, the mixture was transferred to a 50-mL Teflon-lined autoclave, and heated at 180 °C for 24 h. The $NaYF_4$ microplates were prepared under similar conditions except that the amount of OA doubled. It is noted that the total molar composition of rare earth metals remains constant. For example, to prepare $NaYF_4$ doped with 10% Nd^{3+}, 10% Yb^{3+}, 2% Er^{3+}, 5 mL (0.936 mmol) 0.1872 M $Y(NO_3)_3$, 1 mL (0.12 mmol) 0.12 M $Nd(NO_3)_3$, 1 mL (0.12 mmol) 0.12 M $Yb(NO_3)_3$, and 1 mL (0.024 mmol) of 0.024 M $Er(NO_3)_3$ were mixed together instead of 6.41 mL of

0.1872 M $Y(NO_3)_3$. The obtained microcrystals were washed with ethanol and water to remove the oleic acid and other remnants, and then dried in the air at 60 °C for 12 h.

3.2. Characterization

The as-prepared samples were characterized by X-ray powder diffraction (XRD) on a Rigaku D/max-γB diffractometer, which was equipped with a rotating anode and a Cu Kα source (λ = 0.154056 nm). Micrographs of the prepared powders were obtained by using a field emission scanning electron microscope (FESEM, MX2600FE, AIKE Sepp, Oxford, UK). To measure the emitted UC luminescence, the synthesized powders were pressed to form a smooth, flat disk, which was irradiated with a focused 5 W power-controllable 808 nm diode laser (Hi-Tech Optoelectronics Co. Ltd, Beijing, China). The emitted UC luminescence was then collected by a lens-coupled monochromator (Zolix Instruments Co. Ltd, Beijing, China) of 2 nm spectral resolution with an attached photomultiplier tube (Hamamatsu CR131, Hamamatsu Photonics, Hamamatsu, Japan).

4. CONCLUSIONS

In conclusion, we have presented our systematic synthesis results on hexagonal $NaYF_4$:10% Nd^{3+}, 10% Yb^{3+}, 2% Er^{3+} micro- and nanocrystals with various morphologies and size. Morphologies with nanorods, short microrods, and microplates can be successfully achieved by varying the amount of OA coordination ligand in the initial precursor solution. This is ascribed to the fact that the OA is able to modulate the growth rate of (0001) and (1011) crystallographic facets due to their preferential attachment to them, thus governing the formation of the final morphologies. Moreover, through tuning the molar ratio of F^-/Ln^{3+}, sphere-like, octadecahedral, and hexagonal microprisms with protruding centers $NaYF_4$:10% Nd^{3+}, 10% Yb^{3+}, 2% Er^{3+} can be obtained. This shape evolution is a direct cooperative result between the attachment of F^- and OA to the involved crystal facets of (0001), (1011), and (1010). Furthermore, under 808 nm laser excitation and the same measurement conditions, spectroscopic investigations of the resulting particles revealed a morphology-, size-, and crystallinity-dependent upconverted luminescence; the highest luminescence intensity was from $NaYF_4$:10% Nd^{3+}, 10% Yb^{3+}, 2% Er^{3+} microplates. These Nd^{3+}-sensitized hexagonal $NaYF_4$ nanoparticles with various size and shape have important implications for biophotonic applications with minimized heating effects.

ACKNOWLEDGMENTS

This work is supported by the Natural Science Foundation of China (51102066), the Fundamental Research Funds for the Central Universities (Grant No. HIT. NSRIF.2015048 and AUGA5710052614), international scientific and technological cooperation projects (Grant No. 2014DFA50740).

AUTHOR CONTRIBUTIONS

Y.S and S.H conceived and designed the study. Y.S was responsible for the experimental work. The manuscript was written by Y.S and S.H. S.H, C.Y and G.C reviewed and edited the manuscript. All of the authors read and approved the manuscript.

CONFLICTS OF INTEREST

The authors declare no conflict of interest.

REFERENCES

1. Haase, M.; Schäfer, H. Upconverting nanoparticles. *Angew. Chem. Int. Ed.* **2011**, *50*, 5808–5829.

2. Shen, J.; Zhao, L.; Han, G. Lanthanide-doped upconverting luminescent nanoparticle platforms for optical imaging-guided drug delivery and therapy. *Adv. Drug Deliv. Rev.* **2013**, *65*, 744–755.

3. Gu, Z.; Yan, L.; Tian, G.; Li, S.; Chai, Z.; Zhao, Y. Recent advances in design and fabrication of upconversion nanoparticles and their safe theranostic applications. *Adv. Mater.* **2013**, *25*, 3758–3779.

4. Liu, Y.; Tu, D.; Zhu, H.; Ma, E.; Chen, X. Lanthanide-doped luminescent nano-bioprobes: From fundamentals to biodetection. *Nanoscale* **2013**, *5*, 1369–1384.

5. Hao, S.W.; Chen, G.Y.; Yang, C.H. Sensing using rare-earth-doped upconversion nanoparticles.*Theranostics* **2013**, *3*, 331–345.

6. Chen, G.Y.; Shen, J.; Ohulchanskyy, T.Y.; Patel, N.J.; Kutikov, A.; Li, Z.P.; Song, J.; Pandey, R.K.; Agren, H.; Prasad, P.N.; *et al.* (alpha-$NaYbF_4:Tm^{3+}$)/CaF_2 core/shell nanoparticles with efficient near-

infrared to near-infrared upconversion for high-contrast deep tissue bioimaging. *ACS Nano***2012**, *6*, 8280–8287.

7. Chen, G.Y.; Yang, C.H.; Prasad, P.N. Nanophotonics and nanochemistry: Controlling the excitation dynamics for frequency up- and down-conversion. *Acc. Chem. Res.* **2013**, *46*, 1474–1486.

8. Tian, G.; Gu, Z.J.; Zhou, L.J.; Yin, W.Y.; Liu, X.X.; Yan, L.; Jin, S.; Ren, W.L.; Xing, G.M.; Li, S.J.; *et al.* Mn^{2+} dopant-controlled synthesis of $NaYF_4$:Yb/Er upconversion nanoparticles for *in vivo*imaging and drug delivery. *Adv. Mater.* **2012**, *24*, 1226–1231.

9. Yang, D.; Kang, X.; Ma, P.; Dai, Y.; Hou, Z.; Cheng, Z.; Li, C.; Lin, J. Hollow structured upconversion luminescent $NaYF_4:Yb^{3+}$, Er^{3+} nanospheres for cell imaging and targeted anti-cancer drug delivery. *Biomaterials* **2013**, *34*, 1601–1612.

10. Shan, J.N.; Budijono, S.J.; Hu, G.H.; Yao, N.; Kang, Y.B.; Ju, Y.G.; Prud'homme, R.K. Pegylated composite nanoparticles containing upconverting phosphors and meso-tetraphenyl porphine (TPP) for photodynamic therapy. *Adv. Funct. Mater.* **2011**, *21*, 2488–2495.

11. Zhang, P.; Steelant, W.; Kumar, M.; Scholfield, M. Versatile photosensitizers for photodynamic therapy at infrared excitation. *J. Am. Chem. Soc.* **2007**, *129*, 4526–4527.

12. Zou, W.Q.; Visser, C.; Maduro, J.A.; Pshenichnikov, M.S.; Hummelen, J.C. Broadband dye-sensitized upconversion of near-infrared light. *Nat. Photonics* **2012**, *6*, 560–564.

13. Xie, X.; Gao, N.; Deng, R.; Sun, Q.; Xu, Q.-H.; Liu, X. Mechanistic investigation of photon upconversion in Nd^{3+}-sensitized core–shell nanoparticles. *J. Am. Chem. Soc.* **2013**, *135*, 12608–12611.

14. Xie, X.; Liu, X. Photonics upconversion goes broadband. *Nat. Mater.* **2012**, *11*, 842–843.

15. Weber, M.J. Optical properties of Yb^{3+} and Nd^{3+}-Yb^{3+} energy transfer in $YAlO_3$. *Phys. Rev. B* **1971**,*4*, 3153–3159.

16. Kushida, T.; Marcos, H.M.; Geusic, J.E. Laser transition cross section and fluorescence branching ratio for Nd^{3+} in yttrium aluminum garnet. *Phys. Rev.* **1968**, *167*, 289–291.

17. Liégard, F.; Doualan, J.L.; Moncorgé, R.; Bettinelli, M. Nd^{3+}-Yb^{3+} energy transfer in a codoped metaphosphate glass as a model

for Yb^{3+} laser operation around 980 nm. *Appl. Phys. B* **2005**, *80*, 985–991.

18. Wang, Y.-F.; Liu, G.-Y.; Sun, L.-D.; Xiao, J.-W.; Zhou, J.-C.; Yan, C.-H. Nd^{3+}-sensitized upconversion nanophosphors: Efficient *in vivo* bioimaging probes with minimized heating effect.*ACS Nano* **2013**, *7*, 7200–7206.

19. Shen, J.; Chen, G.; Vu, A.-M.; Fan, W.; Bilsel, O.S.; Chang, C.-C.; Han, G. Engineering the upconversion nanoparticle excitation wavelength: Cascade sensitization of tri-doped upconversion colloidal nanoparticles at 800 nm. *Adv. Opt. Mater.* **2013**, *1*, 644–650.

20. Wen, H.; Zhu, H.; Chen, X.; Hung, T.F.; Wang, B.; Zhu, G.; Yu, S.F.; Wang, F. Upconverting near-infrared light through energy management in core-shell-shell nanoparticles. *Angew. Chem. Int. Ed.***2013**, *52*, 13419–13423.

21. Cui, H.; Hong, C.; Ying, A.; Yang, X.; Ren, S. Ultrathin gold nanowire-functionalized carbon nanotubes for hybrid molecular sensing. *ACS Nano* **2013**, *7*, 7805–7811.

22. Chen, G.Y.; Qiu, H.L.; Prasad, P.N.; Chen, X.Y. Upconversion nanoparticles: Design, nanochemistry, and applications in theranostics. *Chem. Rev.* **2014**, *114*, 5161–5214.

23. Dou, Q.; Zhang, Y. Tuning of the structure and emission spectra of upconversion nanocrystals by alkali ion doping. *Langmuir* **2011**, *27*, 13236–13241.

24. Qu, X.; Yang, H.K.; Pan, G.; Chung, J.W.; Moon, B.K.; Choi, B.C.; Jeong, J.H. Controlled fabrication and shape-dependent luminescence properties of hexagonal $NaCeF_4$, $NaCeF_4$:Tb^{3+} nanorods via polyol-mediated solvothermal route. *Inorg. Chem.* **2011**, *50*, 3387–3393.

25. Hao, S.; Shao, W.; Qiu, H.L.; Shang, Y.F.; Fan, R.W.; Guo, X.Y.; Zhao, L.L.; Chen, G.Y.; Yang, C.H. Tuning the size and upconversion emission of $NaYF_4$:Yb^{3+}/Pr^{3+} nanoparticles through Yb^{3+}doping. *RSC Adv.* **2014**, *4*, 56302–56306.

26. Li, C.; Quan, Z.; Yang, P.; Huang, S.; Lian, H.; Lin, J. Shape-controllable synthesis and upconversion properties of lutetium fluoride (doped with Yb^{3+}/Er^{3+}) microcrystals by hydrothermal process. *J. Phys. Chem. C* **2008**, *112*, 13395–13404.

27. Hao, S.; Chen, G.; Qiu, H.; Xu, C.; Fan, R.; Meng, X.; Yang, C. Controlled growth along circumferential edge and upconverting luminescence of β-$NaYF_4$: 20%Yb^{3+}, 1%Er^{3+} microcrystals.*Mater. Chem. Phys.* **2012**, *137*, 97–102.

28. Lin, M.; Zhao, Y.; Liu, M.; Qiu, M.S.; Dong, Y.Q.; Duan, Z.F.; Li, Y.H.; Murphy, B.P.; Lu, T.J.; Xu, F. Synthesis of upconversion $NaYF_4$:Yb^{3+}, Er^{3+} particles with enhanced luminescent intensity through control of morphology and phase. *J. Mater. Chem. C* **2014**, *2*, 3671–3676.

29. Kim, F.; Connor, S.; Song, H.; Kuykendall, T.; Yang, P.D. Platonic gold nanocrystals. *Angew. Chem. Int. Ed.* **2004**, *43*, 3673–3677.

30. Zhu, L.; Qin, L.; Liu, X.D.; Li, J.Y.; Zhang, Y.F.; Meng, J.; Cao, X.Q. Morphological control and luminescent properties of CeF_3 nanocrystals. *J. Phys. Chem. C* **2007**, *111*, 5898–5903.

Chapter 5

Recent Advances in the Optimization and Functionalization of Upconversion Nanomaterials for *in Vivo* Bioapplications

Wei Feng[1], Xingjun Zhu[1] and Fuyou Li[1]

[1]*Department of Chemistry and State Key Laboratory of Molecular Engineering of Polymers, Fudan University, Shanghai, People's Republic of China*

Topof page

ABSTRACT

Upconversion nanophosphors are considered to be important components in advanced materials designed for *in vivo* bioapplications and benefit from their unique anti-Stokes upconversion luminescence properties. The near-infrared light excitation allows for a large penetration depth in biotissues during *in vivo* applications. This review presents recent advances in the optimization and functionalization of upconversion nanomaterials for bioimaging and *in vivo* therapy. The most effective protocols are introduced in detail. Current limitations and future prospects are also included to provide a general summary and suggest future research directions.

Keywords: bioimaging; *in vivo*; therapy; upconversion nanomaterials

INTRODUCTION

Upconversion luminescence (UCL) materials have attracted significant attention in recent years because of their unique ability to generate shorter wavelength emissions under longer wavelength excitations.[1] This anti-Stokes luminescence process requires the luminescent center to subsequently absorb the energy of at least two photons. The mechanism is significantly different from that of other anti-Stokes processes, such as two-photon luminescence or second harmonic generation, in which the absorption of multiple photons occurs simultaneously.

Recently, nanosized upconversion materials have been used for *in vivo*bioapplications.[2, 3] Using near-infrared (NIR) radiation as an excitation source, visible or NIR emissions can be collected as signals for bioimaging. Because of the minimum absorption and scattering effects in biotissues for light in the NIR range, upconversion bioimaging is expected to have a large tissue penetration depth that is suitable for *in vivo* applications.[4, 5] Currently reported results already demonstrate the ability to obtain signals at penetration depths of 3.2 cm.[6] In addition, the unique upconversion process avoids interference from the autofluorescence of biotissues because no biotissues have similar abilities to generate anti-Stokes luminescence with large Stokes shift. In addition, commonly used upconversion nanoprobes are inorganic nanocrystals that have the extremely high stability needed to ensure that the signals are steady and reliable.[7] Beyond the simple upconversion nanomaterials for bioimaging, nanocomposites containing upconversion nanophosphors (UCNPs) can also be designed to act as drug delivery containers or NIR-phototriggered devices.[8] This type of progress has greatly accelerated research into luminescence-based *in vivo* bioapplications.

To date, some researchers have already summarized the synthesis and emission color-tuning processes for the upconversion of nanoparticles and their applications in bioimaging.[3, 9, 10, 11, 12, 13, 14, 15] Therefore, this review concentrates on recent advances made in the optimization and functionalization of upconversion nanomaterials for innovative *in vivo* bioapplication designs. The main discussion focuses on how the designs or protocols ensure that the upconversion nanomaterials are superior to other types of luminescent material for *in vivo* applications.

Topof page

Typical upconversion nanomaterials

To date, the most popular upconversion nanomaterials are based on lanthanide elements that utilize abundant metastable excited states with long lifetimes as the intermediate state in the upconversion process.[1] The host lattices are usually rare earth fluorides that have low phonon energies to minimize the possibility of nonradiative transition. The most commonly used emitters are Er^{3+}, Tm^{3+} and Ho^{3+} that have main upconversion emission bands located in the green and red regions, the blue and NIR regions and the green region, respectively (Figure 1).[16] To date, the most widely investigated upconversion nanosystem is the $NaREF_4$ nanocrystal, where RE stands for rare earth elements, including Y and all the lanthanides (La to Lu). Because of the low solubility of $NaREF_4$ nanocrystals in both aqueous and organic solutions, $NaREF_4$ upconversion nanocrystals are easily synthesized using the coprecipitation method from corresponding precursors to provide the required amount of elements. In addition, carrying out the coprecipitation method in high-boiling-point solvents permits more effective control over the size and shape of the products. This method is now considered the standard method for obtaining upconversion nanocrystals.[17, 18, 19, 20, 21, 22] Therefore, most advances made in recent years are based on these upconversion $NaREF_4$ nanocrystals.

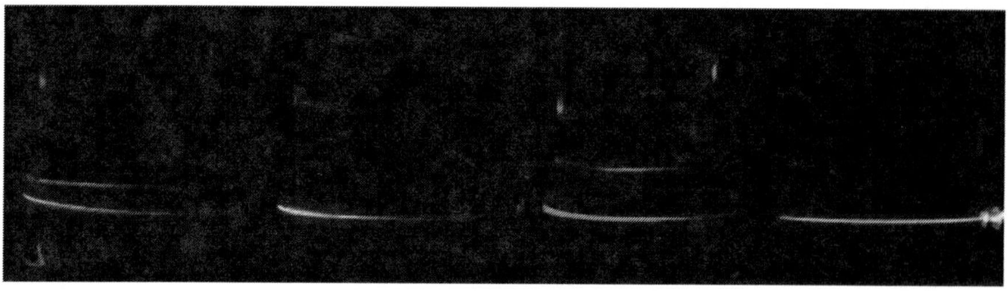

Figure 1. *Typical upconversion emission photographs from upconversion nanophosphors (UCNPs) under excitation with a 980 nm laser, including $NaYbF_4$:Tm, $NaYbF_4$:Ho, $NaYbF_4$:Er and $NaYF_4$:Yb (seen from left to right). The pictures shown here are taken without optical filters and are reproduced with permission from Ehlert et al.*[16] *Copyright 2008 American Chemical Society.*

Topof page

High-efficiency upconversion nanomaterials for *in vivo*bioapplications

Low brightness is one of the main drawbacks of upconversion nanomaterials and limits their range during *in vivo* applications. In a typical application, a laser with a relatively high excitation power density must be used to achieve efficient UCL. For *in vivo* bioapplications in particular, a high excitation power (>500 mW cm^{-2}) leads to a greater heating effect and increases the cost of the instruments.[23] To solve this problem, recent efforts have attempted to improve the upconversion efficiency of UCNPs by reducing the required excitation power density and improving the performance of UCNPs in bioapplications.

The low brightness of UCNPs originates from two main causes. One is the extremely low quantum efficiency. Nanosized lanthanide luminescent materials usually have much lower luminescent efficiencies than the corresponding bulk material because of large surface-to-volume ratios and poor crystallinity that produce many defects that act as quenching centers. This occurrence is an extremely important issue for UCNPs because in the upconversion process, both the intermediate excited state and the final excited state can transfer excitation energy to the quenching centers that significantly reduces the upconversion efficiency.[24] To date, the upconversion quantum efficiency of small-size UCNPs has been reported to be <1%.[25, 26]

To solve these problems, new materials are being designed to reduce the possibility of nonradiative transition, enhance the absorption of the sensitizers or use a combination of these methods to improve the upconversion efficiency of UCNPs.

Topof page

Reduce the nonradiative transition possibility

The selection of the proper host lattice can enhance the upconversion emission intensities of UCNPs. In bulk upconversion materials, hexagonal phase $NaYF_4$ is thought to be an ideal host because of its ultra-low phonon energy.[27] However, in nanoscale upconversion nanocrystals, the existence of a larger number of surface quenching centers and crystal defects originating from the small size effect enables more nonradiative transition routes that reduce the upconversion efficiency of the upconversion nanomaterials. This effect is extremely important for the UCNPs designed for *in vivo* bioapplications because small particles are necessary to enable quick excretion.

Our research group has recently reported that sub-10 nm $NaLuF_4$-based upconversion nanocrystals that are synthesized using the thermal decomposition method have higher upconversion efficiencies than corresponding $NaYF_4$-based upconversion nanocrystals obtained using the same method.[28] The absolute quantum yield of the upconversion emission centered at 800 nm for hexagonal phase $NaLu(Gd)F_4$:Yb,Tm nanocrystals (~8 nm) can reach as high as 0.47%(Figure 2). The increase in the upconversion efficiency of UCNPs also improves their *in vivo* bioimaging performance as a result of the increase in the penetration depth and detection limits.

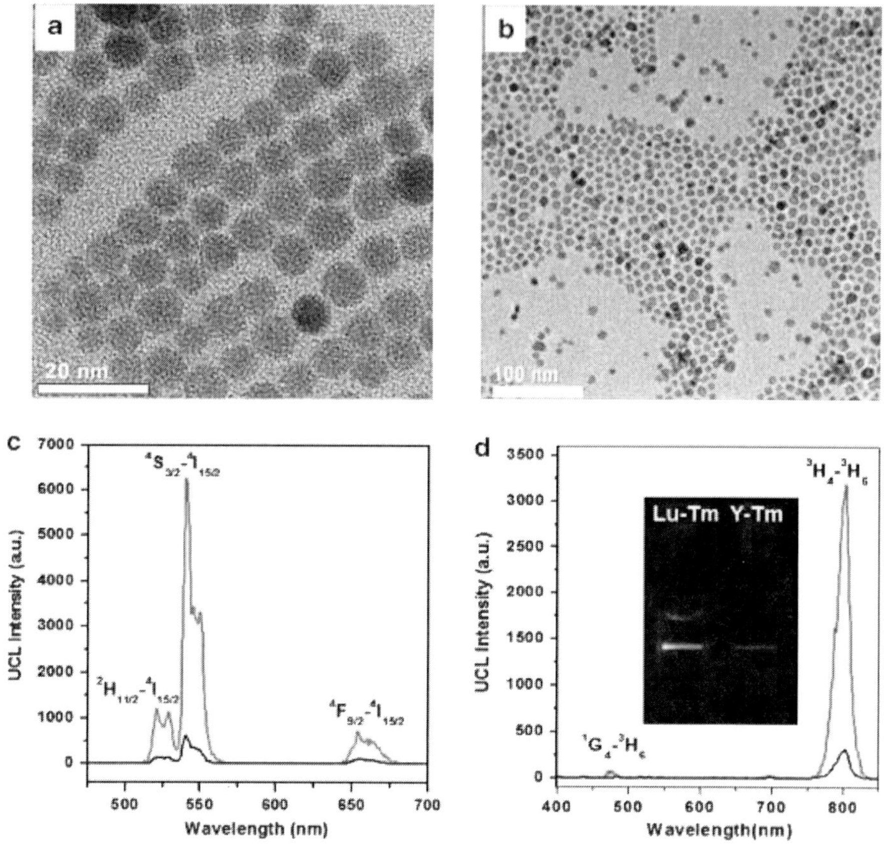

Figure 2. Transmission electron microscopy (TEM) images of (a) sub-10 nm β-phase $NaLuF_4$:24%Gd,20%Yb,2%Er (Lu) and (b) sub-20 nm α-phase $NaLuF_4$:20%Yb,2%Er; also shown are the upconversion luminescence (UCL) spectra of (c) Lu, (red line), $NaYF_4$: 20%Yb,2%Er (Y, black line), (d) $NaLuF_4$:24%Gd,20%Yb,2%Tm (Lu–Tm, red line), and $NaYF_4$: 20%Yb, 1%Tm (Y–Tm black line). The inset shows a photograph of the UCL emission of Lu–Tm and Y–Tm. Reprinted with permission from Liu*et al.*[26] and Yang *et al.*[29] Copyright 2011 American Chemical Society. Copyright 2012 Elsevier B.V.

Li and Yang and coworkers[29] reported that cubic-phase $NaLuF_4$:Yb,Tm UCNPs have higher upconversion efficiencies than $NaYF_4$:Yb,Tm UCNPs with similar dimensions (Figure 2). The as-prepared $NaLuF_4$:Yb,Tm can even be used for *in vivo* bioimaging of fur-rich mice, black mice and rabbits. In bulk materials, using $NaLuF_4$ as the host does not show obvious efficiency enhancements compared with using a $NaYF_4$ host; thus, the increase in the upconversion efficiency is mainly ascribed to the reduction in the number of surface ligands. This reduction may be the result of the different coordination abilities between the surface ligands and the various rare earth ions.

The construction of a core–shell structure is one of the most effective ways to improve the upconversion efficiency. The principle and typical structures of core–shell UCNPs designed to increase the upconversion efficiency are shown in Figure 3. Core-inert shell structures, in which the shell is the undoped host material, are designed to reduce the possibility of nonradiative transition. The introduction of an inert shell usually requires the shell material to have the same or similar lattice parameters as the core structure to minimize any lattice strains. Because there is no sensitizer or activator ion in the inert shell, the shell layer removes the energy transfer route from the activator (or sensitizer) to the surface quenching centers that reduces the possibility of nonradiative transition and increases the upconversion efficiency.

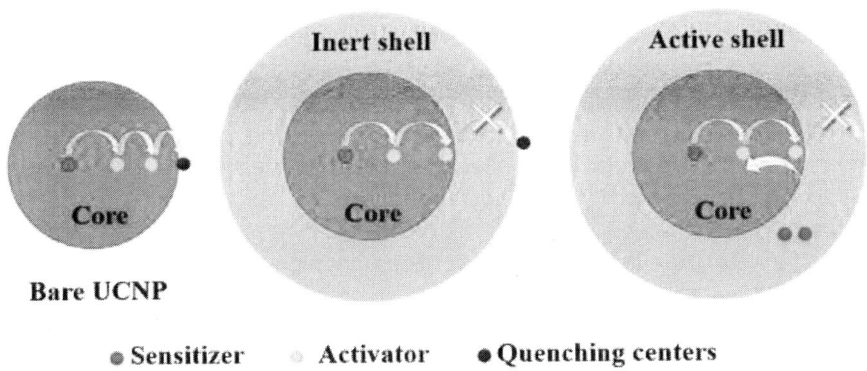

Figure 3. The design and structures of core–shell upconversion nanocomposites for improving upconversion efficiencies. UCNP, upconversion nanophosphor.

Early work in the synthesis of core–shell nanocomposites is based on the epitaxial growth of an additional shell layer with the same lattice as the core material, but without any sensitizer or activator ion. Because rare earth ions have similar properties and ionic radii, the doping of lanthanide ions causes few changes to the lattice parameters of the host material, and the inert layer can be the same as the host lattice. For example, Yan and coworkers[30] synthesized a core–shell α-$NaYF_4$:Yb,Er/Tm@α-$NaYF_4$ nanocomposite using a two-step growth method. The as-prepared α-$NaYF_4$:Yb,Er/Tm nanocrystals were used as crystallization nuclei and placed into the reaction solutions to further the growth of the inert α-$NaYF_4$ shell. Zhang and coworkers[31] also fabricated core–shell β-$NaYF_4$:Yb,Er/Tm@β-$NaYF_4$ nanocomposites to improve the upconversion efficiency. This method is widely used and has been extended to similar upconversion systems, such as rare earth fluorides,[30, 32] oxides[33] and oxyfluorides.[34] In addition, Zhang and Zhao and coworkers[35] controlled the addition of the precursors in the shell layer to precisely control the thickness of the shell with an accuracy at a monolayer level. Furthermore, they investigated the relationship between the thickness of the shell layer and the upconversion emission intensities. An electron energy loss spectroscopy mapping technique was employed at low temperatures to characterize the shell thickness and to avoid any damage originating from electron collisions.

Recently, van Veggel and coworkers[36] demonstrated a new protocol to synthesize core–shell structures. Using α-$NaYF_4$ nanocrystals as precursors, β-$NaYF_4$ shells were grown on the surfaces of β-$NaYF_4$:Yb,Er nanocrystals (core) layer by layer via a recrystallization process. The thickness of the shell layer can be controlled by adjusting the initial ratio of α-$NaYF_4$ to β-$NaYF_4$:Yb,Er in the reaction solution.

In addition to the growth of a shell with the same lattice as the host of core material, other inorganic host materials with similar lattice parameters are also used to construct core–shell upconversion nanocomposites. Recently, Yan and coworkers[37] developed a CaF_2-coated upconversion nanocomposite for improving the UCL efficiency. Because CaF_2 has nearly the same lattice parameter as the cubic $NaYF_4$ host, it is possible to precisely control the epitaxial growth process to obtain the desired thickness in the shell layer. The core–shell structure is easily characterized in a high-angle annular dark-field image because of the wide difference between the atomic numbers of Y and Ca. The upconversion emission intensity can be enhanced by ~300 times in $NaYF_4$:Yb,Er@CaF_2. Furthermore, the CaF_2 shell hinders the effective release of rare earth ions into the biosurroundings, reducing the possible toxicity of UCNPs (Figure 4).

Enhance the absorption ability

Another reason for the low upconversion efficiency in UCNPs is the low absorption cross-section of the lanthanide ions. The absorption cross-section of Yb^{3+}, a commonly used sensitizer, is much lower than for organic dyes, quantum dots or noble metal nanocrystals.[38] Therefore, the absorption efficiency of UCNPs is lower than for the conventional fluorescence process.

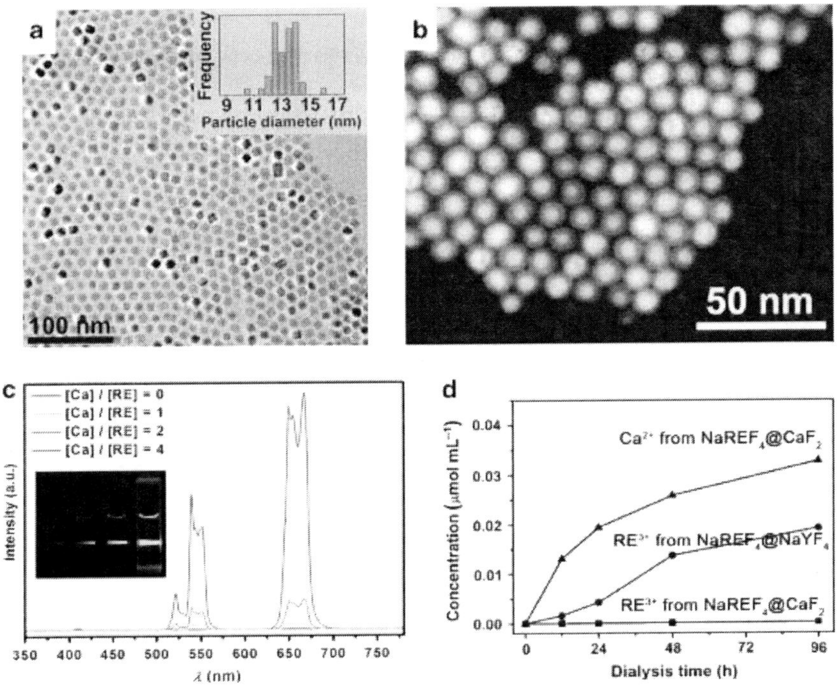

Figure 4. (a) Transmission electron microscopy (TEM) image, (b) HAADF image and (c) upconversion emission spectra of the core–shell $NaYF_4$:Yb,Er@CaF_2 nanocomposites with different Ca/RE (RE=Y+Yb+Er) ratios; (d) the released concentration of rare earth or Ca^{2+} ions from different upconversion nanostructures, indicating the suppressed RE^{3+}leakage upon coating of the CaF_2 shell. Reproduced with permission from Wang *et al.*[37] Copyright 2012 Wiley-VCH Verlag GmbH & Co. KGaA, Weinheim.

To solve this problem, an active shell has recently been designed for upconversion nanocomposites to increase the upconversion efficiency by doping the sensitizer Yb^{3+} ions in the shell layer. The additional sensitizer ions increase the possibility to absorb excitation photons, improving the UCL.

For example, to absorb more excitation photons at 980 nm, Capobianco and coworkers[39] fabricated a core–shell $NaGdF_4$:Yb,Er@$NaGdF_4$:Yb nanocomposite by introducing an active shell containing an appropriate amount of Yb^{3+} ions. Some other core-active shell nanostructures, such as $BaGdF_5$:Yb,Er@$BaGdF_5$:Yb[40] and @$NaYF_4$:Yb,Er@$NaYF_4$:Yb,[41] have also been fabricated to enhance the upconversion emission intensities. In these core–shell structured UCNPs, the amount of Yb^{3+} in the active shell needs to be optimized because the Yb^{3+} can also provide a route to transfer the absorbed excitation energy to the surface quenching centers, which weakens the upconversion emission.

Another route for increasing the absorption of NIR light is by utilizing a well-known metal surface-enhanced fluorescence phenomenon. It has been proven that incident electric fields redistribute on the surface of metal nanocrystals because of the existence of surface plasmons in metals. The intensity of the electric field can be enhanced by several orders of magnitude in this way at some hot spots. Commonly used metals include Gold (Au) and Silver (Ag) for the surface plasmon resonance bands located in the visible range. Early studies have proven that the upconversion emissions from Er^{3+} in solid-state films are enhanced in the presence of Ag nanocrystals.[42] Yan and coworkers[43] reported that the upconversion emissions from $NaYF_4$:Yb,Er nanocrystals are also enhanced in the presence of Ag nanowires, which proves it is possible to construct nanocomposites with noble metals to enhance the upconversion emission intensity. Duan and coworkers[44] demonstrated enhancements to the UCL of $NaYF_4$:Yb,Tm nanocrystals by absorbing Au nanocrystals through electrostatic interactions with the assistance of polyelectrolytes. The emission intensity was enhanced by up to 150% with the proper amount of Au nanoparticles. In addition, a series of similar upconversion nanocomposites with noble metals has been fabricated to enhance the upconversion emissions.[45, 46, 47, 48, 49] Unfortunately, most of this work focuses on observing the enhancement effects rather than describing the detailed mechanism behind them. Therefore, there is no confirmed information available to direct the design of such nanocomposites to further improve the upconversion emissions until now.

The third and newest route to enhance the absorption abilities of UCNPs is to use a NIR organic dye as an energy absorber. Although organic dyes have larger absorption cross-sections than commonly used sensitizer Yb^{3+} ions, it is difficult to use a sensitizer to increase the UCL because of the lack of fundamental studies on the energy transfer process from organic dyes to rare earth ions. Recently, Hummelen and coworkers[50] proved the feasibility of using a NIR dye to increase the absorption of excitation light. A commercial cyanine dye, IR-780, was selected as the antenna molecule. After carboxylic acid functionalization, the IR-780 bound to the surface of a $NaYF_4$:Yb,Er nanocrystal (UCNPs) instead of the original oleylamine through a ligand exchange process. The resulting nanocomposite had an enhanced upconversion emission that is ~1100 times stronger than that of the original oleylamine-coated UCNPs and broadened the excitation peaks to permit the utilization of more excitation light. This experimental result is important because it enables the use of organic dyes to sensitize lanthanide upconversion nanophosphors.

UCNPs based on triplet–triplet annihilation

Another type of UCNP using a triplet–triplet annihilation (TTA) upconversion process has recently been employed for *in vivo* bioapplications using organic molecules as sensitizers. For this upconversion mechanism, the sensitizer and emitter are both organic molecules, and the triplet state of these organic molecules is used as the intermediate state in the upconversion process.[51] TTA-based UCL processes benefit from the high absorption cross-section of organic dyes and the highly efficient TTA process between emitter molecules that usually have much stronger absorption and higher UCL efficiencies than lanthanide-based UCL processes. To date, the highest quantum efficiency reported for TTA-based UCL processes exceeds 40%,[52] and some systems have proven to be functional under excitation from only sunlight.[51]

The problem limiting bioapplications of TTA-based upconversion materials is that the TTA process needs to occur in a medium with proper liquidity to ensure opportunities for collisions between the emitters and sensitizers so that energy transfer takes place. Therefore, TTA-based materials cannot be used for *in vivo*bioapplications until a suitable assembly method is developed to obtain upconversion nanoparticles with fluxible, or at least flexible, media inside. Recently, we reported a silica coating method to wrap the sensitizer and emitter molecules together in an inorganic structure with the assistance of polyethyleneglycol F127 (Figure 5).[53] The as-prepared 10 nm silica nanospheres, containing 2,7,8,12,13,17,18-octaethylporphyrin palladium (PdOEP) and 9,10-diphenylanthracene (DPA) molecules (as shown in Figure 5), were used as sensitizers and emitters, respectively. These nanospheres could be used as upconversion nanoprobes for *in vivo* bioimaging experiments despite the upconversion emission located in the blue spectral region when they were subjected to a 532 nm laser excitation. This was the first demonstration of TTA-based UCNPs being used for *in vivo* bioimaging of the whole body of a small animal. The excitation power density can decrease to 8.5 mW cm^{-2} because the PdOEP sensitizer has a large absorption cross-section. The quantum yield of these TTA UCNPs reached 4.5% (for an excitation power density of 260 mW cm^{-2}) in an aqueous solution.

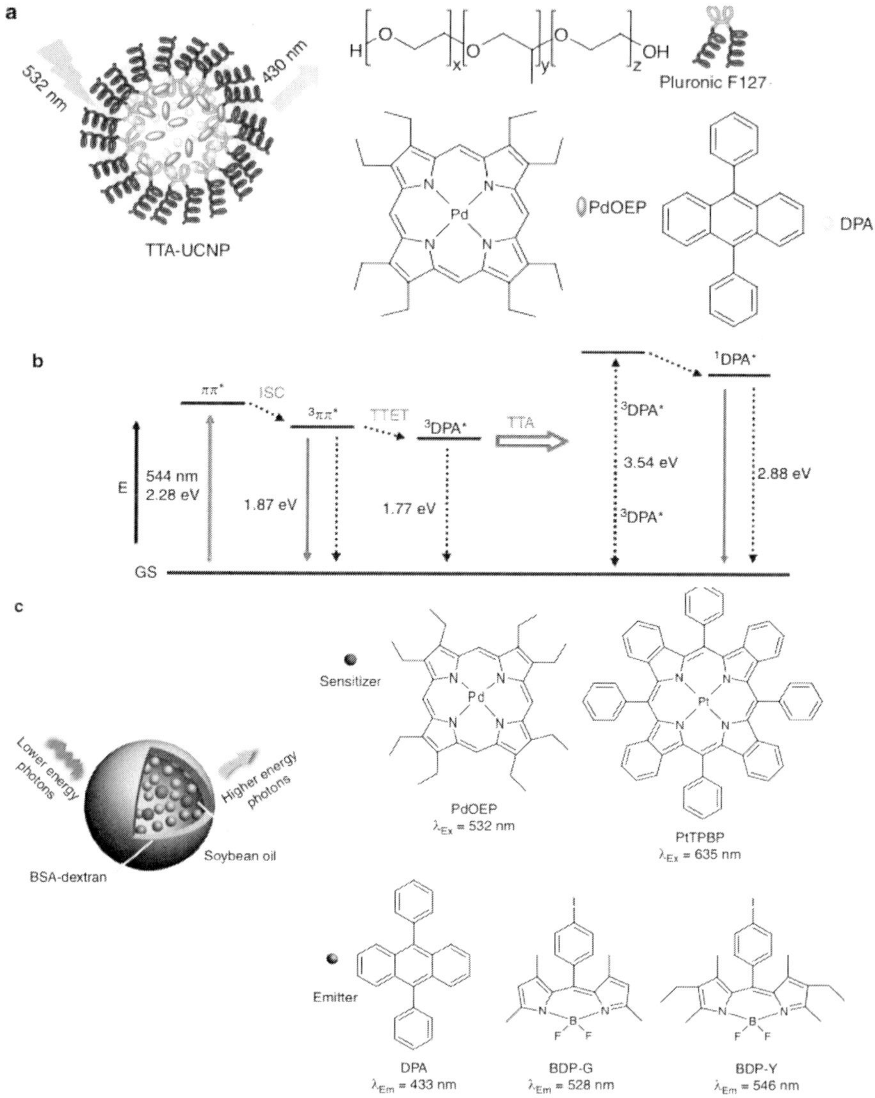

Figure 5. (a) The structure of the hydrophilic triplet–triplet annihilation upconversion nanophosphor (TTA-UCNP) based on silica coating in which the yellow ring represents the silica shell; (b) a schematic illustration of the corresponding upconversion mechanism based on the TTA process and (c) the structure of the bovine serum albumin (BSA)–dextran-based TTA-UCNP, sensitizers and emitters. Reproduced with permission from Liu *et al.*[53, 54] Copyright 2010 American Chemical Society.

Furthermore, we developed a nanocapsule strategy to load TTA-based UCL systems into nanoparticles.[54] The nanocapsules were composed of an oil droplet and a surface bovine serum albumin–dextran layer to protect the oil phase. The TTA sensitizer and emitter molecules were easily loaded into the oil droplets without significantly affecting their mobility and luminescence properties. The high solubility of the sensitizer and emitter molecules in the oil droplets also suppressed possible aggregation-induced quenching of the luminescence. By using platinum(II) tetraphenyltetrabenzoporphyrin (PtTPBP) and boron-dipyrromethene (BODIPY)-based molecules as the sensitizer and emitter (Figure 5), respectively, the quantum yield of the nanocapsule reached 4.8% in an aqueous solution. Because the development of TTA-based upconversion nanocomposites is still in the initial stages, we only provide these examples here. The following chapters will still focus on lanthanide-based upconversion nanomaterials.

Surface modification of UCNPs for biocompatibility

UCNPs for *in vivo* bioapplications must have biocompatible surface properties. Because the majority of high-efficiency and uniform UCNPs are synthesized in organic solvents by the thermal decomposition method or the modified solvothermal method, the surfaces of these UCNPs are capped by hydrophobic organic ligands that hinder the dispersion of UCNPs in biocompatible aqueous solutions.[17, 18, 19, 20] To date, various methods have been developed to modify the surface of UCNPs and are summarized in the published reviews.[3] Almost all types of functional bioimaging applications employing UCNPs use surface modification methods to introduce the required surface ligands to improve biocompatibility.

Silica coatings are generally used to provide biocompatibility to hydrophobic UCNPs. Reverse micelle-based silica coating is a well-developed method for coating a silica layer on hydrophobic UCNPs.[19] Because this method is well investigated and reviewed, it is not discussed in detail here. Alternatively, we focus on a recently developed ligand exchange procedure that is a convenient and effective method for modifying the surface of hydrophobic UNCPs.

In particular, three methods have recently been developed to synthesize organic ligand-free UCNPs. Capobianco and coworkers[55] used HCl to acidify a solution containing OA (oleic acid)-capped UCNPs. OA molecules can be protonated when the pH of the solution is adjusted to a particular value that greatly weakens the coordination ability of the OA molecules to the rare earth ions. Thus, the OA ligands are replaced by other counter-ions, and the hydrophobic UCNPs become hydrophilic. Murray and coworkers[56] developed

$NOBF_4$ as an agent to remove the original capping ligand on the hydrophobic UCNPs. The ion NO^+ can replace surface rare earth ions and coordination ligands. The as-prepared UCNPs are hydrophilic, with NO^+ on the surface and BF_4^- as the counter-ion (Figure 6). Our group reported a similar protocol using Gd^{3+} as cations instead of NO^+ to replace the surface rare earth ions and form a 'naked' surface.[57]

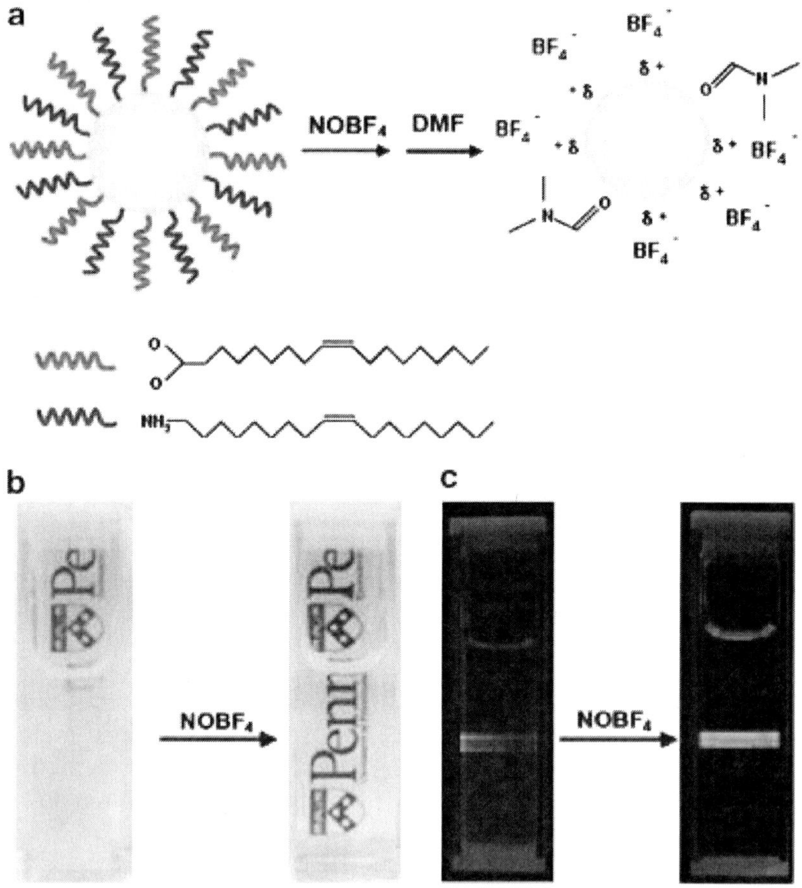

Figure 6. (a) Schematic illustration of the preparation procedure to produce ligand-free upconversion nanophosphors (UCNPs) with the assistance of $NOBF_4$, (b) the improved colloidal stability and (c) the upconversion efficiency after treatment. Reproduced with permission from Dong *et al.*[56] Copyright 2010 American Chemical Society.

Importantly, the interaction between surface rare earth ions and the introduced counter-ions is quite weak; thus, the counter-ions are easily replaced by other additional, functional ligands with good coordination abilities with respect to rare earth ions. Therefore, the so-called ligand-free state is available to bind to the following ligands as required, and this method can be used as a general protocol for the surface modification of OA-capped UCNPs.

Although these ligand exchange protocols have been proven to effectively modify the surface of UCNPs, the detailed state of the newly introduced surface ligands is not clear after the upconversion materials are injected into small animals and needs to be explored in future research.

After causing the hydrophobic UCNPs to become hydrophilic, further surface modifications are applied to prolong the circulation time of the UCNPs in the blood. Typically, after intravenous injection, the UCNPs can only perform simple imaging on the liver, spleen or lungs, through which the UCNPs are filtered or captured by the macrophage system of the animal. Various efforts have been undertaken to solve this problem, and polyethylene glycol (PEG) is now regarded as the most effective surface ligand to prolong the circulation time of UCNPs in blood by suppressing the capture of UCNPs by the macrophage system. The PEG moieties can be introduced onto UCNPs by using amino- or COOH-modified PEG species or by using some amphiphilic polymers containing PEG chains directly. Liu and coworkers[58] investigated the long-term biodistribution of UCNPs with different surface modification species. Compared with the polyacrylic acid-modified UCNPs, PEG-coated UCNPs have longer blood circulation half-lives. This result proves that PEG reduces the nonspecific interactions of UCNPs with the macrophage system and delays uptake by macrophages.

Topof page

Upconversion nanomaterials as functional indicators for bioimaging

Recent developments in upconversion nanomaterial research have focused on the design and fabrication of upconversion probes for functional bioimaging. To date, upconversion probes have been successfully designed for cell tracking and for *in vivo* monitoring of the lymph nodes, tumors and specific analytes. Multimodality bioimaging probes have also been designed for colocalization of the probes.

UCNPs for lymph bioimaging

The lymphatic system holds a very important place in the tumor metastasis process, but it is difficult to study the lymphatic system because of the shortage of imaging techniques with proper sensitivity and temporal–spatial resolution. UCNPs are ideal fluorescence probes used for *in vivo* lymph imaging because the probes injected into the paw of a mouse can easily reach the sentinel lymph nodes through lymphatic vessels and exhibit intense UCL signals.[59, 60] Recently, TTA-based UCNPs have also been successfully used for *in vivo* lymph imaging. Our group has used silica-coated PdOEP and DPA nanoparticles with blue upconversion emission (excited at 532 nm) for the imaging of lymph nodes in living mice.[53] Because the detection signal from the blue upconversion emission becomes poor as the penetration depth increases, imaging can only be obtained after removal of the skin of the mice. If another TTA-based UCNP composed of PtTPBP and BDP, oil and bovine serum albumin–dextran possessing a green (or yellow) upconversion emission (excited at 635 nm) is used as the probe, then lymphatic images can be easily obtained *in vivo* and without removal of the skin.[54]

Topof page

UCNPs designed for tumor-targeted imaging

In addition to basic imaging of the liver or spleen by simply injecting the hydrophilic UCNPs into the mouse via the tail vein, researchers also want to utilize UCNPs for functional imaging experiments. Tumor targeting is one of the most attractive functional imaging applications because of the importance of tumor detection and in-clinic therapy.

Some studies have reported that the enhanced permeability and retention effect of tumors can cause nanosized probes to concentrate in the tumor site. However, the targeting efficiency is relatively low for UCNPs, and additional functionalization is needed for more accurate targeting. Folic acid, peptides and antibodies are the most commonly used targeting groups to help nanoparticles find tumor sites.

Our group has reported using folic acid-modified $NaYF_4$:Yb,Er (FA-UCNPs) for *in vivo* tumor target imaging.[61] Folic acid is joined with the amino group containing UCNPs to target tumor cells with a folic receptor. After intravenous injection of FA-UCNPs into a HeLa tumor-bearing athymic nude mouse for 24 h, a red upconversion emission was detected at the tumor site. Our group also used an Arg-Gly-Asp (RGD) tripeptide to target $\alpha_v\beta_3$-overexpressing U87MG tumors.[62] To accomplish this, RGD tripeptide was covalently bonded to PEG-NH_2-modified $NaYF_4$:Yb,Tm nanocrystals (RGD-UCNPs). The 800 nm upconversion emission of RGD-UCNPs was detected at tumor sites, reaching a maximum emission 4 h after intravenous injection of the UCNPs into a U87MG tumor-bearing mouse.

In addition, antibodies can be joined to the UCNPs for targeted imaging of tumor cells, based on the antibody–antigen interaction. For example, Xu, Mao and coworkers[63] linked rabbit anti-CEA8 antibodies to silica-coated $NaYF_4$:Yb,Er UCNPs with $-NH_2$ groups to target a carcinoembryonic antigen in HeLa cells. However, thus far there has been no mention in the literature of tumor imaging at the animal level using UCNPs designed to focus on the antibody–antigen interaction.

UCNPs for monitoring analytes *in vivo*

UCNPs can be used for upconversion detection in a nanosystem composed of UCNPs and chemodosimeters with an efficient energy transfer process. If this nanosystem is introduced into animals, the detection of upconversion may be used to monitor the concentration of analytes *in vivo*.

Very recently, our group employed a hydrophobic heptamethine cyanine dye (hCy7) as a $MeHg^+$-responsive chemodosimeter to combine with UCNPs to detect $MeHg^+$ *in vivo*.[64] An amphiphilic polymer (P-PEG) was used to coat the $NaYF_4$:Yb,Er,Tm UCNP to provide a hydrophobic layer to load hCy7 and a hydrophilic layer to endow the nanocomposite with biocompatibility. The absorption of the hCy7 corresponded to an upconversion emission at 660 nm from Er^{3+} so that efficient energy transfer would occur to stop any green upconversion emissions in the initial state. After the addition of $MeHg^+$, the hCy7 reacted with $MeHg^+$ and changed to hCy7′, and the absorption band redshifted to 845 nm. This matched the upconversion emission at 800 nm from Tm^{3+}. The ratio of the upconversion emission intensity at 660 nm to that at 800 nm can be used as a detection signal to monitor the concentration of $MeHg^+$. Because of the high penetration depth of the radiation at 800 nm, these nanocomposites were further applied to monitor the presence of $MeHg^+$ in the liver of a living mouse for *in vivo* upconversion detection (Figure 7). Although this is the first example of *in vivo* upconversion detection at the animal level, the concept and protocol shown in this report are broad and can be used to construct similar systems to monitor other functional species *in vivo*, thus providing information about the analytes in a more direct and timely manner.

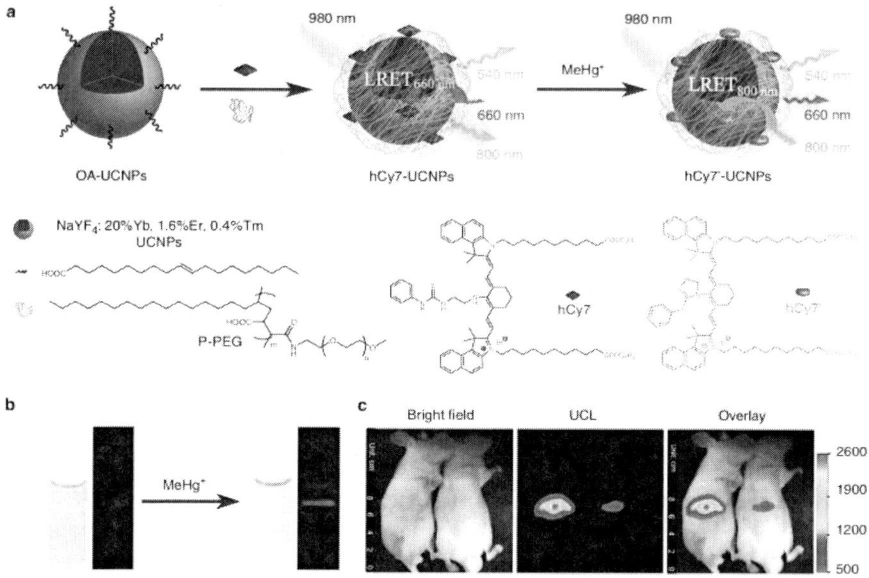

Figure 7. (a) Schematic design of the upconversion probes for *in vivo* monitoring of $MeHg^+$, (b) the detection of $MeHg^+$ in solution and (c) a mouse. Reproduced with permission from Liu *et al.*[64] Copyright 2013 American Chemical Society.

UCNPs for cell tracking

Another bioimaging application of UCNPs is cell tracking. After incubation with UCNPs, specific cells can be stained by the UCNPs and exhibit upconversion emissions during their activity in the biosystem. Because of the ultra-high photostability of UCNPs, cell tracking using UCNPs as probes is expected to allow longer-term tracking of cells than what is presently accomplished with commercial dyes. For example, Liu and coworkers[65] used $NaY_{0.78}Yb_{0.2}Er_{0.02}F_4$, $NaY_{0.69}Yb_{0.3}Er_{0.01}F_4$ and $NaY_{0.78}Yb_{0.2}Tm_{0.02}F_4$ nanocrystals with different emission wavelengths to label mouth epidermal carcinoma (KB) cells for cell tracking. KB cells labeled by UCNPs with different emission wavelengths were injected into different sites of a nude mouse. At 1 week after injection, the upconversion emissions from three types of UCNPs were detectable in the three different tumor sites, as expected, that demonstrates the cell tracking ability of UCNPs (Figure 8).

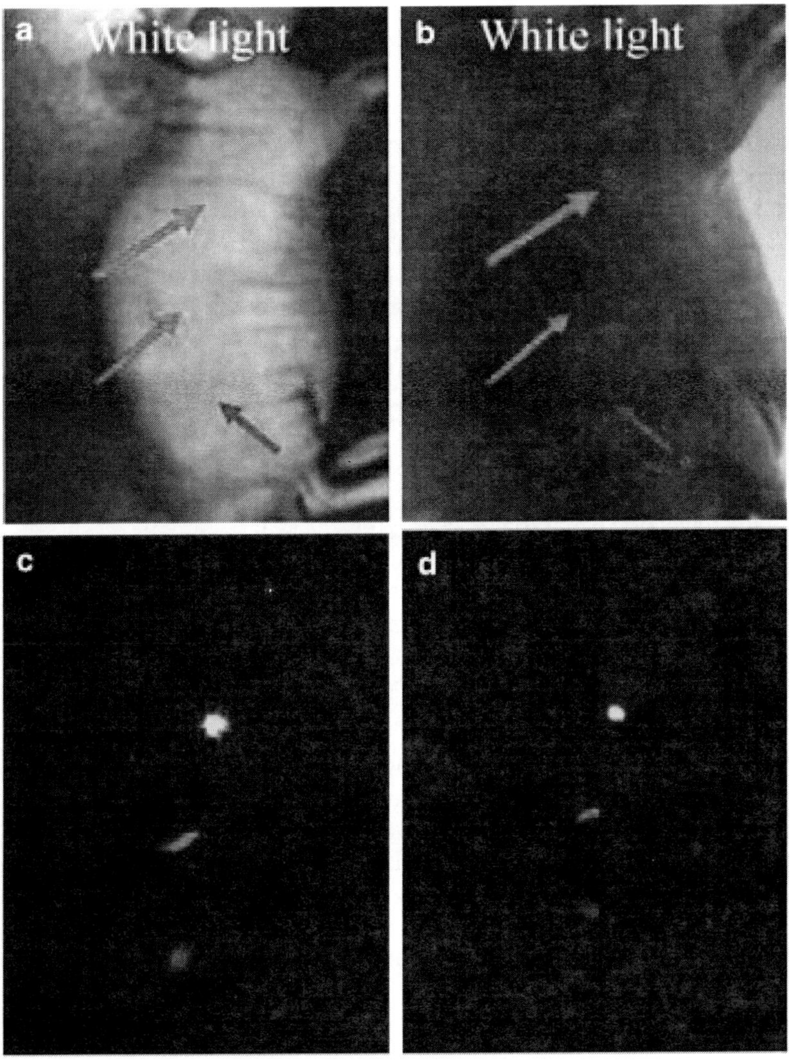

Figure 8. Multicolor *in vivo* cancer cell tracking and imaging with different upconversion nanophosphors (UCNPs). The upper images are (a) white light photos of a nude mouse after subcutaneous injection of the UCNP-labeled KB cells and (b) the same mouse after 1 week with visible tumor tissues. The lower images (c, d) are the corresponding false color upconversion luminescence images. Reproduced with permission from Cheng *et al.*[65] Copyright 2010 The Authors.

Our group also reported on core–shell structured $NaYF_4$:Yb,Er,Tm@$NaLuF_4$ UCNPs with bright upconversion emissions that can be used for cell tracking.[66] The UCNP-labeled JEG-3 cells were then injected intravenously into a nude mouse. Tumors were detected in the lung, and the upconversion emission signal illustrated the ongoing metastasis process. This upconversion signal was still detectable 21 days after the injection. Recently, UCNPs have also been used to track stem cells. Liu and coworkers[67] used a PEG-conjugated Au-Fe_3O_4-UCNP nanocomposite to label marrow mesenchymal stem cells taken from Balb/C mice. The magnetic field induced accumulation of the labeled stem cells in the wound site, proving the potential of the upconversion imaging technique to guide the concentration of stem cells. Han and coworkers[68] also reported the labeling of rat mesenchymal stem cells and tracked the labeled stem cells to determine their osteogenic and adipogenic behaviors.

The upconversion materials used for cell tracking have no specific requirements, but effective cell staining and high photostability must be achieved without hindering the biological activities of the cell.

Multimodality imaging

Another type of upconversion nanomaterial attracting a great deal of attention for *in vivo* bioimaging is the multimodality probe. Multimodality images not only confirm the upconversion imaging results but also combine the advantages of multiple types of imaging modes that bridge gaps in sensitivity, resolution and penetration depth for multilevel molecular imaging from the cellular scale to the whole body. To date, upconversion nanoprobes have been successfully designed for integration with commercial medical imaging devices, such as magnetic resonance imaging, X-ray computer tomography, positron emission tomography (PET) and single-photon emission computed tomography (SPECT).

Typically, the T_1-enhanced magnetic resonance imaging ability is provided by Gd^{3+} doped into the lattice or joined to the surface of UCNPs.[69, 70] The T_2-enhanced magnetic resonance imaging is usually provided by Fe_3O_4 components integrated with the UCNPs.[71, 72] X-ray computer tomography imaging contrast is achieved by using the rare earth elements themselves[73] or by introducing a commercial contrast agent.[74]

Recently, PET and SPECT imaging functionality has also been developed for UCNPs. This development is especially useful for *in vivo* tracking of UCNPs and for providing useful information for biodistribution studies of these nanoprobes. Our group reported an ^{18}F labeling method to integrate PET imaging ability into UCNPs.[57] Because of the strong interaction between F and rare earth elements, $^{18}F^-$ cations were easily bonded to the surface of UCNPs by simply mixing the solution containing $^{18}F^-$ and UCNPs. After the $^{18}F^-$ labeled UCNPs were injected into the paw of a mouse, the PET imaging signal could be detected in the lymph node. Because of the high sensitivity of the PET imaging technique, this PET–UCL dual modality imaging can also be used to monitor the biodistribution of UCNPs. However, the radiative half-life of $^{18}F^-$ is only 2 h, which is too short to monitor the long-term biodistribution of UCNPs. SPECT imaging provides a robust technique for long-term observation of the probes by utilizing ^{153}Sm as the radioactive isotope with a half-life as long as 46.3 h.

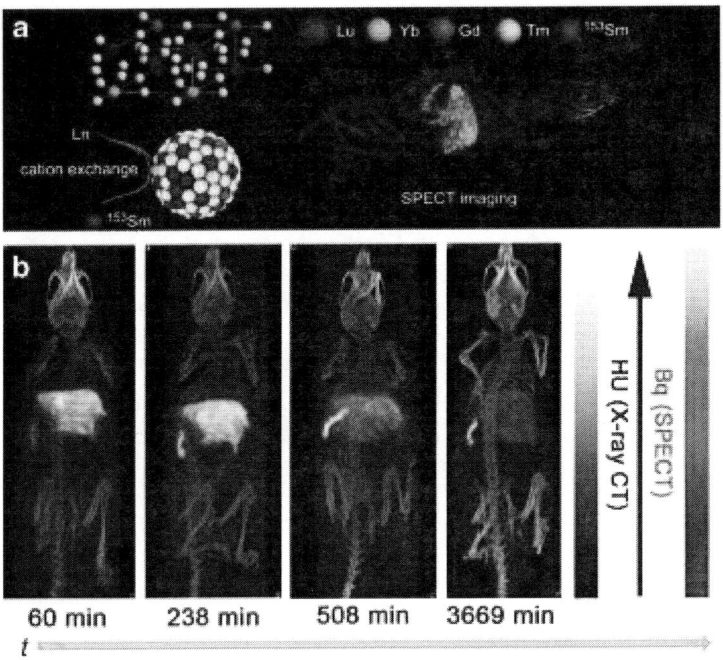

Figure 9. (a) The design and structure of the nanoparticles used to integrate radioactive ^{153}Sm into the lattice of the upconversion nanophosphors (UCNPs) and (b) an *in vivo* single-photon emission computed tomography (SPECT) image of a mouse obtained at different times after intravenous injection of ^{153}Sm-labeled UCNPs. Reproduced with permission from Sun *et al.*[75] Copyright 2012, Elsevier Ltd.

Our group also developed a cation exchange method to introduce ^{153}Sm into the lattice of UCNPs to replace the original rare earth ions.[75] The labeling procedure is fast and convenient and is based on simply mixing a solution containing $^{153}Sm^{3+}$ and UCNPs. The ^{153}Sm-labeled UCNPs can then be used to investigate the long-term *in vivo* biodistribution of UCNPs. For example, the redistribution process of UCNPs from the liver to the spleen after intravenous injection can be clearly observed in SPECT images (Figure 9). Moreover, by taking advantage of the long half-life, ^{153}Sm can also be doped into the lattice of UCNPs during the synthesis procedure.[76] The long-term biodistribution of ^{153}Sm-labeled UCNPs has been successfully studied using quantitative SPECT imaging.

Because these multimodality contrast agents can be integrated into a nanocomposite to achieve *in vivo* multimodality imaging, the combination of different imaging techniques can be easily achieved to provide *in vivo* images with high accuracy, sensitivity and spatial resolution.

Upconversion nanomaterials for therapy

In addition to imaging applications, UCNPs have recently been introduced as important components in advanced nanocomposites designed for therapy. In these nanocomposites, UCNPs can be used as radiation sources to convert the wavelength of incident light to make functional components work under NIR excitation or as luminescent indicators to help locate the position or monitor the state of nanocomposites for therapy.

UCNPs as NIR indicators for photothermal therapy

UCNPs are widely used to trace other drugs for therapy by simply integrating the drugs and UCNPs together without any interaction. In this type of nanostructure, UCNPs are only used to trace the drugs by relying on their UCL emissions. For example, photothermal therapy (PTT) is a recently developed and convenient method to kill cancer cells using heating caused by the absorption of the incident light. During the therapy process, PTT drugs only need to remain at the tumor site to absorb the incident light and do not need to be released as chemotherapy drugs. Therefore, the UCNPs integrated with the PTT drugs only need to indicate the position of the nanostructure.

Au and Ag nanoparticles with surface plasmon resonance absorption in the visible to NIR regions are the most common PTT drugs. Song and coworkers[77] synthesized core–shell structured $NaYF_4$:Yb,Er@Ag for therapeutic applications. Ag nanocrystals were grown on the surface of UCNPs with the assistance of thiol groups grafted onto the UCNPs. The as-prepared nanostructures could then be used for upconversion cell imaging. The cancer cells incubated with the $NaYF_4$:Yb,Er@Ag were killed by applying 808 nm laser radiation to heat the cells. Liu and coworkers[78] also reported on a nanostructure that combined PTT drugs and UCNPs. An Au shell was chosen as the PTT drug to be grown outside the $NaYF_4$:Yb,Er nanocrystal with the assistance of an Fe_3O_4 intermediate layer. The nanocomposites were heated by absorbing 808 nm incident radiation, and the $NaYF_4$:Yb,Er provided UCL signals. The intermediate Fe_3O_4 layer endowed the nanostructures with magnetic resonance imaging contrast abilities as well as the potential to be manipulated by an external magnetic field.

UCNPs as NIR indicators for drug delivery

Chemotherapy drugs often need to be loaded into a drug delivery system to control the release of drugs or to protect them before reaching the afflicted region. In a drug delivery system, UCNPs can first be used as simple luminescent indicators to help locate the position of the nanocomposites. The simplest system utilizes silica-coated UCNPs that have upconversion properties and a suitable cavity in the silica layer to load the drugs. Drugs can be loaded into the nanocavity by simple physical adsorption or by hydrophobic-hydrophobic interactions inside the porous silica layer. For example, Lin and coworkers[79]coated a mesoporous silica layer onto β-$NaYF_4$:Yb,Er nanocrystals to load drug molecules. The mesoporous structure of the silica layer provided a large cavity volume to increase the loading capacity of the upconversion nanocomposite. The upconversion emission was used to track the distribution of the drug carriers.

Furthermore, UCNPs can be used as indicators to monitor the release of drugs in the delivery system based on the energy transfer process. Lin and coworkers[80]loaded ibuprofen molecules into core–shell structured multifunctional Fe_3O_4@SiO_2@m-SiO_2@$NaYF_4$:Yb,Er nanocomposites. The magnetic core allowed the drug carrier to be manipulated by a external magnetic field. Because of the spectral overlap between the absorption of the ibuprofen molecules and the green upconversion emission of $NaYF_4$:Yb,Er, the effective energy transfer reduced the green upconversion emission intensity. Along with the release of ibuprofen drugs, the green upconversion emissions were detected and used as a signal for monitoring the quantity of drugs released.

Another recently developed nanostructure for drug delivery uses a yolk-shell silica layer outside the upconversion nanocrystal.[81] The cavity of the yolk-shell silica layer is much larger than the common mesoporous silica layer and allows the opportunity to load more drugs into the nanocomposites.

UCNPs as NIR phototriggers for drug delivery

In these drug delivery nanocomposites, drug molecules can only be released by the diffusion resulting from a concentration gradient that cannot be controlled during the metabolism process. Recently, phototrigger systems have been used to allow the remote-controlled release of drugs. The physical or chemical properties of the phototrigger system are modified by absorbing radiation of a specific wavelength. If a change in properties can lead to the release of drugs, then a controllable drug carrier system can be constructed. Conventional phototrigger systems work under ultraviolet (UV) or visible light excitation that cannot be used to control drug release because of their absorption by skin and biotissues. UCNPs allow the use of low-power NIR radiation, which possesses a large penetration depth, as the excitation radiation for phototrigger systems.

Branda and Zhao and coworkers[82] reported on an upconversion phototriggered micelle system composed of poly(ethylene oxide)-block-poly(4,5-dimethoxy-2-nitrobenzyl methacrylate) and $NaYF_4$:Yb,Tm@$NaYF_4$ UCNPs. The mechanism behind this system involves UCNPs converting the 980 nm excitation radiation to UV upconversion emissions and transferring the energy to the *o*-nitrobenzyl group on the micelle core-forming block that induces a photo-cleavage reaction and destroys the micelle structure. A hydrophobic compound, Nile Red, can be used as the model for drug molecules to be loaded inside the micelle. Under 980 nm laser radiation, the micelle was broken to release Nile Red, proving the viability of this NIR-triggered drug release system.

The photo-induced isomerization of azobenzene can also be used as a phototrigger switch to control the release of drugs at a cellular level.[83] In a recently reported example, the anticancer drug doxorubicin was loaded in the mesoporous silica coating on $NaYF_4$:Yb,Tm@$NaYF_4$. Azobenzene groups were covalently bonded to the walls in the mesoporous silica shell. Under NIR excitation, the UCNPs generated visible and UV emissions to drive the isomerization of azobenzene. The corresponding wagging motion acted as a molecular impeller to help release the drug molecules (Figure 10).

Subsequent research has been performed on a similar structure for *in vivo*applications. Xing and coworkers[84] demonstrated use of the 1-(2-nitrophenyl)ethyl group as a phototrigger to control the release of molecular model D-luciferin. The photo-caged D-luciferin molecules were covalently bonded to the surface of $NaYF_4$:Yb,Tm@$NaYF_4$@SiO_2 nanoparticles. Under irradiation from a 980 nm laser, UCNPs generated upconversion emissions at 365 nm and transferred energy to the 1-(2-nitrophenyl)ethyl group to release the D-luciferin molecules. The NIR radiation-driven photorelease of D-luciferin has been demonstrated in MCF-7 cells and living mice.

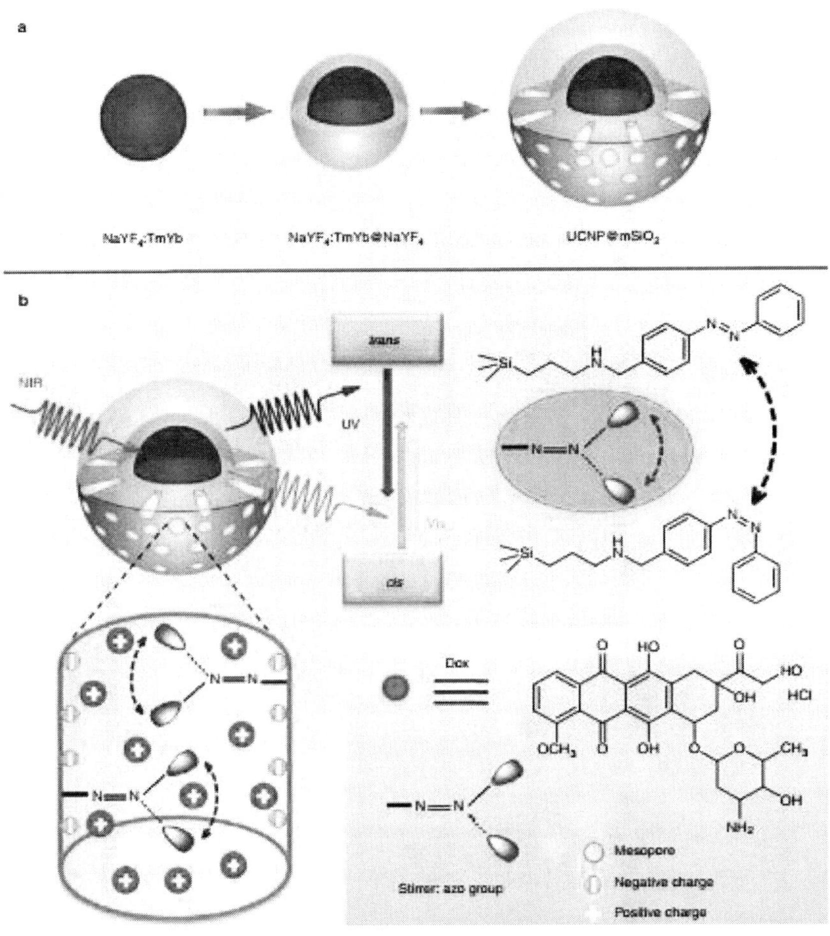

Figure 10. (a) An illustration of the structure of the near-infrared (NIR)-triggered drug delivery system and (b) the drug release mechanism driven by NIR irradiation. Reproduced with permission from Liu *et al.*[83] Copyright 2013 Wiley-VCH Verlag GmbH & Co. KGaA, Weinheim.

Because of the excellent tissue penetration depth of NIR light and high UCL energy, phototrigger systems have successfully been used with NIR light to control the delivery of small interfering RNA both *in vitro* and *in vivo*.[85, 86] As a result of UV emissions from UCNPs generated by 980 nm light exposure, the small interfering RNA loaded on the UCNPs can be released at similar levels of bioactivity by silencing the gene expression in a cell or even in deep tissue.

Different types of drug delivery systems can be fabricated with UCNPs. Future research will be carried out to enable the controllable, precise release of drugs via the upconversion process. The phototrigger groups are key factors that need to be explored to increase the quantum efficiency.

UCNPs as wavelength converter for photodynamic therapy (PDT)

Another typical therapeutic use for UCNP-based nanosystems involves employing UCNPs as wavelength converters to enable a commercial PDT agent to be activated under NIR excitation. Generally, commercial PDT drugs are excited by UV or visible light to generate reactive oxygen species and kill the target cells. The poor penetration ability of UV and visible light limits the application of current PDT drugs to the treatment of superficial diseases. The integration of UCNPs and PDT drugs utilizes the unique upconversion process to modify the NIR excitation light to UV or visible emissions. The subsequent energy transfer process from the UCNP to the PDT drugs generates reactive oxygen species and activates PDT drugs (Figure 11).[87] PDT drugs can be loaded by adsorption in the mesoporous silica layer that is coated onto UCNP surfaces to effectively receive energy from UCNPs.[88] Further research is needed to covalently bond PDT drugs to the silica layer to permit more stable PDT drugs to function in the NIR range.[32, 89]

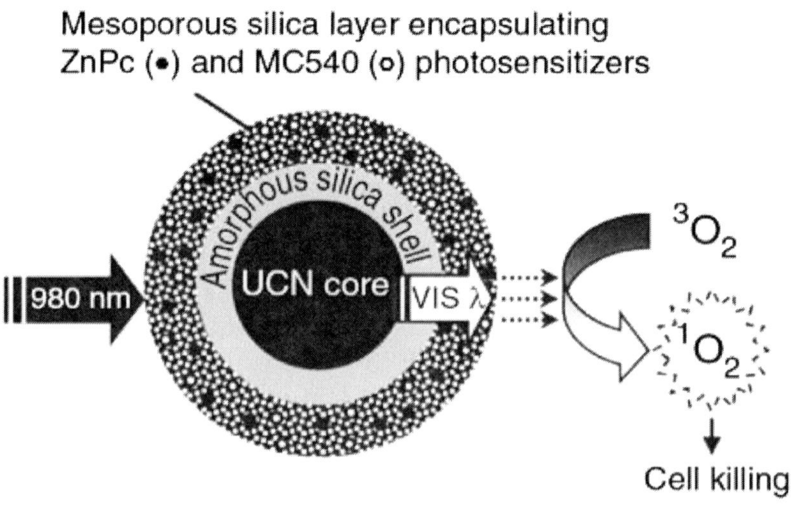

Figure 11. Schematic illustration of the upconversion nanostructure used to load drugs for near-infrared (NIR)-triggered photodynamic therapy using an energy transfer process. Reproduced with permission from Idris *et al.*[90] Copyright 2013 Wiley-VCH Verlag GmbH & Co. KGaA, Weinheim.

All of these nanosystems have proven to have significant therapeutic effects at both the cellular and animal levels. For example, Zhang and coworkers[90] loaded photosensitizer merocyanine 540 (MC540) and zinc(II) phthalocyanine (ZnPc) into a mesoporous silica layer that was coated onto $NaYF_4$:Yb,Er nanocrystals. The UCNPs generated green and red upconversion emissions to stimulate MC540 and ZnPc, respectively, to produce singlet oxygen and destroy nearby cells. The dual PDT approach based on excitation by a single wavelength laser is more effective than using a PDT agent. Liu and coworkers[91] developed a simple strategy to load the PDT agent chlorin e6 (Ce6) directly into the hydrophobic layer outside UCNPs and subsequently coat an amphiphilic polymer outside the UCNPs to make them water dispersible. The as-prepared nanocomposites achieved effective NIR light-induced PDT in the animal experiment. Hyeon and coworkers[92] also reported similar nanocomposite with Ce6 covalently bonded to amine-functionalized UCNPs for PDT applications. More recently, Liu and coworkers[93] constructed a charge-reversible nanocomposite with pH-sensitive surface properties containing Ce6 for better PDT performance against acidic tumors.

Conclusions and future prospects

The advantages of UCNPs make them useful for *in vivo* bioapplications. Rapid developments in this area have proven that UCNPs can be used as luminescent probes with high penetration depths, the absence of autofluorescence and high photostability for bioimaging. In addition to other functional components, UCNPs in nanocomposites exhibit wavelength conversion abilities. Because of the small size of UCNPs and the unique UCL process, most of the potential future applications of UCNPs are still based on *in vivo* bioapplications, where their advantages can be fully utilized. However, there are still several problems that need to be solved before clinical experiments can be initiated.

First, the luminescent efficiency of UCNPs is still low, and lasers need to be used as the excitation source for the imaging process and for other applications. The high power density of lasers leads to possible tissue damage and causes heating effects. The core–shell structure currently functions well to improve the quantum efficiency. However, the low absorption cross-section of the sensitizer Yb^{3+} still poses a major problem for future research and limits the utilization of excitation light. Some protocols discussed here attempt to solve this problem, but none have exhibited sufficient improvement to make UCNPs operational at low excitation energy densities (~10 mW cm^{-2}). Organic dye-sensitized lanthanide upconversion nanosystems seem to provide a potential route to reduce the required excitation energy density, but the detailed mechanism and upconversion efficiencies need to be further investigated. Moreover, the photostability of such a nanocomposite also becomes problematic because of the introduction of organic NIR dyes.

Recently, TTA-based upconversion materials have been reported for bioprobes as an alternative to lanthanide upconversion phosphors.[53, 54] Because of their relatively high absorption coefficient and upconversion efficiency, these TTA-based upconversion materials can be excited by low-power density radiation. However, the synthesis of such TTA-based upconversion nanomaterials through NIR emissions, as well as related issues such as long-term stability, biodistribution and toxicity, still need to be addressed.

Second, the relatively strong absorption of 980 nm light by water causes noticeable overheating effects. Improving the upconversion efficiency certainly reduces the required excitation power density and thus suppresses the effects of overheating. However, new efforts have been recently undertaken to design new structures to solve this problem. For example, Zhan and coworkers[23] reported using a 915 nm laser as an excitation source to achieve *in vivo* bioimaging with lower overheating effects. More recently, Nd^{3+} ions have been introduced as antenna to absorb the excitation light and further transfer energy to Yb^{3+} ions to sensitize activators to generate UCL at excitations of 808 nm.[94, 95, 96] This new strategy has been proven to mitigate overheating effects.

The third problem is the possible toxicity of UCNPs. To date, only a few reports have considered the biodistribution and possible excretion of UCNPs. The current data show that most UCNPs accumulate in the liver and spleen after injection and require a very long time to be cleared. Further research needs to focus on this issue to determine the possible applications for UCNPs. If UCNPs do not cause long-term effects in animals, they can be used directly by injection into the blood circulation system. Otherwise, they can only be used for applications in conjunction with a proven method to clear the UCNPs from the body. Once these concerns are addressed, UCNPs are destined for use in a new generation of probes for a wide range of *in vivo* applications.

ACKNOWLEDGEMENTS

We thank the National Science Foundation of China (21231004, 21201038), MOST of China (2011AA03A407 and 2012CB932403) and Shanghai Sci. Tech. Comm. (12JC1401300) for their financial support.

REFERENCES

1. Auzel, F. Upconversion and anti-stokes processes with f and d ions in solids. *Chem. Rev.* **104**, 139–173 (2004).

2. Chatterjee, D. K., Gnanasammandhan, M. K. & Zhang, Y. Small upconverting fluorescent nanoparticles for biomedical applications. *Small***6**, 2781–2795 (2010).

3. Zhou, J., Liu, Z. & Li, F. Y. Upconversion nanophosphors for small-animal imaging. *Chem. Soc. Rev.* **41**, 1323–1349 (2012).

4. Nyk, M., Kumar, R., Ohulchanskyy, T. Y., Bergey, E. J. & Prasad, P. N. High contrast *in vitro* and *in vivo* photoluminescence bioimaging using near infrared to near infrared up-conversion in Tm^{3+} and Yb^{3+} doped fluoride nanophosphors. *Nano Lett.* **8**, 3834–3838 (2008).

5. Pichaandi, J., Boyer, J. C., Delaney, K. R. & van Veggel, F. C. J. M. Two-photon upconversion laser (scanning and wide-field) microscopy using Ln^{3+}-doped $NaYF_4$ upconverting nanocrystals: a critical evaluation of their performance and potential in bioimaging. *J. Phys. Chem. C* **115**, 19054–19064 (2011).

6. Chen, G. Y., Shen, J., Ohulchanskyy, T. Y., Patel, N. J., Kutikov, A., Li, Z. P., Song, J., Pandey, R. K., Ågren, H., Prasad, P. N. & Han, G. (α-$NaYbF_4$:Tm^{3+})/CaF_2 core/shell nanoparticles with efficient near-infrared to near-infrared upconversion for high-contrast deep tissue bioimaging. *ACS Nano* **6**, 8280–8287 (2012).

7. Liu, Q., Feng, W., Yang, T. S., Yi, T. & Li, F. Y. Upconversion luminescence imaging of cells and small animals. *Nat. Protoc.* **8**, 2033–2044 (2013).

8. Feng, W., Han, C. M. & Li, F. Y. Upconversion-nanophosphor-based functional nanocomposites. *Adv. Mater.* **25**, 5287–5303 (2013).

9. Haase, M. & Schäfer, H. Upconverting nanoparticles. *Angew. Chem. Int. Ed.***50**, 5808–5829 (2011).

10. Wang, F., Banerjee, D., Liu, Y. S., Chen, X. Y. & Liu, X. G. Upconversion nanoparticles in biological labeling, imaging, and therapy. *Analyst* **135**, 1839–1854 (2010).

11. Mader, H. S., Kele, P., Saleh, S. M. & Wolfbeis, O. S. Upconverting luminescent nanoparticles for use in bioconjugation and bioimaging. *Curr. Opin. Chem. Biol.* **14**, 582–596 (2010).

12. Hilderbrand, S. A. & Weissleder, R. Near-infrared fluorescence: application to *in vivo* molecular imaging. *Curr. Opin. Chem. Biol.* **14**, 71–79 (2010).

13. Xu, C. T., Zhan, Q. Q., Liu, H. C., Somesfalean, G., Qian, J., He, S. L. & Andersson-Engels, S. Upconverting nanoparticles for pre-clinical diffuse optical imaging, microscopy and sensing: current trends and future challenges. *Laser Photonics Rev.* **7**, 663–697 (2013).

14. Gorris, H. H. & Wolfbeis, O. S. Photon-upconverting nanoparticles for optical encoding and multiplexing of cells, biomolecules, and microspheres.*Angew. Chem. Int. Ed.* **52**, 3584–3600 (2013).

15. Gnach, A. & Bednarkiewicz, A. Lanthanide-doped up-converting nanoparticles: merits and challenges. *Nano Today* **7**, 532–563 (2012).

16. Ehlert, O., Thomann, R., Darbandi, M. & Nann, T. A four-color colloidal multiplexing nanoparticle system. *ACS Nano* **2**, 120–124 (2008).

17. Mai, H. X., Zhang, Y. W., Si, R., Yan, Z. G., Sun, L. D., You, L. P. & Yan, C. H. High-quality sodium rare-earth fluoride nanocrystals: controlled synthesis and optical properties. *J. Am. Chem. Soc.* **128**, 6426–6436 (2006).

18. Boyer, J. C., Vetrone, F., Cuccia, L. A. & Capobianco, J. A. Synthesis of colloidal upconverting $NaYF_4$ nanocrystals doped with Er^{3+}, Yb^{3+} and Tm^{3+}, Yb^{3+} via thermal decomposition of lanthanide trifluoroacetate precursors. *J. Am. Chem. Soc.* **128**, 7444–7445 (2006).

19. Jalil, R. A. & Zhang, Y. Biocompatibility of silica coated $NaYF_4$ upconversion fluorescent nanocrystals. *Biomaterials* **29**, 4122–4128 (2008).

20. Yi, G. S. & Chow, G. M. Synthesis of hexagonal-phase $NaYF_4$:Yb,Er and $NaYF_4$:Yb,Tm nanocrystals with efficient up-conversion fluorescence. *Adv. Funct. Mater.* **16**, 2324–2329 (2006).

21. Zeng, J. H., Su, J., Li, Z. H., Yan, R. X. & Li, Y. D. Synthesis and upconversion luminescence of hexagonal-phase $NaYF_4$:Yb, Er, phosphors of controlled size and morphology. *Adv. Mater.* **17**, 2119–2123 (2005).

22. Wang, G. F., Peng, Q. & Li, Y. D. Lanthanide-doped nanocrystals: synthesis, optical-magnetic properties, and applications. *Acc. Chem. Res.* **44**, 322–332 (2011).

23. Zhan, Q. Q., Qian, J., Liang, H. J., Somesfalean, G., Wang, D., He, S. L., Zhang, Z. G. & Andersson-Engels, S. Using 915 nm laser excited $Tm^{3+}/Er^{3+}/Ho^{3+}$-doped $NaYbF_4$ upconversion nanoparticles for *in vitro* and deeper *in vivo* bioimaging without overheating irradiation. *ACS Nano* **5**, 3744–3757 (2011).

24. Wang, F., Wang, J. & Liu, X. G. Direct evidence of a surface quenching effect on size-dependent luminescence of upconversion nanoparticles.*Angew. Chem. Int. Ed.* **49**, 7456–7460 (2010).

25. Boyer, J. C. & van Veggel, F. C. J. M. Absolute quantum yield measurements of colloidal $NaYF_4:Er^{3+},Yb^{3+}$ upconverting nanoparticles.*Nanoscale* **2**, 1417–1419 (2010).

26. Liu, Q., Sun, Y., Yang, T. S., Feng, W., Li, C. G. & Li, F. Y. Sub-10 nm hexagonal lanthanide-doped $NaLuF_4$ upconversion nanocyrstals for sensitive bioimaging *in vivo*. *J. Am. Chem. Soc.* **133**, 17122–17125 (2011).

27. Krämer, K. W., Biner, D., Frei, G., Güdel, H. U., Hehlen, M. P. & Lüthi, S. R. Hexagonal sodium yttrium fluoride based green and blue emitting upconversion phosphors. *Chem. Mater.* **16**, 1244–1251 (2004).

28. Liu, Q., Sun, Y., Yang, T., Feng, W., Li, C. & Li, F. Sub-10 nm hexagonal lanthanide-doped $NaLuF_4$ upconversion nanocrystals for sensitive bioimaging *in vivo*. *J. Am. Chem. Soc.* **133**, 17122–17125 (2011).

29. Yang, T. S., Sun, Y., Liu, Q., Feng, W., Yang, P. Y. & Li, F. Y. Cubic sub-20 nm $NaLuF_4$-based upconversion nanophosphors for high-contrast bioimaging in different animal species. *Biomaterials* **33**, 3733–3742 (2012).

30. Mai, H. X., Zhang, Y. W., Sun, L. D. & Yan, C. H. Highly efficient multicolor up-conversion emissions and their mechanisms of monodisperse $NaYF_4$:Yb,Er core and core/shell-structured nanocrystals. *J. Phys. Chem. C***111**, 13721–13729 (2007).

31. Wang, Y., Tu, L. P., Zhao, J. W., Sun, Y. J., Kong, X. G. & Zhang, H. Upconversion luminescence of beta-$NaYF_4:Yb^{3+},Er^{3+}$@beta-$NaYF_4$ core/shell nanoparticles: excitation power, density and surface dependence. *J. Phys. Chem. C* **113**, 7164–7169 (2009).

32. Zhao, Z. X., Han, Y. N., Lin, C. H., Hu, D., Wang, F., Chen, X. L., Chen, Z. & Zheng, N. F. Multifunctional core–shell upconverting nanoparticles for imaging and photodynamic therapy of liver cancer cells. *Chem. Asian J.* **7**, 830–837 (2012).

33. Zhang, F., Haushalter, R. C., Haushalter, R. W., Shi, Y. F., Zhang, Y. C., Ding, K. L., Zhao, D. Y. & Stucky, G. D. Rare-earth upconverting nanobarcodes for multiplexed biological detection. *Small.* **7**, 1972–1976 (2011).

34. Yi, G. S., Peng, Y. F. & Gao, Z. Q. Strong red-emitting near-infrared-to-visible upconversion fluorescent nanoparticles. *Chem. Mater.* **23**, 2729–2734 (2011).

35. Zhang, F., Che, R. C., Li, X. M., Yao, C., Yang, J. P., Shen, D. K., Hu, P., Li, W. & Zhao, D. Y. Direct imaging the upconversion nanocrystal core/shell structure at the subnanometer level: shell thickness dependence in upconverting optical properties. *Nano Lett.* **12**, 2852–2858 (2012).

36. Johnson, N. J., Korinek, A., Dong, C. & van Veggel, F. C. J. M. Self-focusing by ostwald ripening: a strategy for layer-by-layer epitaxial growth on upconverting nanocrystals. *J. Am. Chem. Soc.* **134**, 11068–11071 (2012).

37. Wang, Y. F., Sun, L. D., Xiao, J. W., Feng, W., Zhou, J. C., Shen, J. & Yan, C. H. Rare-Earth nanoparticles with enhanced upconversion emission and suppressed rare-Earth-ion leakage. *Chem. Eur. J.* **18**, 5558–5564 (2012).

38. Li, S. F., Zhang, M., Peng, Y., Zhang, Q. Y. & Zhao, M. S. Rate equation model analysis on the infrared and upconversion emission of Er/Yb co-doped borate-silicate glass. *J. Rare Earths* **28**, 237–242 (2010).

39. Vetrone, F., Naccache, R., Mahalingam, V., Morgan, C. G. & Capobianco, J. A. The active-core/active-shell approach: a strategy to enhance the upconversion luminescence in lanthanide-doped nanoparticles. *Adv. Funct. Mater.* **19**, 2924–2929 (2009).

40. Yang, D. M., Li, C. X., Li, G. G., Shang, M. M., Kang, X. J. & Lin, J. Colloidal synthesis and remarkable enhancement of the upconversion luminescence of $BaGdF_5$:Yb^{3+}/Er^{3+} nanoparticles by active-shell modification. *J. Mater. Chem.* **21**, 5923–5927 (2011).

41. Liu, X. M., Kong, X. G., Zhang, Y. L., Tu, L. P., Wang, Y., Zeng, Q. H., Li, C. G., Shi, Z. & Zhang, H. Breakthrough in concentration quenching threshold of upconversion luminescence via spatial separation of the

emitter doping area for bio-applications. *Chem. Commun.* **47**, 11957 (2011).

42. Mertens, H. & Polman, A. Plasmon-enhanced erbium luminescence. *Appl. Phys. Lett.* **89**, 211107 (2006).

43. Feng, W., Sun, L. D. & Yan, C. H. Ag nanowires enhanced upconversion emission of $NaYF_4$:Yb,Er nanocrystals via a direct assembly method. *Chem. Commun.* **29**, 4393–4395 (2009).

44. Zhang, H., Li, Y. J., Ivanov, I. A., Qu, Y. Q., Huang, Y. & Duan, X. F. Plasmonic modulation of the upconversion fluorescence in $NaYF_4$:Yb/Tm hexaplate nanocrystals using gold nanoparticles or nanoshells. *Angew. Chem. Int. Ed.* **49**, 2865–2868 (2010).

45. Schietinger, S., Aichele, T., Wang, H. Q., Nann, T. & Benson, O. Plasmon-enhanced upconversion in single $NaYF_4:Yb^{3+}/Er^{3+}$ codoped nanocrystals.*Nano Lett.* **10**, 134–138 (2010).

46. Liu, N., Qin, W. P., Qin, G. S., Jiang, T. & Zhao, D. Highly plasmon-enhanced upconversion emissions from Au@beta-NaYF4:Yb,Tm hybrid nanostructures. *Chem. Commun.* **47**, 7671–7673 (2011).

47. Yuan, P. Y., Lee, Y. H., Gnanasammandhan, M. K., Guan, Z. P., Zhang, Y. & Xu, Q. H. Plasmon enhanced upconversion luminescence of $NaYF_4$:Yb, Er@SiO_2@Ag core-shell nanocomposites for cell imaging. *Nanoscale* **4**, 5132–5137 (2012).

48. Zhang, F., Braun, G. B., Shi, Y. F., Zhang, Y. C., Sun, X. H., Reich, N. O., Zhao, D. Y. & Stucky, G. Fabrication of Ag@SiO_2@Y_2O_3:Er nanostructures for bioimaging: tuning of the upconversion fluorescence with silver nanoparticles. *J. Am. Chem. Soc.* **132**, 2850–2851 (2010).

49. Fujii, M., Nakano, T., Imakita, K. & Hayashi, S. Upconversion luminescence of Er and Yb codoped $NaYF_4$ nanoparticles with metal shells. *J. Phys. Chem. C* **117**, 1113–1120 (2013).

50. Zou, W. Q., Visser, C., Maduro, J. A., Pshenichnikov, M. S. & Hummelen, J. C. Broadband dye-sensitized upconversion of near-infrared light. *Nat. Photonics* **6**, 560–564 (2012).

51. Singh-Rachford, T. N. & Castellano, F. N. Photon upconversion based on sensitized triplet-triplet annihilation. *Coordin. Chem. Rev.* **254**, 2560–2573 (2010).

52. Cheng, Y. Y., Fückel, B., Khoury, T., Clady, R. G. C. R., Tayebjee, M. J. Y., Ekins-Daukes, N. J., Crossley, M. J. & Schmidt, T. W. Kinetic analysis of photochemical upconversion by triplet–triplet annihilation: beyond any spin statistical limit. *J. Phys. Chem. Lett.* **1**, 1795–1799 (2010).

53. Liu, Q., Yang, T. S., Feng, W. & Li, F. Y. Blue-emissive upconversion nanoparticles for low-power-excited bioimaging *in vivo*. *J. Am. Chem. Soc.***134**, 5390–5397 (2012).

54. Liu, Q., Yin, B. R., Yang, T. S., Yang, Y. C., Shen, Z., Yao, P. & Li, F. Y. A general strategy for biocompatible, high-effective upconversion nanocapsules based on triplet–triplet annihilation. *J. Am. Chem. Soc.* **135**, 5029–5037 (2013).

55. Bogdan, N., Vetrone, F., Ozin, G. A. & Capobianco, J. A. Synthesis of ligand-free colloidally stable water dispersible brightly luminescent lanthanide-doped upconverting nanoparticles. *Nano Lett.* **11**, 835–840 (2011).

56. Dong, A. G., Ye, X. C., Chen, J., Kang, Y. J., Gordon, T., Kikkawa, J. M. & Murray, C. B. A generalized ligand-exchange strategy enabling sequential surface functionalization of colloidal nanocrystals. *J. Am. Chem. Soc.* **133**, 998–1006 (2010).

57. Liu, Q., Sun, Y., Li, C. G., Zhou, J., Li, C. Y., Yang, T. S., Zhang, X. Z., Yi, T., Wu, D. M. & Li, F. Y. ^{18}F-Labeled magnetic-upconversion nanophosphors via rare-Earth cation-assisted ligand assembly. *ACS Nano* **5**, 3146–3157 (2011).

58. Cheng, L., Yang, K., Shao, M. W., Lu, X. H. & Liu, Z. *In vivo*pharmacokinetics, long-term biodistribution and toxicology study of functionalized upconversion nanoparticles in mice. *Nanomedicine* **6**, 1327–1340 (2011).

59. Kobayashi, H., Kosaka, N., Ogawa, M., Morgan, N. Y., Smith, P. D., Murray, C. B., Ye, X. C., Collins, J., Kumar, G. A., Bell, H. & Choyke, P. L.*In vivo* multiple color lymphatic imaging using upconverting nanocrystals.*J. Mater. Chem.* **19**, 6481–6484 (2009).

60. Cao, T. Y., Yang, Y., Gao, Y., Zhou, J., Li, Z. Q. & Li, F. Y. High-quality water-soluble and surface-functionalized upconversion nanocrystals as luminescent probes for bioimaging. *Biomaterials* **32**, 2959–2968 (2011).

61. Xiong, L. Q., Chen, Z. G., Yu, M. X., Li, F. Y., Liu, C. & Huang, C. H. Synthesis, characterization, and *in vivo* targeted imaging of amine-

functionalized rare-earth up-converting nanophosphors. *Biomaterials* **30**, 5592–5600 (2009).

62. Xiong, L. Q., Chen, Z. G., Tian, Q. W., Cao, T. Y., Xu, C. J. & Li, F. Y. High contrast upconversion luminescence targeted imaging *in vivo* using peptide-labeled nanophosphors. *Anal. Chem.* **81**, 8687–8694 (2009).

63. Wang, M., Mi, C. C., Wang, W. X., Liu, C. H., Wu, Y. F., Xu, Z. R., Mao, C. B. & Xu, S. K. Immunolabeling and NIR-excited fluorescent imaging of HeLa cells by using $NaYF_4$:Yb,Er upconversion nanoparticles. *ACS Nano* **3**, 1580–1586 (2009).

64. Liu, Y., Chen, M., Cao, T. Y., Sun, Y., Li, C. Y., Liu, Q., Yang, T. S., Yao, L. M., Feng, W. & Li, F. Y. A cyanine-modified nanosystem for *in vivo*upconversion luminescence bioimaging of methylmercury. *J. Am. Chem. Soc.* **135**, 9869–9876 (2013).

65. Cheng, L., Yang, K., Zhang, S., Shao, M. W., Lee, S. T. & Liu, Z. Highly-sensitive multiplexed *in vivo* imaging using PEGylated upconversion nanoparticles. *Nano Res.* **3**, 722–732 (2010).

66. Sun, Y. Multifunctionalized Rare Earth Upconversion Nanophosphors for *in Vivo* Imaging. Doctoral thesis, Fudan Univ. (2012).

67. Cheng, L., Wang, C., Ma, X. X., Wang, Q. L., Cheng, Y., Wang, H., Li, Y. G. & Liu, Z. Multifunctional upconversion nanoparticles for dual-modal imaging-guided stem cell therapy under remote magnetic control. *Adv. Funct. Mater.* **23**, 272–280 (2013).

68. Zhao, L., Kutikov, A., Shen, J., Duan, C. Y., Song, J. & Han, G. Stem cell labeling using polyethylenimine conjugated α-$NaYbF_4$:Tm^{3+}/CaF_2upconversion nanoparticles. *Theranostics* **3**, 249–257 (2013).

69. Chen, F., Bu, W., Zhang, S., Liu, J., Fan, W., Zhou, L., Peng, W. & Shi, J. Gd^{3+}-ion-doped upconversion nanoprobes: relaxivity mechanism probing and sensitivity optimization. *Adv. Funct. Mater.* **23**, 298–307 (2013).

70. Xia, A., Chen, M., Gao, Y., Wu, D. M., Feng, W. & Li, F. Y. Gd^{3+} complex-modified $NaLuF_4$-based upconversion nanophosphors for trimodality imaging of NIR-to-NIR upconversion luminescence, X-Ray computed tomography and magnetic resonance. *Biomaterials* **33**, 5394–5405 (2012).

71. Xia, A., Gao, Y., Zhou, J., Li, C. Y., Yang, T. S., Wu, D. M., Wu, L. M. & Li, F. Y. Core-shell $NaYF_4$:Yb^{3+},Tm^{3+}@Fe_xO_y nanocrystals for dual-modality

T_2-enhanced magnetic resonance and NIR-to-NIR upconversion luminescent imaging of small-animal lymphatic node. *Biomaterials* **32**, 7200–7208 (2011).

72. Chen, F., Zhang, S. J., Bu, W. B., Liu, X. H., Chen, Y., He, Q. J., Zhu, M., Zhang, L. X., Zhou, L. P., Peng, W. J. & Shi, J. L. A "neck-formation" strategy for an antiquenching magnetic/upconversion fluorescent bimodal cancer probe. *Chem. Eur. J* **16**, 11254–11260 (2010).

73. Liu, Y. L., Ai, K. L., Liu, J. H., Yuan, Q. H., He, Y. Y. & Lu, L. H. A High-performance ytterbium-based nanoparticulate contrast agent for *in vivo* X-Ray computed tomography imaging. *Angew. Chem. Int. Ed.* **51**, 1437–1442 (2011).

74. Xiao, Q. F., Bu, W. B., Ren, Q. G., Zhang, S. J., Xing, H. Y., Chen, F., Li, M., Zheng, X. P., Hua, Y. Q. & Zhou, L. P. Radiopaque fluorescence-transparent TaO_x decorated upconversion nanophosphors for *in vivo*CT/MR/UCL trimodal imaging. *Biomaterials* **33**, 7530–7539 (2012).

75. Sun, Y., Liu, Q., Peng, J. J., Feng, W., Zhang, Y. J., Yang, P. Y. & Li, F. Y. Radioisotope post-labeling upconversion nanophosphors for *in vivo*quantitative tracking. *Biomaterials* **34**, 2289–2295 (2013).

76. Yang, Y., Sun, Y., Cao, T. Y., Peng, J. J., Liu, Y., Wu, Y. Q., Feng, W., Zhang, Y. J. & Li, F. Y. Hydrothermal synthesis of $NaLuF_4$:^{153}Sm,Yb,Tm nanoparticles and their application in dual-modality upconversion luminescence and SPECT bioimaging. *Biomaterials* **34**, 774–783 (2013).

77. Dong, B., Xu, S., Sun, J., Bi, S., Li, D., Bai, X., Wang, Y., Wang, L. P. & Song, H. W. Multifunctional $NaYF_4$: Yb^{3+},Er^{3+}@Ag core/shell nanocomposites: integration of upconversion imaging and photothermal therapy. *J. Mater. Chem.* **21**, 6193–6200 (2011).

78. Cheng, L., Yang, K., Li, Y. G., Zeng, X., Shao, M. W., Lee, S. T. & Liu, Z. Multifunctional nanoparticles for upconversion luminescence/MR multimodal imaging and magnetically targeted photothermal therapy. *Biomaterials* **33**, 2215–2222 (2012).

79. Kang, X. J., Cheng, Z. Y., Li, C. X., Yang, D. M., Shang, M. M., Ma, P. A., Li, G. G., Liu, N. & Lin, J. Core–shell structured up-conversion luminescent and mesoporous $NaYF_4$:Yb^{3+}/Er^{3+}@nSiO_2@mSiO_2 nanospheres as carriers for drug delivery. *J. Phys. Chem. C* **115**, 15801–15811 (2011).

80. Gai, S. L., Yang, P. P., Li, C. X., Wang, W. X., Dai, Y. L., Niu, N. & Lin, J. Synthesis of magnetic, up-conversion luminescent, and mesoporous

core-shell-structured nanocomposites as drug carriers. *Adv. Funct. Mater.* **20**, 1166–1172 (2010).

81. Fan, W. P., Shen, B., Bu, W. B., Chen, F., Zhao, K. L., Zhang, S. J., Zhou, L. P., Peng, W. J., Xiao, Q. F., Xing, H. Y., Liu, J. N., Ni, D. L., He, Q. J. & Shi, J. L. Rattle-structured multifunctional nanotheranostics for synergetic chemo-/radiotherapy and simultaneous magnetic/luminescent dual-mode imaging. *J. Am. Chem. Soc.* **135**, 6494–6503 (2013).

82. Yan, B., Boyer, J. C., Branda, N. R. & Zhao, Y. Near-infrared light-triggered dissociation of block copolymer micelles using upconverting nanoparticles.*J. Am. Chem. Soc.* **133**, 19714–19717 (2011).

83. Liu, J. N., Bu, W. B., Pan, L. M. & Shi, J. L. NIR-triggered anticancer drug delivery by upconverting nanoparticles with integrated azobenzene-modified mesoporous silica. *Angew. Chem. Int. Ed.* **52**, 4375–4379 (2013).

84. Yang, Y. M., Shao, Q., Deng, R. R., Wang, C., Teng, X., Cheng, K., Cheng, Z., Huang, L., Liu, Z., Liu, X. G. & Xing, B. G. *In vitro* and *in vivo* uncaging and bioluminescence imaging by using photocaged upconversion nanoparticles. *Angew. Chem. Int. Ed.* **51**, 3125–3129 (2012).

85. Yang, Y. M., Liu, F., Liu, X. G. & Xing, B. G. NIR light controlled photorelease of siRNA and its targeted intracellular delivery based on upconversion nanoparticles. *Nanoscale* **5**, 231–238 (2013).

86. Jayakumar, M. K. G., Idris, N. M. & Zhang, Y. Remote activation of biomolecules in deep tissues using near-infrared-to-UV upconversion nanotransducers. *Proc. Natl Acad. Sci. USA* **109**, 8483–8488 (2012).

87. Chatterjee, D. K., Fong, L. S. & Zhang, Y. Nanoparticles in photodynamic therapy: an emerging paradigm. *Adv. Drug Delivery Rev.* **60**, 1627–1637 (2008).

88. Qian, H. S., Guo, H. C., Ho, P. C. L., Mahendran, R. & Zhang, Y. Mesoporous-silica-coated up-conversion fluorescent nanoparticles for photodynamic therapy. *Small* **5**, 2285–2290 (2009).

89. Qiao, X. F., Zhou, J. C., Xiao, J. W., Wang, Y. F., Sun, L. D. & Yan, C. H. Triple-functional core-shell structured upconversion luminescent nanoparticles covalently grafted by photosensitizer for luminescent, magnetic resonance imaging and photodynamic therapy *in vitro*. *Nanoscale* **4**, 4611–4623 (2012).

90. Idris, N. M., Gnanasammandhan, M. K., Zhang, J., Ho, P. C., Mahendran, R. & Zhang, Y. *In vivo* photodynamic therapy using upconversion nanoparticles as remote-controlled nanotransducers. *Nat. Med.* **18**, 1580–1585 (2012).

91. Wang, C., Tao, H. Q., Cheng, L. & Liu, Z. Near-infrared light induced *in vivo* photodynamic therapy of cancer based on upconversion nanoparticles.*Biomaterials* **32**, 6145–6154 (2011).

92. Park, Y. I., Kim, H. M., Kim, J. H., Moon, K. C., Yoo, B., Lee, K. T., Lee, N., Choi, Y., Park, W., Ling, D., Na, K., Moon, W. K., Choi, S. H., Park, H. S., Yoon, S.-Y., Suh, Y. D., Lee, S. H. & Hyeon, T. Theranostic probe based on lanthanide-doped nanoparticles for simultaneous *in vivo* dual-modal imaging and photodynamic therapy. *Adv. Mater.* **24**, 5755–5761 (2012).

93. Wang, C., Cheng, L., Liu, Y. M., Wang, X. J., Ma, X. X., Deng, Z. Y., Li, Y. G. & Liu, Z. Imaging-guided pH-sensitive photodynamic therapy using charge reversible upconversion nanoparticles under near-infrared light.*Adv. Funct. Mater.* **23**, 3077–3086 (2013).

94. Wang, Y. F., Liu, G. Y., Sun, L. D., Xiao, J. W., Zhou, J. C. & Yan, C. H. Nd^{3+}-sensitized upconversion nanophosphors: efficient *in vivo* bioimaging probes with minimized heating effect. *ACS Nano* **7**, 7200–7206 (2013).

95. Xie, X. J., Gao, N. Y., Deng, R. R., Sun, Q., Xu, Q. H. & Liu, X. G. Mechanistic investigation of photon upconversion in Nd^{3+}-sensitized core-shell nanoparticles. *J. Am. Chem. Soc.* **135**, 12608–12611 (2013).

96. Shen, J., Chen, G. Y., Vu, A.-M., Fan, W., Bilsel, O. S., Chang, C. C. & Han, G. Engineering the upconversion nanoparticle excitation wavelength: cascade sensitization of tri-doped upconversion colloidal nanoparticles at 800 nm. *Adv. Opt. Mater.* **1**, 644–650 (2013).

Chapter 6

Recent Advance of Biological Molecular Imaging Based on Lanthanide-Doped Upconversion-Luminescent Nanomaterials

Yuanzeng Min [1], Jinming Li [1], Fang Liu [1], Parasuraman Padmanabhan [2,*], Edwin K. L. Yeow [1,*] and Bengang Xing [1,*]

[1]Division of Chemistry and Biological Chemistry, School of Physical and Mathematical Sciences, Nanyang Technological University, Singapore 637371, Singapore

[2]The Lee Kong Chian School of Medicine, Nanyang Technological University, 50 Nanyang Drive, Research Techno Plaza, Singapore 637553, Singapore

ABSTRACT

Lanthanide-doped upconversion-luminescent nanoparticles (UCNPs), which can be excited by near-infrared (NIR) laser irradiation to emit multiplex light, have been proven to be very useful for *in vitro* and *in vivo* molecular imaging studies. In comparison with the conventionally used down-conversion fluorescence imaging strategies, the NIR light excited luminescence of UCNPs displays high photostability, low cytotoxicity, little background auto-fluorescence, which allows for deep tissue penetration, making them attractive as contrast agents for biomedical imaging applications. In this review, we will mainly focus on the latest development of a new type of lanthanide-doped UCNP material and its main applications for *in vitro* and *in vivo* molecular imaging and we will also discuss the challenges and future perspectives.

Keywords: upconversion nanoparticles; molecular imaging; luminescent nanomaterials

1. INTRODUCTION

Currently, molecular imaging is a brand new and currently upcoming technology, which allows people to non-invasively monitor the cellular functions or the biological processes in complicated living systems. Usually, the observation of physiological events under single cell and/or molecular levels can provide unique insights into the function (or dysfunction) of living organisms that are normally associated with many diseases like cancer, heart disease and neurological disorders [1,2,3,4,5,6,7,8]. Therefore, molecular imaging will be a very powerful tool for its potential biomedical applications in early stage diseases diagnosis or monitoring the therapy intervention [1,2,3,4,5]. Generally, molecular imaging can be classified into different specialties that contain optical imaging, computed tomography (CT), magnetic resonance imaging (MRI), single photon emission tomography (SPECT), positron emission tomography (PET) and ultrasound imaging based on the imaging mechanism, detection instrumentation or the applied imaging contrast agents [1,2,3,4,5,6,7,8].

Among the different imaging modalities, optical imaging techniques including the fluorescence and bioluminescence imaging which utilize light sources under various wavelengths for image generation represents a simple and direct observation of specific molecular targets or biological pathways *in vitro*and *in vivo*. In general, bioluminescent imaging mainly relies on the native light emission from several organisms such as marine bacterial luciferases, the eukaryotic firefly luciferases and Renilla (sea pansy) luciferases [9,10]. Fluorescent imaging can be extensively visualized upon the light illumination of various fluorochromes with appropriate light sources and the emission signals are collected at a shifted wavelength.

Currently, quite a number of commonly used fluorochromes such as coumarin, boron-dipyrromethene (BODIPY), fluorescein, rhodamine and their analogs have been successfully applied for *in vitro* cell culture fluorescent observation and bio-sensing [11,12,13,14,15,16,17,18,19,20,21]. Some other organic fluorophores including cyanine dyes, squaraine dyes and phthalocyanines derivatives that function in the near infrared (NIR) region have also been extensively used for the non-invasively monitoring of gene expression [11], *in vivo* cell trafficking [12], enzyme activities identification [13,14,15,16,17,18,19,20,21], early stage of disease screening and new drug development [22,23] *in vitro*and in living subjects, mostly attributed to their promising advantages of reduced light scattering, high tissue penetration and relatively minimal auto-fluorescence from biological samples. However, despite the widely accepted utility of these NIR organic fluorophores in fluorescent imaging, most of the dye molecules indicated limited quantum yield, poor photostability, a short lifetime and less Stokes shift in aqueous solution [17,18,19,20,21]. Therefore, the development of alternative long emission materials with improved photo-physical properties will be still highly required.

In recent decades, the disciplines of nanomedicine and diagnostic imaging have attracted considerable academic and industry attentions, in which real-time imaging has shown to be particularly valuable for investigating the aspects, including target specificity, pharmacokinetic profiles and bio-distribution, as well as to assess the therapeutic effects of nanoparticle conjugates [15]. One of the most extensively studied nanomaterials for optical imaging applications are quantum dots (QDs) [24,25,26,27,28,29,30,31]. QDs are mono-dispersing inorganic fluorescent nano-crystalline particles made from semiconductor materials with numerous superior optical properties when compared to those of organic fluorophores [11,12,13,14,15,16,17,18,19,20,21], such as high quantum yields, high molar extinction coefficients, long fluorescent lifetimes (> 10 ns), narrow emission spectra, continuous absorption spectra ranging from the UV to NIR area and large effective Stokes shifts [24,25,26,27,28,29,30,31]. So far, QDs have attracted considerable attention in bionanotechnology and biomedical fields, particularly, multiplexed bio-sensing and labeling, targeted drug or gene delivery, and in vitro or *in vivo* biological imaging [24,25,26,27,28,29,30,31]. In spite of their significant success in principle, recent advances also highlighted the possible toxicity of Cd based QD nanomaterials in biological systems. The toxicity of the Cd related compounds would be one potential concern that limits the extensive use of these visible or NIR emitting nano-crystals, especially for their applications directly in human healthcare [27,28,29,30,31]. Therefore, searching for non-toxic, stable and more importantly, longer wavelength excited and deeper tissue penetration substitutes with high quantum yields is highly desirable and it remains a big challenge in the relevant fields.

Recently, lanthanide-doped upconversion nanoparticles (UCNPs) have been used as a primary and potentially promising vector for molecular imaging, due to their unique photo-luminescent properties and good biocompatibility. The UCNPs can absorb near-infrared (NIR) light and convert it into multiplexed emissions that span over a broad range (*i.e.*, from the UV to NIR region) [32,33], resulting in a nonlinear optical upconversion process (Figure 1), which is different from those of traditionally used down-conversion fluorescent organic fluorophores and QDs [34,35,36,37]. In general, the UCNPs indicate many advantages, including narrow-bandwidth emission, a long lifetime, tunable emission, high photostability and high resistance to photo-bleaching [38,39,40,41,42]. Furthermore, the less background auto-fluorescence and light scattering that allows deep tissue light penetration will also make UCNPs work as attractive optical contrast agents for *in vitro* and *in vivo* bio-imaging applications. More importantly, upon the doping of different elements such as Gd and Sm etc, into UCNPs structures during the process of their synthesis, the prepared UCNP can also be endowed with more promising functions for multimodality imaging by the combination of fluorescent imaging with other imaging techniques including computed tomography (CT), magnetic resonance imaging (MRI), single photon emission tomography (SPECT), positron emission tomography (PET), *etc.* [43].

So far, there are some excellent review articles that have summarized the different emphasis on the synthesis or bio-applications of UCNPs. Some mainly discussed the emphasis on synthesis [34,35], while others analyzed their applications [38,43], including the photo-physical properties of UCNPs [13,14,15,16,17,18,19,20,21,22,23,24,25,26,27,28,29,30,31]. However, reviews on the applications of molecular imaging using UCNPs are still very limited. In this review paper, we summarize the significance and recent advances in UCNPs for molecular imaging, and we also discuss the future challenges and promises of using UCNPs for molecular imaging *in vitro* and *in vivo*.

2. SYNTHESIS, SURFACE FUNCTIONALIZATION AND BIOCOMPATIBILITY OF LANTHANIDE UCNPS

In general, UCNPs are composed of a host matrix doped with a sensitizer (light absorbers) and an activator (light-emitting ions) (Figure 1) [34,35]. For activators, rare earth ions especially Ln^{3+} ions have mainly been used for activators, because of their ladder-like arranged energy level that promotes upconversion by the absorption of multiple photons. The host materials usually require close lattice matches to dopant ions, which have low phonon energies. Based on this, currently, rare earth fluorides such as binary REF_3 and complex $AREF_4$ (RE = rare earth, A = alkali), including $NaYF_4$, $NaGdF_4$, $NaLuF_4$, KYF_4, $NaYbF_4$, LaF_3, CaF_2, $KMnF_3$, YF_3, $KGdF_4$ and a few others are the mostly used host components for UCNP preparation [34,35,38].

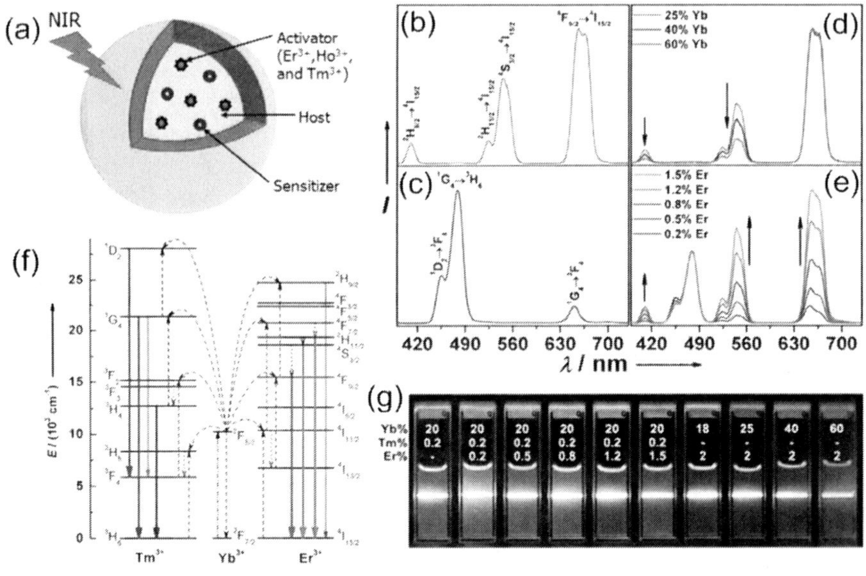

Figure 1. The structure and optical properties of upconversion nanoparticles (UCNPs): (**a**) Scheme illustration of the structure and components of the UCNPs; (**b–e**) Upconversion multicolor tuning in Ln^{3+}-doped cubic $NaYF_4$ UCNPs. Room temperature UCNP emission spectra of (**b**) $NaYF_4$:Yb/Er (18/2 mol%), (**c**) $NaYF_4$:Yb/Tm (20/0.2 mol%), (**d**) $NaYF_4$:Yb/Er (25–60/2 mol%), and (**e**) $NaYF_4$:Yb/Tm/Er (20/0.2/0.2–1.5 mol%) particles in ethanol solutions (10 mM). The spectra in (**d**) and (**e**) were normalized to Er^{3+} 660 nm and Tm^{3+} 480 nm emissions, respectively; (**f**) The proposed energy transfer mechanisms showing the upconversion processes in Er^{3+}, Tm^{3+}, and Yb^{3+} doped crystals under 980 nm diode laser excitation. The dashed-dotted, dashed, dotted and full arrows represent photon excitation, energy transfer, multiphoton relaxation, and emission processes, respectively. Only visible and NIR emissions are shown here. (**g**) Upconversion multicolor fine-tuning through the use of lanthanide-doped $NaYF_4$ nanoparticles with varied dopant ratios. Adapted with permission from references [33,34,35] and [39], respectively. NIR, near-infrared. Copyright: American Chemical Society, 2008; Royal Society of Chemistry, 2009; and John Wiley and Sons, 2013.

Normally, the suitable size and uniform shape are the vital factors in UCNPs towards effective bio-imaging applications. There are several methods that have been developed to prepare UCNPs with appropriate size (from sub-10 nm to sub micrometer) and uniform shape (nanospheres, nanorods, nanocubes and nanoplates, *etc.*). Most of them have been prepared mainly based on techniques including thermal decomposition

[33,44,45,46,47,48,49] hydrothermal synthesis [50,51,52,53,54], co-precipitation [55], sol-gel processing, *etc.* [56,57,58,59,60]. So far, among the various techniques, thermal decomposition and hydrothermal synthesis have been proven to be the two most popular methods for fabricating the UCNPs [34,35,40,41,42,43].

Besides the size-dependent properties, the biological applications of UCNPs may also be affected by their surface properties. Currently, most of the synthesized UCNPs are usually capped with hydrophobic organic ligands, such as oleic acid (OA), oleylamine (OM) and adamantine-acetic acid (Ad), which may potentially influence their utilities in molecular imaging, mostly because of their limited solubility in aqueous solution or in biological buffers. In order to obtain biocompatible UCNPs for potential biomedical applications, it is very necessary to conduct surface modification. Therefore, water solubilization and surface functionalization are the critical steps to allow UCNP to serve as reliable platforms in biological applications. Toward this direction, a variety of strategies have been extensively reported, which mostly employ functional polymers (e.g., especially polyethylene glycol (PEG), *etc.*) [61], silica [62], peptide sequences [63] and other biomolecules [64], to functionalize UCNPs with methods of ligand exchange [65,66], ligand removal [67,68], ligand oxidation [69,70], layer-by-layer assembly [71,72], silica coating [62,73,74] or amphiphilic polymer coating, *etc.* [34,35,38,75,76]. These approaches have been successfully applied to functionalize UCNPs structures and to make them water soluble, chemical stable, biocompatible, and importantly, to have optimal reaction sites for subsequent bio-conjugation. Figure 2 illustrates the general strategies for the surface functionalization of UCNPs. It is obvious that most UCNPs have been modified with $-N_3$, $-NH_2$, $-SH$ and $-COOH$ groups, which will greatly facilitate the convenient and effective bio-conjugation on the surface of UCNPs to enhance their biocompatibility and further biomedical applications.

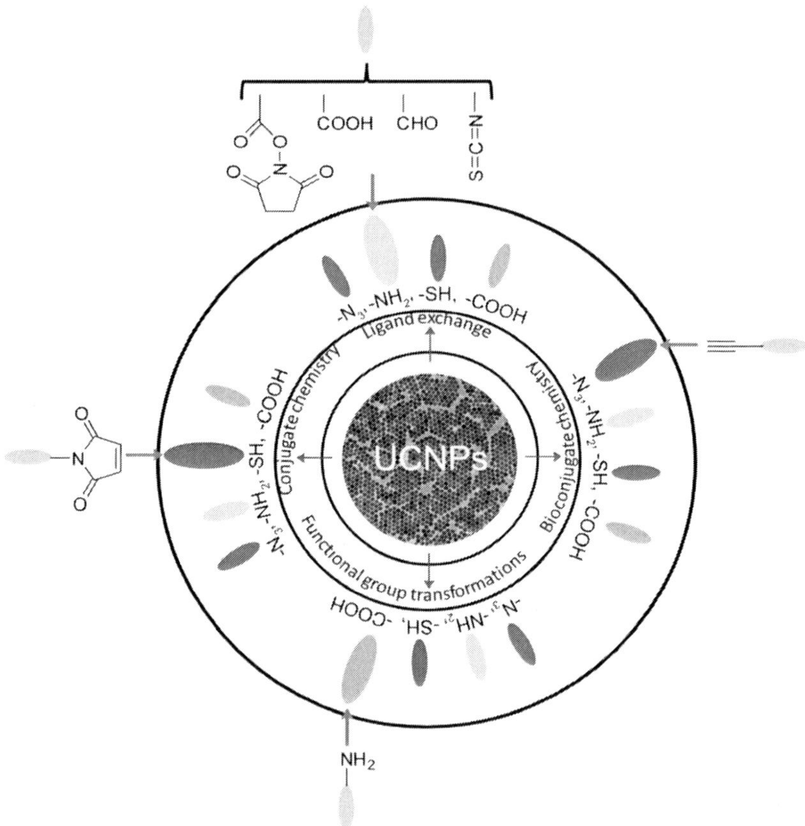

Figure 2. Surface functionalization of UCNPs and their functional groups for biological applications. Adapted with permission and modified from references [27,28,29,30,31]. Copyright: Royal Society of Chemistry, 2013.

The biocompatibility of UCNPs is of great importance for their subsequent biological applications [77]. Therefore, the biocompatibility and potential toxicity of UCNPs need to be fully addressed before their extensive bio-applications *in vitro* and *in vivo*. So far, there has been a large amount of *in vitro* and*in vivo* experiments evaluating the cytotoxicity of UCNPs by using methods, including the MTT (methyl thiazolyl tetrazolium) assay, MTS assay (3-(4,5-dimethylthiazol-2-yl)-5-(3-carboxymethoxyphenyl)-2-(4-sulfophenyl)-2H-tetrazolium, sodium salts) and CKK-8 assays (Cell Counting Kit-8 mitochondrial metabolic activity assays) [64,78,79]. Most of these results suggested the negligible cytotoxicity of UCNPs upon the functionalization of the particle surface with different types of ligands or the aforementioned biocompatible coating molecules at cellular levels.

Furthermore, extensive *in vivo* experiments were also conducted recently to evaluate the long-term toxicity of UCNPs in different subjects such as *Caenorhabditis elegans* (*C. elegans*) worms [80], zebrafish embryos [81] and mice [64], respectively. For example, the *in vivo* results on *C. elegans* demonstrated that UCNPs prepared by homogenous precipitation in an aqueous solution of $Y(NO_3)_3{\cdot}6H_2O$ (50 mM), $Yb(NO_3)_3{\cdot}5H_2O$ (1 mM), $Er(NO_3)_3{\cdot}5H_2O$ (0.5 mM) and urea (15 mM) did not display notable toxicity when the concentrations are up to 10 mg/mL or above [80]. The potential toxicity evaluation in mice was also conducted through behavior observation, body weight measurement, histology and hematology analysis, and serum biochemistry assay. Similar tests based on polyacrylic acid (PAA)-coated $NaYF_4$:Yb/Tm UCNPs further confirmed that there was no any apparent toxicity with an administration of a dose of 15 mg/kg of PAA-UCNPs in living mice for about three months [64]. Very recently, good biocompatibility was also observed for the Gd/Mn-doped $NaYF_4$:Yb/Er UCNPs which displayed very low toxicity in mice as the survival rate was 100% for thirty days after injection. Upon the analysis of inductively coupled plasma mass spectrometry (ICP-MS), it has been confirmed that Gd^{3+}/Mn^{2+} ions have been confined in a rigid matrix of host UCNPs nano-crystals, which was not easily released into the living environments and, thus, greatly minimized their potential cytotoxicity [54].

3. LANTHANIDE UCNPS FOR OPTICAL IMAGING

In terms of superior biocompatibility and less cytotoxicity upon the surface modification, UCNPs can be used for promising biomedical applications, especially in the areas of molecular imaging, mostly owing to their unique optical properties, which can be excited with long wavelength NIR light from 795 to 980 nm, and such long wavelength light illumination can be converted into multiplexed emissions that span over a broad range from the UV to NIR region [32,33,34,35]. Such promising optical functions allow UCNP to serve as effective platforms for non-invasively visualizing biological distribution and real-time imaging of biological processes with deep tissue and light penetration.

As potential agents for molecular imaging *in vitro* and *in vivo*, UCNPs indicated many unique advantages over conventionally used organic dyes, such as narrow-bandwidth emission, a long lifetime, tunable emission, high photo-stability and high resistance to photo-bleaching. With the excitation of NIR light, the emitted light from the UV to NIR region can be applied for multicolor imaging [38,39]. Among the various converted emissions, NIR-to-NIR (using NIR excited light to emit NIR light)-based UCNPs luminescence imaging has gained increasing attention, as both the excitation and emission

NIR light can penetrate the "optical transmission window" of tissues (700–1000 nm) [82], thus providing a direct powerful tool for visualizing the biodistribution of UCNPs *in vivo*.

Generally, in the process of the development of lanthanide-based UCNP nanomaterials for their applications in biological imaging, three main aspects have been classified, which include non-specific targeting imaging, active targeting imaging, and activatable imaging, mostly based on the reaction mechanism of the UCNP-based imaging probes and their related biological targets *in vitro* and *in vivo*.

For example, early studies of non-specific targeting imaging based on UCNPs demonstrated that 50 nm of $NaYF_4$:Yb/Er UCNPs could be employed for *in vivo* imaging in anesthetized Wistar rats (Figure 3) when the rats were injected with the nanoparticles directly under the skin in the groin and upper leg regions [83]. After excitation of a 980 nm NIR laser, the nanoparticles could be detected up to 10 mm beneath the skin, far deeper than depth through the use of QDs. As one of the pioneering animal imaging studies based on UCNPs, these results provided promising examples for *in vivo* optical imaging at a deeper tissue level for performing minimal invasive imaging in real-time. With the superiority of the deep penetration of their emitted light, the $NaYF_4$:Yb/Er UCNPs doped with 30 mol % Mn^{2+} ions and $KMnF_3$ UCNPs doped with Yb/Er, Yb/Ho or Yb/Tm have also been applied for NIR-to-NIR single-band luminescence imaging [54]. Up to now, Tm^{3+} doped $NaYF_4$, $NaGdF_4$, $NaLuF_4$ and $NaYbF_4$ UCNPs have been extensively utilized for non-specific imaging in living mice [84,85,86]. Recently, Prasad *et al.*[78] also provided a new approach for *in vivo* imaging by using $NaYF_4$:Yb/Tm-based UCNPs as high contrast photoluminescence imaging probes, with their promising light penetration depth up to ~20 mm in mouse (Figure 4).

By employing sub-10 nm citric acid-capped $NaLuF_4$-based nano-crystals co-doped with Gd^{3+}, $Yb^{3+}$and Er^{3+} (or Tm^{3+}), the high contrast imaging in living mouse would allow an excellent detection limit down to 50 and 1000 UCNP-labeled cells upon the subcutaneous and intravenous injections of these nano-crystal based probes, respectively [87] (Figure 5a,b). More recently, the non-specific whole-body imaging based on polyacrylic acid (PAA) coated NaYF4:Yb/Tm (PAA-UCNPs) as luminescence probes for long-term *in vivo* distribution also showed that most of the PAA-UCNPs in the liver and the spleen would be cleared very slowly from the body through hepatobiliary excretion. Similarly, PAA-$NaLuF_4$:Yb/Tm UCNPs as an optical probe have also been used in *in vivo* imaging of black mouse and even living rabbit, with an excellent signal-to-noise ratio [88]. Moreover, oligo-arginine modified UCNPs have been applied towards whole-body imaging at the single cell level using ultra-sensitive stem cell labeling for long-term stem cell tracking. It was reported that as few as ~10 cells in a mouse could be detected with an ultra-high sensitivity when compared with the currently used exogenous stem cell labeling nanoagents, such as QDs [89] (Figure 5).

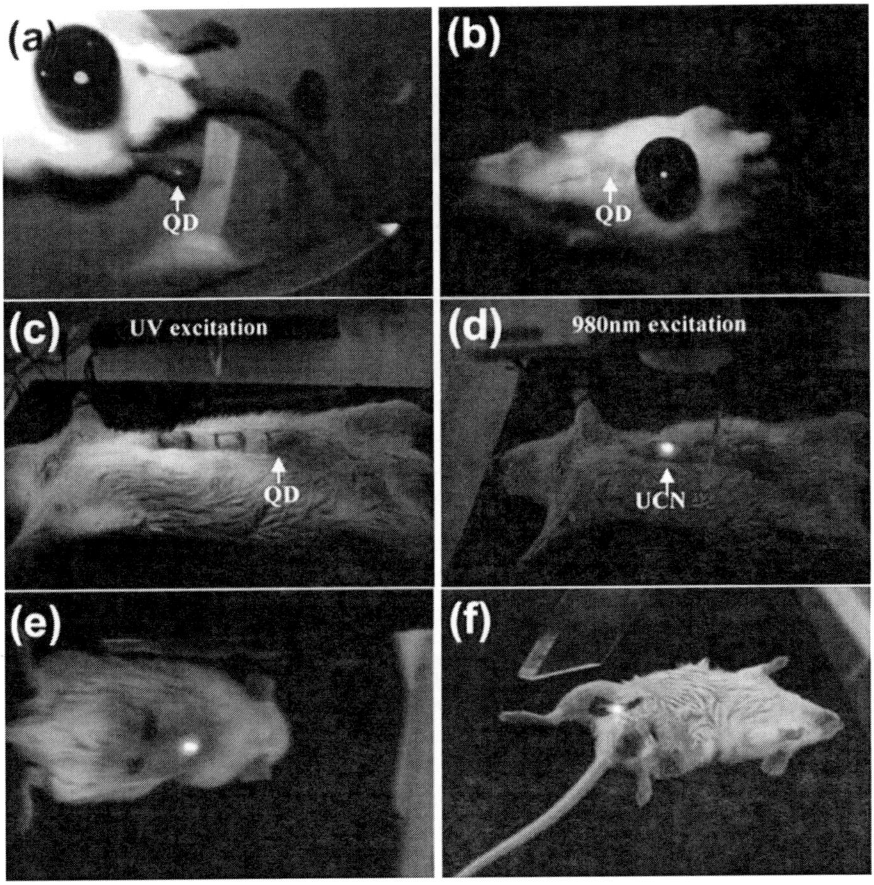

Figure 3. *In vivo* imaging of living rats with quantum dots (QDs) injected into the translucent skin of the foot (**a**) showing fluorescence, but not through thicker skin of the back (**b**) or abdomen (**c**), $NaYF_4$:Yb/Er nanoparticles injected below the abdominal skin (**d**), thigh muscles (**e**) or below the skin of the back (**f**) show luminescence. QDs on a black disk in (**a**,**b**) are used as the control. Adapted with permission from reference [83]. Copyright: Elsevier, 2008.

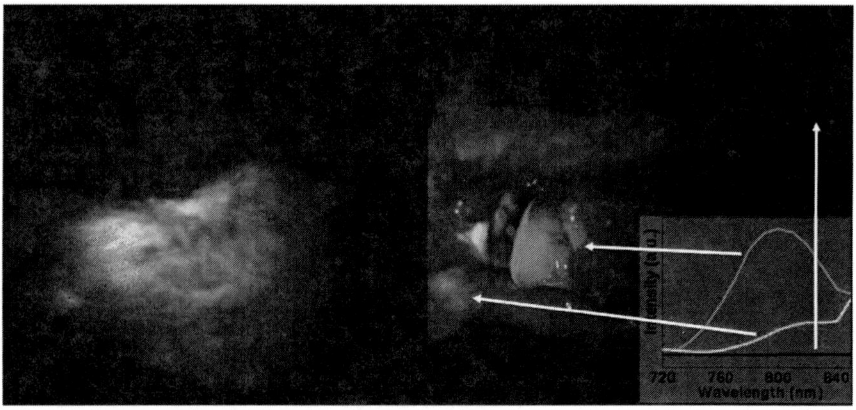

Figure 4. *In vivo* imaging of a mouse with the injection of UCNPs: intact mouse (**left**); the same mouse after dissection (**right**). The red color indicates emission from UCNPs; green and black show the background, as indicated by the arrows. The inset presents the photoluminescence spectra corresponding to the spectrally unmixed components of the multispectral image obtained with the Maestro system. Adapted with permission from reference [78]. Copyright: American Chemical Society, 2008.

Despite the promising applications of UCNPs in living animal imaging, however, as non-specific contrast agents, the biocompatible UCNPs lack reactive groups for selective recognition and further effective conjugation, which may potentially limit their extensive applications in biological imaging. Therefore, one simple and effective strategy to overcome the nonspecific targeting and to improve their selectivity towards the target site in a living system is to conjugate the bioactive ligand moieties with the UCNPs, which would exhibit strong affinity to specific molecular targets. Thus, the ligand modified UCNP conjugates can recognize the targets and can be easily trapped at the targeting region for a sufficient length of time. In the meantime, the unbound UCNPs lack affinity within the specific region and will thereby result in limited accumulation during the circulation process. In general, most of the applied high-affinity ligand moieties include small organic molecules, peptides, proteins, antibodies or their fragments [90]. Following are some representative examples for active targeting optical imaging on the basis of ligand modified UCNPs.

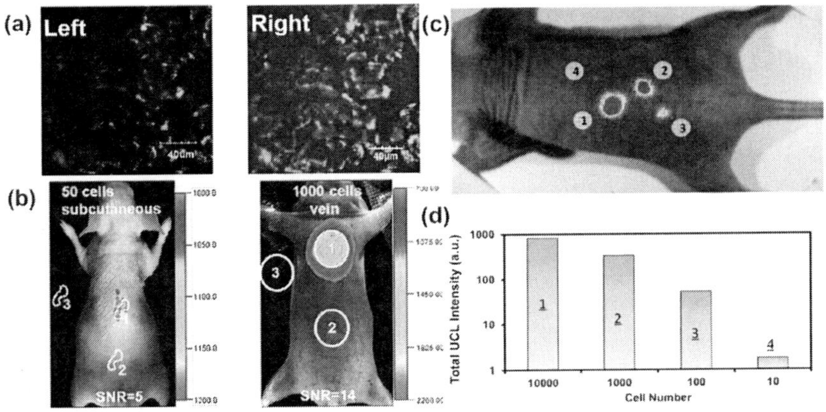

Figure 5. UCNPs for cellular labeling and *in vivo* tracking analysis. (**a**) Confocal UCNP imaging (**left**) and its overlay with a bright field image (**right**) of cells stained with 200 μg mL^{-1} $NaLuF_4$ UCNPs for 3 h at 37 °C. (**b**) *In vivo* UCNPs imaging of athymic nude mice after subcutaneous injection of 50 human nasopharyngeal epidermal carcinoma KB cells (**left**) and tail-vein injection of 1000 KB cells (**right**). The KB cells were pre-incubated with 200μg mL^{-1} NaLuF4 UCNPs for 3 h at 37 °C before injection. (**c** and **d**) *In vivo* detection of UCNP-labeled mMSCs (an exogenous contast agent to track mouse Mesenchymal Stem Cells). (**c**) An upconversion luminescence image of a mouse subcutaneously injected with various numbers of mouse mesenchymal stem cells (1×10^5) labeled with UCNPs. (**d**) Quantification of UCNPs luminescence signals in (**c**). Adapted with permission from references [87,89], respectively. UCL, upconversion luminescence. Copyright: American Chemical Society, 2011; and Elsevier, 2012.

For instance, folic acid modified UCNPs and arginine-glycine-aspartic (RGD) peptide modified UCNPs have been used for *in vivo* tumor targeting [91,92]. As indicated in Figure 6, after intravenous injection of folic acid modified UCNPs into living mouse in which tumors have been implanted by the injection of Hela cells, the Hela tumor cell lines were known to over-express folate receptors. The obvious luminescence signal at 600–700 nm was observed in the tumor, whereas no significant luminescence signal could be detected in the control group. Similarly, as revealed by other *in vivo*imaging in living animals, the RGD peptide conjugated UCNPs could also effectively target the U87 MG human glioblastoma tumor. It has been well known that U87 MG (Malignant Glioma) human glioblastoma tumors could over-express $\alpha_v\beta_3$ integrin receptors, which plays an important role in tumor angiogenesis and could be easily recognized by RGD peptide sequences with high affinity [93,94,95,96]. The further *in vivo* imaging results also confirmed that the neurotoxin-mediated upconversion nanoprobes for tumor targeting and visualization in living mice could produce a high-contrast image through their highly specific binding activity toward tumors [97].

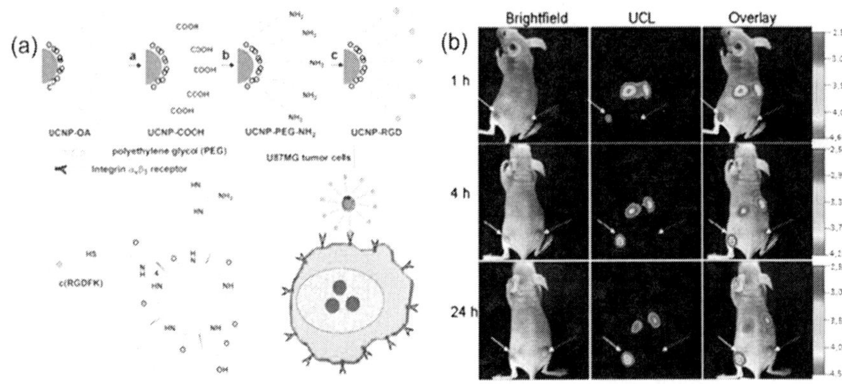

Figure 6. (**a**) Scheme of the synthesis of UCNP-Arginine-Glycine-Asparatic (RGD). (**b**) Time-dependent *in vivo* upconversion luminescence imaging of subcutaneous U87 MG (left hind leg, indicated by short arrows) and MCF-7 (Michigan Cancer Foundation-7) tumor (right hind leg, indicated by long arrows) borne by athymic nude mice after intravenous injection of UCNP-RGD over a 24 h period. UCNP-RGD conjugate was prepared from UCNP-OA complex (OA: Oleylamine). All images were acquired under the same instrumental conditions (power ≈ 80 mW cm^{-2} and temperature ≈ 21.5 °C on the surface of the mouse). Adapted with permission from reference [92]. Copyright: American Chemical Society, 2009.

Similarly, upon surface modification with folic acid ligands, UCNPs could be also used as an effective platform towards targeted drug release and site specific imaging of the targeted drug release into folic acid receptor expressed HeLa cells [73] (Figure 7). By capping the mesoporous silica-UCNPs with the cross-linked *o*-nitrobenzyl photoactivatable linker, the modified silica-UCNP complex could serve as photo-caged nanocarriers to encapsulate the drug payload molecules within the mesopores silica structure. After loading with the antitumor drug, Dox (doxorubicin) followed by NIR light irradiation, the cross-linked *o*-nitrobenzyl photoactive linker on the particle surface could be efficiently cleaved by the converted UV light from UCNPs, which could effectively trigger the photo-controlled drug release into the living cells. Moreover, upon the functionalization of the photo-caged nano-carriers with folic acid units, selective drug delivery can be easily achieved in the targeted tumor cell lines in which the folate receptor has been highly expressed. This novel and effective drug loaded photocaged nanocarrier may demonstrate a new possibility for selective cell imaging and controlled drug release in the living system with less photo damage and deeper light penetration.

In line with the selective recognition of the biological imaging targets in living system, one more alternative strategy to effectively impart the molecular specificity into activatable imaging contrast agents is to develop environmentally responsive targeting probes, which would rely on the controlled manipulation of the fluorescent output of UCNP platforms by changing their local chemical environments or structural conformations. On the basis of this process, the fluorescent signal could be generated or significantly amplified only after the particle-based nano-probes have been specifically activated by an environmental stimulation.

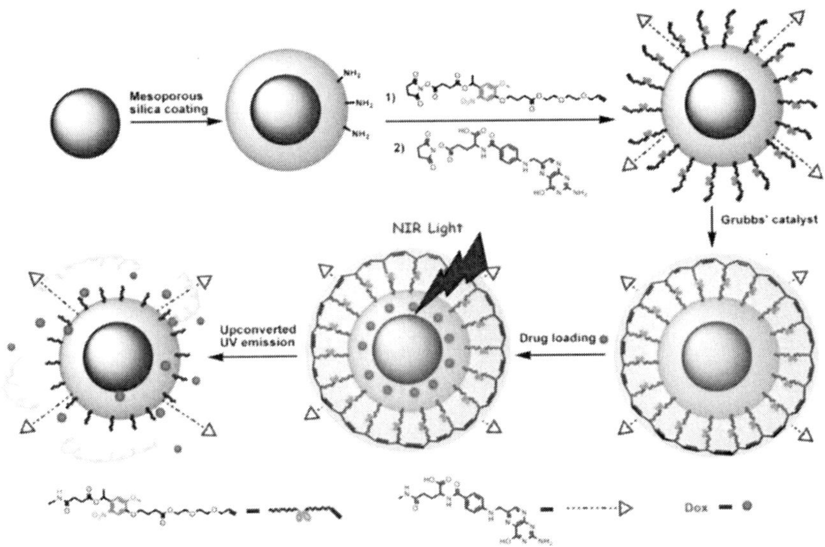

Figure 7. The design for photo-controlled Dox delivery through mesoporous silica coated UCNPs conjugated with folic acid. Adapted with permission from reference [73]. Copyright: John Wiley and Sons, 2013.

Among the different strategies to respond to environmental stimulation, photolysis of photoactivatable or "caged" molecules has been well proven to be one effective strategy for the non-invasive regulation of biological activities and processes in living systems [98,99,100,101], by which the activation process could be easily modulated by a beam of light with high spatial and temporal precision. Generally, in this process, the active sites in targeted biological molecules were covalently modified with photo-labile protecting groups (e.g., 2-nitrobenzyl derivatives, spiropyran or diazobenzene moieties, *etc.*), which would keep these molecules at inactive stages before photolysis. Upon brief light irradiation, these photo-active protecting groups could be cleaved and the photo-caged biological molecules would be released, thus recovering the intrinsic bioactivity [98,99,100,101]. So far, the photo-caged activation strategies have been successfully applied as powerful tools to monitor biological processes and study their native functions in well-resolved space and time resolutions. However, up to now, most of photoactivatable compounds have been designed to mainly respond to UV light irradiation and they could not be activated by visible or NIR light. In general, there were significant shortcomings associated with the utilities of the short wavelength of UV or visible light in the process of photoactivation. For example, excessive exposure to UV light would easily lead to photo cross-linking reactions in nucleic acids or proteins, which could induce significant cellular damage. Meanwhile, short-wavelength UV or visible light irradiation could not penetrate into deep tissue, which greatly limited their extensive applications for living animal imaging and the activation of the photo-caged biomolecules in living conditions. Therefore, the rational design of new photo-caged strategies or alternative materials for high depth of tissue or light penetration but with limited biological damage would be of utmost importance. Herein, by taking the advantage of the unique properties of UCNPs to transmit NIR to UV or visible light emission, lanthanide-doped particle systems have received considerable attention and they could work as promising functional materials towards photo-activatable imaging *in vitro* and *in vivo* without worrying about the issues caused by direct UV or visible light irradiation.

As proof-of-concept, Xing's group [62] recently reported a simple and effective method for the controlled uncaging of D-luciferin and bioluminescence imaging on the basis of photo-caged UCNPs, which takes advantage of the photon upconversion of NIR light to UV light to trigger the uncaging of D-luciferin from D-luciferin-conjugated UCNPs (Figure 8). As the first example of photo-caged imaging through UCNPs, the released D-luciferin effectively conferred enhanced fluorescence and bioluminescence signals *in vitro* and *in vivo* with deep light penetration and low cellular damage.

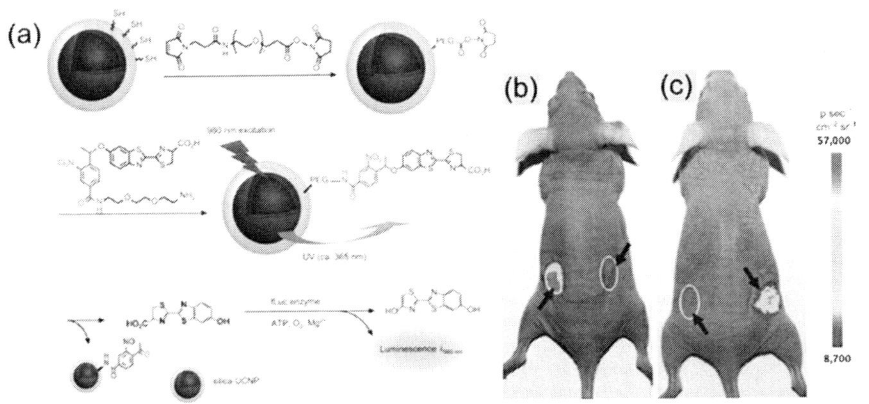

Figure 8. Bioluminescent images of firefly luciferase (fLuc) activity in living mice that were treated with D-luciferin. (**a**) Experimental design for uncaging D-luciferin and subsequent bioluminescence through the use of photo-caged core-shell upconversion nanoparticles. (**b**) **Left**: injection with D-luciferin (20 μM, 20 μL);**right**: injection with photo-caged nanoparticles without NIR light irradiation. (**c**) **Left**: injection with photo-caged nanoparticles and irradiation with UV light for 10 min; **right**: injection with photo-caged nanoparticles and irradiation with NIR light for 1 h. Adapted with permission from reference [62]. Copyright: John Wiley and Sons, 2012.

More recently, the same group also demonstrated a novel and personalized near-infrared (NIR) light activated prodrug nano-platform through the combination of photoactivatable platinum(IV) prodrug and caspase imaging peptide conjugate with silica coated upconversion nanoparticles (UCNPs) for the remote control of antitumor platinum prodrug activation and simultaneously, for the real-time imaging of apoptosis induced by the activated cytotoxicity [102] (Figure 9). Upon NIR light illumination, the Pt(IV) prodrug complex could be activated from the surface of nanoparticles and the selective release of active component could exhibit potent cytotoxicities against human ovarian carcinoma A2780 cells and their cisplatin resistant variant A2780cis cells. More importantly, the systematic programmed cell death was also found to be involved in the NIR responsive tumor inactivation, and the triggered caspases enzymes would effectively cleave the recognition peptide sequence with a flanking fluorescent resonance energy transfer (FRET) pair consisting of NIR fluorophore, Cy5 and relevant quencher Qsy21, thereby allowing the direct imaging of apoptosis by separating Cy5 from Qsy21 quencher *in vitro* and in living cells. Such specific and novel rational design exhibited the promising opportunity to remotely control the activation of platinum drug in the targeted tumors; meanwhile, it could also evaluate in real time the antitumor activities by imaging of apoptosis triggered from the photoactivated drug release, thus providing great potentials for personalized tumor therapeutic and diagnostic applications on the basis of UCNPs in the future.

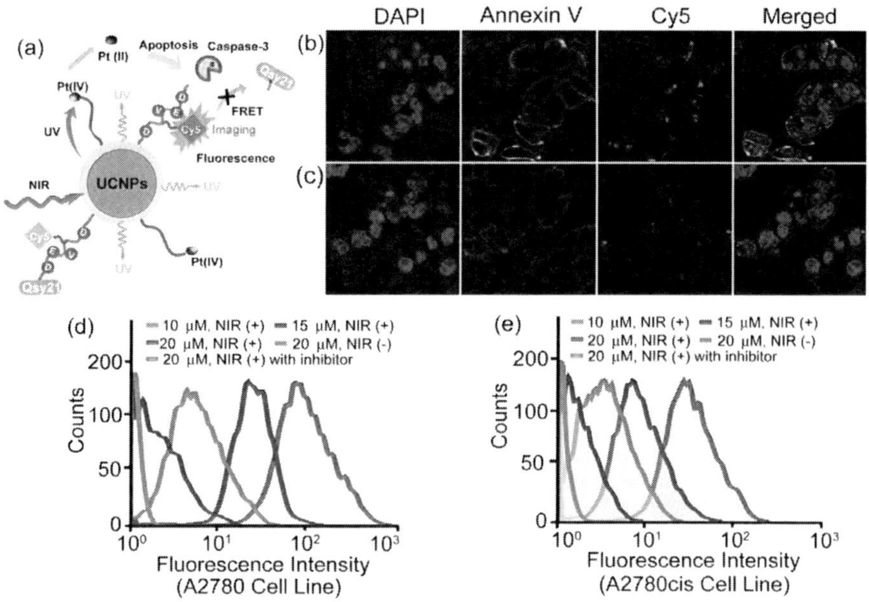

Figure 9. Live-cell apoptosis imaging for NIR irradiation of Pt(IV)-probe UCNPs@SiO_2 (15 μM) incubated cells: (**a**) schematic illustration of NIR light activation of platinum(IV) prodrug and intracellular apoptosis imaging through upconversion nanoparticles; (**b**) A2780 cells; and (**c**) cisplatin resistant A2780cis cells. (blue: DAPI (4,6-Diamidino-2-phenylindole); green: Annexin V; red: Cy5.) Quantitative flow cytometric analysis of (**d**) A2780 and (**e**) A2780cis cells treated with different concentrations of Pt(IV)-probe UCNPs@SiO_2 (10, 15 and 20 μM, respectively) and 1 h NIR irradiation. Cells treated with Ac-DEVD (Aspartic acid-Glutamic acid-Valine-Aspartic acid)-CHO (20 μM) inhibitor and NIR irradiation of cells without Pt(IV)-probe UCNP incubation were used as controls. Adapted with permission from reference [102]. Copyright: John Wiley and Sons, 2013.

4. DOPED LANTHANIDE UCNPS FOR MULTIMODALITY IMAGING

Apart from applications in optical imaging, UCNPs could be further developed to incorporate multiple modality imaging moieties towards integrated identification of specific biological pathways and anatomic structures under high-resolution and three dimensional resolutions, as well as real-time monitoring of physiological functions in living tissues. Upon the structural doping or labeling of bio-medically applied elements, such as Gd, ^{18}F, *etc.* UCNPs could be extensively devised to expand their further applications in multifunctional imaging not only through individual MRI (Magnetic Resonance Imaging), PET (Photoinduced Electron Transfer), CT (Computed Tomography), and SPECT (Single-Photon Emission Tomography) imaging but also through processes based on dual-modal or multi-modal imaging techniques.

For example, Li *et al.* [84] reported Gd-doped UCNPs that could offer both T_1-weighted MRI contrast and optical imaging for *in vivo* dual-modal imaging in living mice. Besides the Gd-doped UCNPs, super-paramagnetic iron oxide nanoparticles (IONP) have been also modified to integrate into hetero-UCNPs structures by surface shielding with a silica shell to achieve dual-modal T_2-weighted MRI and fluorescent imaging through one unified platform. Recently, Liu *et al.* [103] also developed a novel type of multifunctional nanoparticles based on UCNP@IONP@Au with combined optical and magnetic properties useful in dual-modal imaging and targeted photo-thermal tumor therapy in living subjects.

Dual-modal CT/optical imaging *in vivo* was also reported using lanthanide-doped $NaGdF_4$ UCNPs as effective contrast agents with both of the signals co-located well in the subcutaneous injection site in living mice [104]. Very recently, Li *et al.* [105] reported ^{18}F-labeled $NaYF_4$:Yb/Tm UCNPs for PET and luminescent imaging of sentinel lymph node in different living animals. In this study, the UCNPs could be tracked in real time with high sensitivity *in vivo* from mice to large animals by the designed dual modality X-ray and fluorescent imaging strategies. Moreover, considering the potential issue of the short half-life of ^{18}F, the same group also developed another ^{153}Sm-labeled $NaLuF_4$:Yb/Tm UCNP through the hydrothermal synthesis strategy for improved SPECT and fluorescent imaging of living animals. In this new system, the half-life of ^{153}Sm (46.3 h) was found to be greatly enhanced [106], and the biological distribution of the ^{153}Sm-labeled $NaLuF_4$:Yb/Tm UCNPs had been well quantified *in vivo*using the SPECT imaging technique which was consistent with the method done by fluorescent imaging (Figure 10).

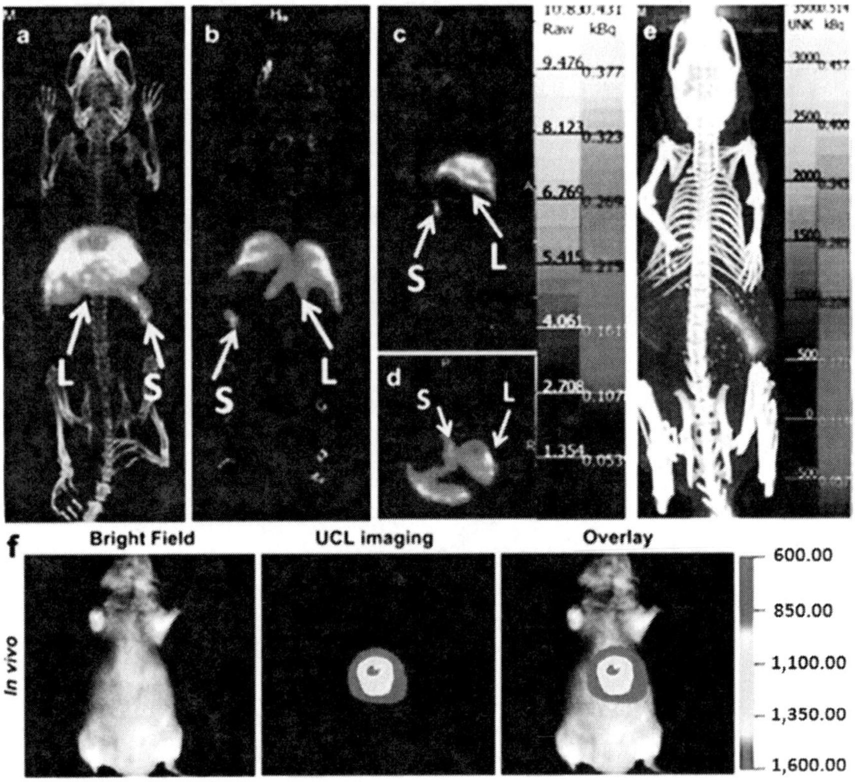

Figure 10. *In vivo* Single Photon Emission Tomography (SPECT)/Optical imaging study after intravenous injection of ^{153}Sm-UCNPs. (**a**) Whole-body three-dimensional projection, (**b**) coronal, (**c**) sagittal and (**d**) transversal images acquired at 1 h and (**e**) whole-body three-dimensional projection images acquired at 24 h are shown respectively. The arrows in the inset point to the liver (L) and spleen (S). (**f**) *In vivo* upconversion luminescence imaging of the Kunming mouse 1 h after tail vein injection of the ^{153}Sm-UCNPs (20 mg/kg). Adapted with permission from reference [105]. Copyright: Elsevier, 2013.

Furthermore, besides the dual-modal imaging based on doped UCNPs, multi-modal imaging through the doping of the different elements into lanthanide-based UCNPs structures could be also achieved. For example, the fabrication of ^{18}F-labeled $NaGdF_4$:Yb/Er UCNPs have been reported which have been further used for PET/MRI/optical triple-modal imaging *in vitro* and *in vivo* [107]. Similarly, the combination of Gd-doped UCNPs with gold nanoparticles based on a simple electrostatic adsorption mechanism was

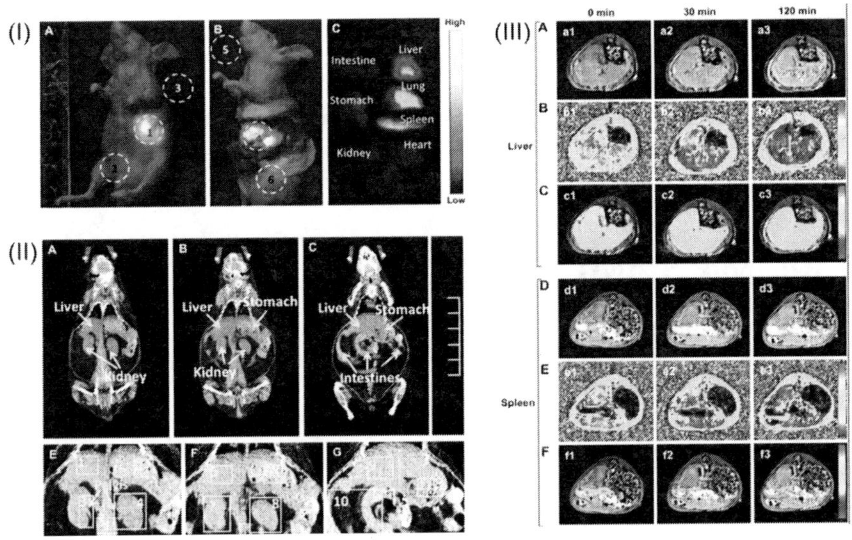

Figure 11. (**I**) A whole-body imaging of UCNPs@SiO_2-GdDTPA (Diethylenetriaminepentaacetic Acid) for 10 min. (**I-A**) *In vivo*imaging of the sacrificed nude mouse after injection with UCNPs@SiO_2-GdDTPA for 10 min. (**I-B**) *Ex vivo* imaging of nude mouse. (**I-C**) *Ex vivo* imaging of viscera. All images were acquired under the same instrumental conditions, and the power density of the 980 nm laser is 150 mW cm^{-2}. (**II**) The application of *in vivo* CT imaging in Kunming mice. (**II-A**, **B** and **C**) serials coronal CT images of Kunming mouse at different layer after injection with UCNPs@SiO_2-GdDTPA. (**II-E**, **F** and **G**) partial enlarged CT view of abdomen. (**III**) The application of *in vivo* MRI imaging of the Kunming mice. (**III-A**) T_1-weighted MR images of liver after injection with UCNPs@SiO_2-GdDTPA for 0, 30 and 120 min (**III-B**) T_1 distribution images of liver after injection with UCNPs@SiO_2-GdDTPA for 0, 30 and 120 min. (**III-C**) Local colorized T_1-weighted MR images of liver after injection with UCNPs@SiO_2-GdDTPA for 0, 30 and 120 min. (**III-D**) T_1-weighted MR images of spleen after injection with UCNPs@SiO_2-GdDTPA for 0, 30 and 120 min. (**III-E**) T_1 distribution images of spleen after injection with UCNPs@SiO_2-GdDTPA for 0, 30 and 120 min. (**III-F**) Local colorized T_1-weighted MR images of spleen after injection with UCNPs@SiO_2-GdDTPA for 0, 30 and 120 min. Adapted with permission from reference [85]. Copyright: Elsevier, 2012.

also reported recently and such a sub-50 nm sized multifunctional nanostructure has been proven to be useful for CT, MRI and fluorescent imaging in living mice [108]. A similar triple-modal imaging application was also reported on the basis of Gd^{3+} complex-modified $NaLuF_4$ core-shelled upconversion nanophosphors for CT, T_1-enhanced MRI and NIR-to-NIR fluorescent imaging of tumor status in living subjects [85] (Figure 11); moreover, the MTT assay and histological analysis of viscera sections further confirmed the good biocompatibility and minimum biological toxicity of mixed UCNPs@SiO_2-Gd-DTPA (Diethylenetriaminepentaacetic Acid) nanoparticle components *in vitro* and in the living animals.

5. SUMMARY AND PERSPECTIVES

This review summarizes the latest advance of lanthanide doped UCNPs within the past few years, mainly concentrating on their bioapplication in molecular imaging *in vitro* and in living subjects. Indeed, upon effective surface modification and light irradiation under long wavelength (mostly at 980 nm), UCNPs particles have been proven to be highly beneficial for real-time imaging of biological processes with a high depth of tissue penetration and, meanwhile, with minimum cellular damage in living systems. Moreover, by taking advantages of the unique luminescent and chemical properties of UCNPs, such promising functional materials will also open new possibilities for more extensive applications, including medical diagnostics and imaging, targeted drug or gene delivery, and light controlled therapy in the future. So far, multifunctional nanoprobes based on UCNPs are emerging as a new type of theranostic tool to achieve more accurate diagnostics and more effective treatments through the integrated multimodal imaging moieties towards MRI, CT, SPECT, PET and photoacoustic screening [84,103,104,105,106,107,108]. Similarly, novel multifunctional materials by the combination of photo-activated therapeutic agents with UCNP nanostructures have also been proposed recently to investigate the possibility for improved *in vivo* therapeutics based on the different strategies including siRNA treatment, photodynamic therapy (PDT), and photothermal therapy (PTT), *etc.* [32,42,109,110,111,112]. Toward these ends, by right, outstanding and promising progress of great scope has been made in designing new types of functional UCNPs and these studies definitely enhance the collaborative efforts from scientists with expertise in diverse research areas. However, despite the great potentials of UCNPs in biomedical applications, many challenges certainly remain to be fully addressed and substantial efforts still need to be made to solve the problems in the applications of lanthanide-based UCNPs.

For example, one major challenge of the utilities of UCNPs is to develop new classes of nano-materials with higher stability and converting optical properties under relatively shorter wavelength light illumination. Currently, most of the irradiation wavelength used for the excitation of the existing UCNPs systems relies on NIR laser at 980 nm, which overlaps the absorption band of the water molecule, one of the most abundant and significant NIR absorber in living systems [82]. Therefore, the 980 nm light irradiation for exciting UCNPs luminescence would be significantly attenuated in the process of penetrating biological tissues. Moreover, upon overexposure with 980 nm irradiation, the biological tissues would easily undergo overheating, and could be potentially damaged by the inevitable thermal effects [74,113] which may greatly limit their extensive applications in living systems. Therefore, new types of UCNP materials with light irradiation at shorter wavelengths would be promising alternatives to overcome the overheating effect [113] (Figure 12). Toward this direction, recently, Hummelen *et al.* [114] reported that some dye-sensitized UCNPs could be excited by 800 nm light irradiation when the lanthanide based UCNPs materials were doped with Nd. Similarly, Han *et al.* [115] also independently presented a new generation of $Nd^{3+}/Yb^{3+}/Er^{3+}$ (Tm^{3+}) cascade sensitized tri-doped core/shell β-$NaYF_4$ colloidal UCNPs with a constitutional excitation peak at 800 nm. With a similar design, core-shell UCNPs with doping of Nd^{3+} were also reported, which were found to be effectively excited at 795 nm [116]. The cellular experiments demonstrated that the Nd^{3+}-doped UCNPs with 795 nm irradiation indicated better biocompatibility compared with the conventional UCNPs under 980 nm laser irradiation. Very recently, Yan *et al.* [110] reported a new set of Nd^{3+}-sensitized upconversion nanophosphors by introducing Nd^{3+} as the sensitizer to build up the core/shell structure to ensure successive $Nd^{3+} \rightarrow Yb^{3+} \rightarrow$activator energy transfer. The *in vivo* imaging showed that 808 nm laser excitation would be efficient for *in vivo* imaging applications. Meanwhile, the established Er@Nd UCNPs and Tm@Nd UCNPs could indicate promising multiplexing imaging effects upon laser irradiation at 808 nm. The *in vivo* overheating effect experiments confirmed that the 808 nm laser-induced local overheating effect could be greatly minimized when compared with that of the 980 nm laser (Figure 13).

In terms of the efforts to minimize the overheating effects caused by long wavelength laser irradiation, the biocompatibility and potential long-term toxicity of UCNPs would be other concerns to be addressed. Although emerging experimental results showed negligible cytotoxicity *in vitro* and *in vivo*when the particle surface was modified with biocompatible ligand moieties, longer time effects for immune response or the reproductive system still require further investigation. Furthermore, the physicochemical properties of UCNPs regarding their surface functionalization, shape, size distribution and their further functions towards reticuloendothelial system (RES) uptake such as in liver, spleen, *etc.*, would also be factors to potentially influence their extensive biomedical applications in living systems.

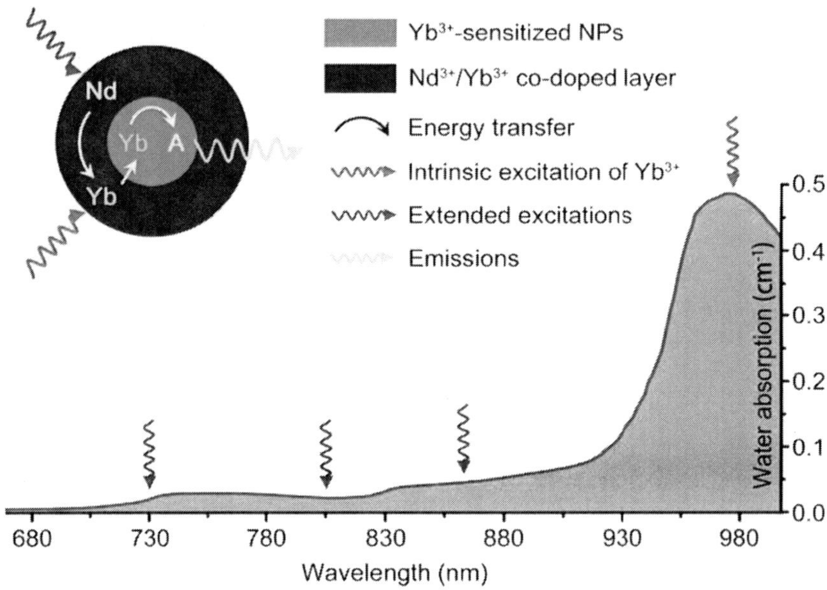

Figure 12. Absorption of water in the NIR and the integration scheme of the Nd^{3+}→Yb^{3+} energy transfer (ET) process by introducing the Nd^{3+}/Yb^{3+} co-doped shell. The resulting Nd^{3+}→Yb^{3+}→activator ET could extend the effective excitation bands for conventional Yb^{3+}-sensitized UCNPs. Featuring lower water absorptions, these alternative excitation bands are expected to minimize the tissue overheating effect caused by NIR laser exposure (the blue line represents the absorption spectrum of water). Adapted with permission from reference [113]. Copyright: American Chemical Society, 2013.

Last, but not least, despite the fact that the UCNP particles have provided unique optical properties that make them successful for various imaging applications in biological systems, the development of alternative strategies or new sets of functional UCNPs with higher converting efficiency, enhanced photostability and quantum yields still remains a big challenge in the related fields. For instance, most existing UCNPs heavily rely on the key component of $NaYF_4$ as the most efficient host materials. The significant enhancement of upconversion luminescence (UCL) efficiency is still difficult to be further achieved. Although some near-infrared dye molecules have been chosen recently as antennas to broaden and enhance the excitation of UCNPs, which could greatly improve the conversion efficiency (e.g., >3000 times) due to the increased absorptivity of the particle complexes [117], new host functional materials with more intense UCL properties to replace $NaYF_4$ are still highly desired. Furthermore, the suitable size distribution of the UCNPs

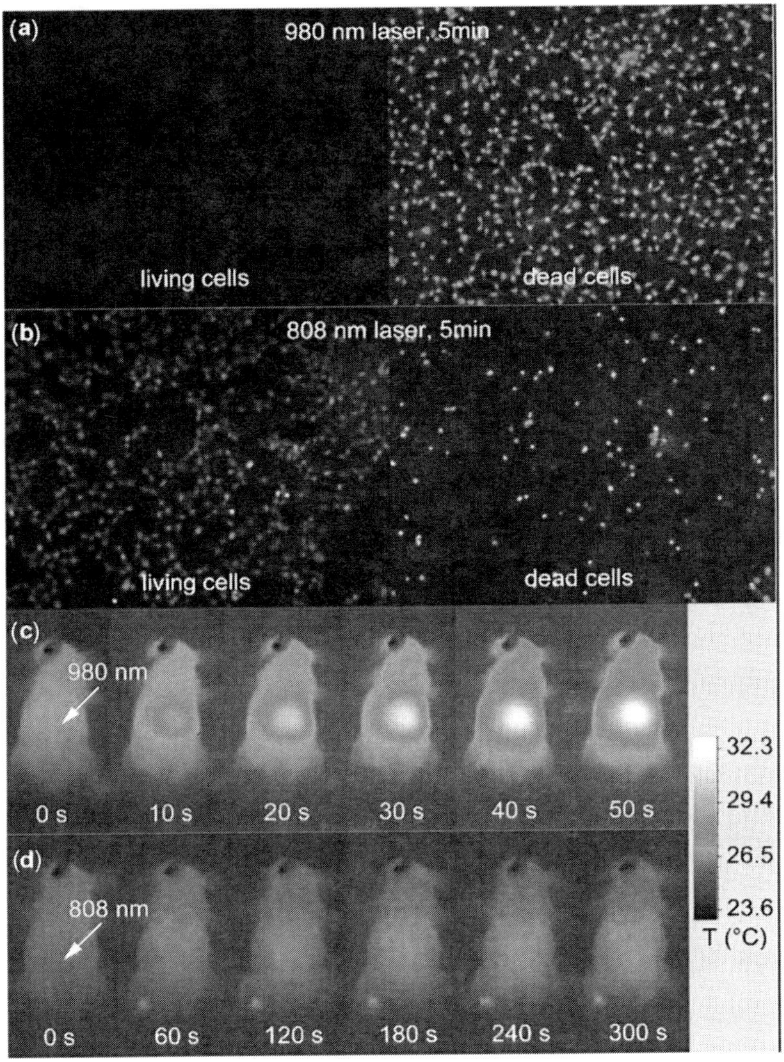

Figure 13. *In vitro* and *in vivo* heating effect induced by laser irradiation. (**a**,**b**) HEK (Human Embryonic Kidney) 293T cells after 5 min irradiation of 980 nm (**a**) and a 808 nm laser (**b**). Living cells and dead cells were stained with calcein AM (Acetomethoxy) and propidium iodide, respectively. (**c**,**d**) Infrared thermal image of a nude mouse during continuous (**c**) 980 nm laser irradiation for 50 s and (**d**) 808 nm laser irradiation for 300 s. Irradiation spots are denoted with the white arrows. Adapted with permission from reference [113]. Copyright: American Chemical Society, 2013.

would be another factor to greatly affect the efficiency of upconversion luminescence. It is known that a smaller size of UCNPs would lead to a weaker luminescence emission, although smaller sized nanoparticles may indicate obvious advantages to facilitate the effective cellular or tissue uptake in the process of imaging applications. However, unfortunately, there is still a lack of satisfactory strategies to prepare the well-defined sub-10 nm UCNPs with a potent luminescence emission so far, which remains an important area with extensive efforts desperately required. Therefore, in the near future, new synthetic methods and alternative functional UCNP materials with more unique optical properties, a higher quantum yield, improved efficiency and a promising excitation wavelength will continue to be developed, which could significantly address the potential issues observed in most of the currently existing lanthanide-based UCNPs. With more innovations in material chemistry, photo-physics and biology, functional UCNPs nanostructures will continually expand their applications, and they will undoubtedly supply more exciting new insights into bio-sensing, non-invasive molecular imaging, effective theranostics, targeted drug or gene delivery and light-controlled personal diseases treatments.

ACKNOWLEDGMENTS

This work was supported by Start-Up Grant (SUG), A*Star (SERC1121202008), Ministry of Education (MOE), Tier 1 (RG 64/10), COS research collaboration award in Nanyang Technological University, Singapore.

CONFLICTS OF INTEREST

The authors declare no conflict of interest.

REFERENCES

1. Achilefu, S. Introduction to concepts and strategies for molecular imaging. *Chem. Rev.* **2010**, *110*, 2575–2578.

2. He, X.; Gao, J.; Gambhir, S.S.; Cheng, Z. Near-infrared fluorescent nanoprobes for cancer molecular imaging: Status and challenges. *Trends Mol. Med.* **2010**, *12*, 574–583.

3. Hildebrandt, I.J.; Gambhir, S.S. Molecular imaging applications for immunology. *Clin. Immunol.***2004**, *2*, 210–224.

4. Weissleder, R.; Pittet, M.J. Imaging in the era of molecular oncology. *Nature* **2008**, *452*, 580–589.

5. Jiang, T.T.; Xing, B.G; Rao, J.H. Recent development of biological reporter technology for detecting gene expression. *Biotech. Genet. Eng. Rev.* **2008**, *25*, 41–75.

6. Massoud, T.F.; Gambhir, S.S. Molecular imaging in living subjects: Seeing fundamental biological processes in a new light. *Genes Dev.* **2003**, *17*, 545–580.

7. Willmann, J.K.; van Bruggen, N.; Dinkelborg, L.M.; Gambhir, S.S. Molecular imaging in drug development. *Nat. Rev. Drug Discov.* **2008**, *7*, 591–607.

8. Mullard, A. Molecular imaging as a de-risking tool: Coming into focus? *Nat. Rev. Drug Discov.* **2013**, *12*, 251–252.

9. Hastings, J.W. Chemistries and colors of bioluminescent reactions: A review. *Gene* **1996**, *173*, 5–11.

10. Leoning, A.M.; Wu, A.M.; Gambhir, S.S. Red-shifted *Renilla reniformis* luciferase variants for imaging in living subjects. *Nat. Methods* **2007**, *4*, 641–643.

11. Contag, C.H.; Bachmann, M.H. Advances in *in vivo* bioluminescence imaging of gene expression. *Annu. Rev. Biomed. Eng.* **2002**, *4*, 235–260.

12. Ray, P.; Gambhir, S.S. Noninvasive imaging of molecular events with bioluminescent reporter genes in living subjects. *Methods Mol. Biol.* **2007**, *411*, 131–144.

13. Wehrman, T.S.; von Degenfeld, G.; Krutzik, P.O.; Nolan, G.P.; Blau, H.M. Luminescent imaging of beta-galactosidase activity in living subjects using sequential reporter-enzyme luminescence. *Nature methods* **2006**, *3*, 295–301.

14. Tung, C.H.; Zeng, Q.; Shah, K.; Kim, D.E.; Schellingerhout, D.; Weissleder, R. *In vivo* imaging of beta-galactosidase activity using far red fluorescent switch. *Cancer Res.* **2004**, *6*, 1579–1583.

15. Shah, K.; Tung, C.H.; Breakefield, X.O.; Weissleder, R. *In vivo* imaging of S-TRAIL-mediated tumor regression and apoptosis. *Mol. Ther.* **2005**, *11*, 926–931.

16. Zhou, W.; Valley, M.P.; Shultz, J.; Hawkins, E.M.; Bernad, L.; Good, T.; Good, D.; Riss, T.L.; Klaubert, D.H.; Wood, K.V. New bioluminogenic substrates for monoamine oxidase assays. *J. Am. Chem. Soc.* **2006**, *128*, 3122–3123.

17. Rao, J.; Dragulescu-Andrasi, A.; Yao, H. Fluorescent imaging *in vivo*: Recent advances. *Curr. Opin. Biotech.* **2007**, *18*, 17–25.

18. Licha, K.; Olbrich, C. Optical imaging in drug discovery and diagnostic applications. *Adv. Drug Deliv. Rev.* **2005**, *57*, 1087–1108.

19. Stefflova, K.; Chen, J.; Zheng, G. Using molecular beacons for cancer imaging and treatment. *Front Biosci.* **2007**, *12*, 4709–4721.

20. Escobedo, J.O.; Rusin, O.; Lim, S.; Strongin, R.M. NIR dyes for bioimaging applications. *Curr. Opin. Chem. Biol.* **2010**, *184*, 64–71.

21. Shao, Q.; Yang, Y.M.; Xing, B.G. Chemistry of Optical Imaging Probes. In *Molecular Imaging Probes for Cancer Research*; World Science: British Columbia, Canada, 2010.

22. Rothman, D.M.; Shults, M.D.; Imperiali, B. Chemical approaches for investigating phosphorylation in signal transduction networks. *Trends Cell Biol.* **2005**, *15*, 502–510.

23. Lawrence, D.S. The preparation and *in vivo* applications of caged peptides and proteins. *Curr. Opin. Chem. Biol.* **2005**, *9*, 570–575.

24. Erathodiyil, N.; Ying, J.Y. Functionalization of inorganic nanoparticles for bioimaging applications. *Acc. Chem. Res.* **2011**, *44*, 925–935.

25. Gao, J.H.; Chen, X.Y.; Cheng, Z. Near-infrared quantum dots as optical probes for tumor imaging.*Curr. Top. Med. Chem.* **2010**, *10*, 1147–1157.

26. Cai, W.B.; Chen, X.Y. Nanoplatforms for targeted molecular imaging in living subjects. *Small* **2007**,*3*, 1840–1854.

27. Biju, V. Chemical modifications and bioconjugate reactions of nanomaterials for sensing, imaging, drug delivery and therapy. *Chem. Soc. Rev.* **2014**, *43*, 744–764.

28. Michalet, X.; Pinaud, F.F.; Bentolila, L.A.; Tsay, J.M.; Doose, S.; Li, J.; Sundaresan, G.; Wu, A.; Gambhir, S.S.; Weiss, S. Quantum dots for live cells, *in vivo* imaging, and diagnostics. *Science* **2005**,*307*, 538–544.

29. Probst, C.E.; Zrazhevskiy, P.; Bagalkot, V.; Gao, X.H. Quantum dots as a platform for nanoparticle drug delivery vehicle design. *Adv. Drug Deliv. Rev.* **2013**, *65*, 703–718.

30. Biju, V.; Itoh, T.; Ishikawa, M. Delivering quantum dots to cells: Bioconjugated quantum dots for targeted and nonspecific extracellular and intracellular imaging. *Chem. Soc. Rev.* **2010**, *39*, 3031–3056.

31. Chi, X.; Huang, D.; Zhao, Z.; Zhou, Z.; Yin, Z.; Gao, J. Nanoprobes for *in vitro* diagnostics of cancer and infectious diseases. *Biomaterials* **2012**, *33*, 189–206.

32. Jayakumar, M.K.; Idris, N.M.; Zhang, Y. Remote activation of biomolecules in deep tissues using near-infrared-to-UV upconversion nanotransducers. *Proc. Natl. Acad. Sci. USA* **2012**, *109*, 8483–8488.

33. Wang, F.; Liu, X. Upconversion multicolor fine-tuning: Visible to near-infrared emission from lanthanide-doped $NaYF_4$ nanoparticles. *J. Am. Chem. Soc.* **2008**, *130*, 5642–5643.

34. Wang, F.; Liu, X. Recent advances in the chemistry of lanthanide-doped upconversion nanocrystals. *Chem. Soc. Rev.* **2009**, *38*, 976–989.

35. Haase, H.; Schafer, H. Upconverting nanoparticles. *Angew. Chem. Int. Ed.* **2011**, *50*, 5808–5829.

36. Mader, H.S.; Kele, P.; Saleh, S.M.; Wolfbeis, O.S. Upconverting luminescent nanoparticles for use in bioconjugation and bioimaging. *Curr. Opin. Chem. Biol.* **2010**, *14*, 582–596.

37. Feng, W.; Sun, L.D.; Zhang, Y.W.; Yan, C.H. Synthesis and assembly of rare earth nanostructures directed by the principle of coordination chemistry in solution-based process. *Coordin. Chem. Rev.***2010**, *254*, 1038–1053.

38. Cheng, L.; Wang, C.; Liu, Z. Upconversion nanoparticles and their composite nanostructures for biomedical imaging and cancer therapy. *Nanoscale* **2013**, *5*, 23–37.

39. Gu, Z.; Yan, L.; Tian, G.; Li, S.; Chai, Z.; Zhao, Y. Recent advances in design and fabrication of upconversion nanoparticles and their safe theranostic applications. *Adv. Mater.* **2013**, *25*, 3758–3779.

40. Liu, Y.; Tu, D.; Zhu, H.; Chen, X. Lanthanide-doped luminescent nanoprobes: Controlled synthesis, optical spectroscopy, and bioapplications. *Chem. Soc. Rev.* **2013**, *42*, 6924–6958.

41. Wang, F.; Banerjee, D.; Liu, Y.; Chen, X.; Liu, X. Upconversion nanoparticles in biological labeling, imaging, and therapy. *Analyst* **2010**, *135*, 1839–1854.

42. Cheng, L.; Yang, K.; Li, Y.; Chen, J.; Wang, C.; Shao, M.; Lee, S.T.; Liu, Z. Facile preparation of multifunctional upconversion nanoprobes for multimodal imaging and dual-targeted photothermal therapy. *Angew. Chem. Int. Ed.* **2011**, *50*, 7385–7390.

43. Zhou, J.; Liu, Z.; Li, F. Upconversion nanophosphors for small-animal imaging. *Chem. Soc. Rev.***2012**, *41*, 1323–1349.

44. Yi, G.S.; Chow, G.M. Synthesis of hexagonal-phase $NaYF_4$:Yb,Er and $NaYF_4$:Yb,Tm nanocrystals with efficient up-conversion fluorescence. *Adv. Funct. Mater.* **2006**, *16*, 2324–2329.

45. Wang, F.; Han, Y.; Lim, C.S.; Lu, Y.; Wang, J.; Xu, J.; Chen, H.; Zhang, C.; Hong, M.; Liu, X. Simultaneous phase and size control of upconversion nanocrystals through lanthanide doping.*Nature* **2010**, *463*, 1061–1065.

46. Mai, H.X.; Zhang, Y.W.; Si, R.; Yan, Z.G.; Sun, L.D.; You, L.P.; Yan, C.H. High-quality sodium rare-earth fluoride nanocrystals: Controlled synthesis and optical properties. *J. Am. Chem. Soc.* **2006**,*128*, 6426–6436.

47. Zhang, Y.W.; Sun, X.; Si, R.; You, L.P.; Yan, C.H. Single-crystalline and monodisperse LaF_3triangular nanoplates from a single-source precursor. *J. Am. Chem. Soc.* **2005**, *127*, 3260–3261.

48. Mai, H.X.; Zhang, Y.W.; Sun, L.D.; Yan, C.H. Highly efficient multicolor up-conversion emissions and their mechanisms of monodisperse $NaYF_4$:Yb,Er core and core/shell-structured nanocrystals.*J. Phys. Chem. C* **2007**, *111*, 13721–13729.

49. Mai, H.X.; Zhang, Y.W.; Sun, L.D.; Yan, C.R. Size- and phase-controlled synthesis of monodisperse $NaYF_4$:Yb,Er nanocrystals from a unique delayed nucleation pathway monitored with upconversion spectroscopy. *J. Phys. Chem. C* **2007**, *111*, 13730–13739.

50. Wang, X.; Zhuang, J.; Peng, Q.; Li, Y. A general strategy for nanocrystal synthesis. *Nature* **2005**,*437*, 121–124.

51. Wang, L.; Li, P.; Zhuang, J.; Bai, F.; Feng, J.; Yan, X.; Li, Y. Carboxylic acid enriched nanospheres of semiconductor nanorods for cell imaging. *Angew. Chem. Int. Ed.* **2008**, *47*, 1054–1057.

52. Wang, M.; Liu, J.L.; Zhang, Y.X.; Hou, W.; Wu, X.L.; Xu, S.K. Two-phase solvothermal synthesis of rare-earth doped $NaYF_4$ upconversion fluorescent nanocrystals. *Mater. Lett.* **2009**, *63*, 325–327.

53. Sun, Y.J.; Chen, Y.; Tian, L.J.; Yu, Y.; Kong, X.G.; Zhao, J.W.; Zhang, H. Controlled synthesis and morphology dependent upconversion luminescence of $NaYF_4$:Yb,Er nanocrystals. *Nanotechnology***2007**, *18*, 275609–275617.

54. Tian, G.; Gu, Z.J.; Zhou, L.J.; Yin, W.Y.; Liu, X.X.; Yan, L.; Jin, S.; Ren, W.L.; Xing, G.M.; Li, S.J.; *et al.* Mn^{2+} Dopant-controlled synthesis of $NaYF_4$:Yb/Er upconversion nanoparticles for *in vivo*imaging and drug delivery. *Adv. Mater.* **2012**, *24*, 1226–1231.

55. Stouwdam, J.W.; van Veggel, F.C.J.M. Near-infrared emission of redispersible Er^{3+}, Nd^{3+}, and Ho^{3+} doped LaF_3 nanoparticles. *Nano. Lett.* **2002**, *2*, 733–737.

56. Patra, A.; Friend, C.S.; Kapoor, R.; Prasad, P.N. Upconversion in Er^{3+}:ZrO_2 nanocrystals. *J. Phys. Chem. B* **2002**, *106*, 1909–1912.

57. Vetrone, F.; Boyer, J.C.; Capobianco, J.A.; Speghini, A.; Bettinelli, M. Significance of Yb^{3+}concentration on the upconversion mechanisms in codoped Y_2O_3:Er^{3+}, Yb^{3+} nanocrystals. *J. Appl. Phys.* **2004**, *96*, 661–667.

58. Xu, L.L.; Yu, Y.N.; Li, X.G.; Somesfalean, G.; Zhang, Y.G.; Gao, H.; Zhang, Z.G. Synthesis and upconversion properties of monoclinic Gd_2O_3:Er^{3+} nanocrystals. *Opt. Mater.* **2008**, *30*, 1284–1288.

59. Kong, W.J.; Shan, J.; Ju, Y.G. Flame synthesis and effects of host materials on Yb^{3+}/Er^{3+} co-doped upconversion nanophosphors. *Mater. Lett.* **2010**, *64*, 688–691.

60. Qin, X.; Yokomori, T.; Ju, Y.G. Flame synthesis and characterization of rare-earth (Er^{3+}, Ho^{3+}, and Tm^{3+}) doped upconversion nanophosphors. *Appl. Phys. Lett.* **2007**, *90*, 073104.

61. Ma, P.; Xiao, H.; Li, X.; Li, C.; Dai, Y.; Cheng, Z.; Jing, X.; Lin, J. Rational design of multifunctional upconversion nanocrystals/polymer nanocomposites for cisplatin(IV) delivery and biomedical imaging. *Adv. Mater.* **2013**, *25*, 4898–4905.

62. Yang, Y.; Shao, Q.; Deng, R.; Wang, C.; Teng, X.; Cheng, K.; Cheng, Z.; Huang, L.; Liu, Z.; Liu, X.;*et al. In vitro* and *in vivo* uncaging and bioluminescence imaging by using photocaged upconversion nanoparticles. *Angew. Chem. Int. Ed.* **2012**, *51*, 3125–3129.

63. Liu, J.N.; Bu, W.; Pan, L.M.; Zhang, S.; Chen, F.; Zhou, L.; Zhao, K.L.; Peng, W.; Shi, J. Simultaneous nuclear imaging and intranuclear drug delivery by nuclear-targeted multifunctional upconversion nanoprobes. *Biomaterials* **2012**, *33*, 7282–7290.

64. Xiong, L.; Yang, T.; Yang, Y.; Xu, C.; Li, F. Long-term *in vivo* biodistribution imaging and toxicity of polyacrylic acid-coated upconversion nanophosphors. *Biomaterials* **2010**, *31*, 7078–7085.

65. Rantanen, T.; Jarvenpaa, M.L.; Vuojola, J.; Kuningas, K.; Soukka, T. Fluorescence-quenching-based enzyme-activity assay by using photon upconversion. *Angew. Chem. Int. Ed.* **2008**, *47*, 3811–3813.

66. Naccache, R.; Vetrone, F.; Mahalingam, V.; Cuccia, L.A.; Capobianco, J.A. Controlled synthesis and water dispersibility of hexagonal phase $NaGdF_4:Ho^{3+}/Yb^{3+}$ nanoparticles. *Chem. Mater.* **2009**,*21*, 717–723.

67. Bogdan, N.; Vetrone, F.; Ozin, G.A.; Capobianco, J.A. Synthesis of ligand-free colloidally stable water dispersible brightly luminescent lanthanide-doped upconverting nanoparticles. *Nano. Lett.***2011**, *11*, 835–840.

68. Chen, J.; Guo, C.R.; Wang, M.; Huang, L.; Wang, L.P.; Mi, C.C.; Li, J.; Fang, X.X.; Mao, C.B.; *et al.* Controllable synthesis of $NaYF^4$:Yb,Er upconversion nanophosphors and their application to *in vivo* imaging of Caenorhabditis elegans. *J. Mater. Chem.* **2011**, *21*, 2632–2638.

69. Zhou, H.P.; Xu, C.H.; Sun, W.; Yan, C.H. Clean and flexible modification strategy for carboxyl/aldehyde-functionalized upconversion nanoparticles and their optical applications. *Adv. Funct. Mater.* **2009**, *19*, 3892–3900.

70. Chen, Z.; Chen, H.; Hu, H.; Yu, M.; Li, F.; Zhang, Q.; Zhou, Z.; Yi, T.; Huang, C. Versatile synthesis strategy for carboxylic acid-functionalized upconverting nanophosphors as biological labels. *J. Am. Chem. Soc.* **2008**, *130*, 3023–3029.

71. Wang, L.Y.; Yan, R.X.; Hao, Z.Y.; Wang, L.; Zeng, J.H.; Bao, J.; Wang, X.; Peng, Q.; Li, Y.D. Fluorescence resonant energy transfer biosensor based on upconversion-luminescent nanoparticles. *Angew. Chem. Int. Edit.* **2005**, *44*, 6054–6057.

72. Bao, Y.; Luu, Q.A.N.; Lin, C.K.; Schloss, J.M.; May, P.S.; Jiang, C.Y. Layer-by-layer assembly of freestanding thin films with homogeneously distributed upconversion nanocrystals. *J. Mater. Chem.* **2010**, *20*, 8356–8361.

73. Yang, Y.M.; Velmurugan, B.; Liu, X.; Xing, B. NIR photoresponsive crosslinked upconverting nanocarriers toward selective intracellular drug release. *Small* **2013**, *9*, 2937–2944.

74. Yang, Y.M.; Liu, F.; Liu, X.G.; Xing, B.G. NIR light controlled photorelease of siRNA and its targeted intracellular delivery based on upconversion nanoparticles. *Nanoscale* **2013**, *5*, 231–238.

75. Cao, T.; Yang, Y.; Gao, Y.; Zhou, J.; Li, Z.; Li, F. High-quality water-soluble and surface-functionalized upconversion nanocrystals as luminescent probes for bioimaging. *Biomaterials***2011**, *32*, 2959–2968.

76. Nichkova, M.; Dosev, D.; Gee, S.J.; Hammock, B.D.; Kennedy, I.M. Microarray immunoassay for phenoxybenzoic acid using polymer encapsulated Eu:Gd_2O_3 nanoparticles as fluorescent labels.*Anal. Chem.* **2005**, *77*, 6864–6873.

77. Peer, D.; Karp, J.M.; Hong, S.; FaroKHzad, O.C.; Margalit, R.; Langer, R. Nanocarriers as an emerging platform for cancer therapy. *Nat. Nanotechnol.* **2007**, *2*, 751–760.

78. Nyk, M.; Kumar, R.; Ohulchanskyy, T.Y.; Bergey, E.J.; Prasad, P.N. High contrast *in vitro* and *in vivo* photoluminescence bioimaging using near infrared to near infrared up-conversion in $Tm^{3+}$and Yb^{3+} doped fluoride nanophosphors. *Nano Lett.* **2008**, *8*, 3834–3838.

79. Xing, H.Y.; Zheng, X.P.; Ren, Q.G.; Bu, W.B.; Ge, W.Q.; Xiao, Q.F.; Zhang, S.J.; Wei, C.Y.; Qu, H.Y.; Wang, Z.; *et al.* Computed tomography imaging-guided radiotherapy by targeting upconversion nanocubes with significant imaging and radiosensitization enhancements. *Sci. Rep.-UK* **2013**, *3*, 1–9.

80. Lim, S.F.; Riehn, R.; Ryu, W.S.; Khanarian, N.; Tung, C.K.; Tank, D.; Austin, R.H. *In vivo* and scanning electron microscopy imaging of up-converting nanophosphors in *Caenorhabditis elegans.Nano Lett.* **2006**, *6*, 169–174.

81. Wang, K.; Ma, J.B.; He, M.; Gao, G.; Xu, H.; Sang, J.; Wang, Y.X.; Zhao, B.Q.; Cui, D.X. Toxicity assessments of near-infrared upconversion luminescent LaF_3:Yb,Er in early development of zebrafish embryos. *Theranostics* **2013**, *3*, 258–266.

82. Mitsunaga, M.; Ogawa, M.; Kosaka, N.; Rosenblum, L.T.; Choyke, P.L.; Kobayashi, H. Cancer cell-selective *in vivo* near infrared photoimmunotherapy targeting specific membrane molecules. *Nat. Med.* **2011**, *17*, 1685–1691.

83. Chatterjee, D.K.; Rufaihah, A.J.; Zhang, Y. Upconversion fluorescence imaging of cells and small animals using lanthanide doped nanocrystals. *Biomaterials* **2008**, *29*, 937–943.

84. Zhou, J.; Sun, Y.; Du, X.; Xiong, L.; Hu, H.; Li, F. Dual-modality *in vivo* imaging using rare-earth nanocrystals with near-infrared to near-infrared (NIR-to-NIR) upconversion luminescence and magnetic resonance properties. *Biomaterials* **2010**, *31*, 3287–3295.

85. Xia, A.; Chen, M.; Gao, Y.; Wu, D.; Feng, W.; Li, F. Gd^{3+} complex-modified NaLuF4-based upconversion nanophosphors for trimodality imaging of NIR-to-NIR upconversion luminescence, X-ray computed tomography and magnetic resonance. *Biomaterials* **2012**, *33*, 5394–5405.

86. Xing, H.; Bu, W.; Ren, Q.; Zheng, X.; Li, M.; Zhang, S.; Qu, H.; Wang, Z.; Hua, Y.; Zhao, K.; *et al.* A $NaYbF_4$:Tm^{3+} nanoprobe for CT and NIR-to-NIR fluorescent bimodal imaging. *Biomaterials* **2012**,*33*, 5384–5393.

87. Liu, Q.; Sun, Y.; Yang, T.; Feng, W.; Li, C.; Li, F. Sub-10 nm hexagonal lanthanide-doped $NaLuF_4$upconversion nanocrystals for sensitive bioimaging *in vivo*. *J. Am. Chem. Soc.* **2011**, *133*, 17122–17125.

88. Yang, T.; Sun, Y.; Liu, Q.; Feng, W.; Yang, P.; Li, F. Cubic sub-20 nm $NaLuF_4$-based upconversion nanophosphors for high-contrast bioimaging in different animal species. *Biomaterials* **2012**, *33*, 3733–3742.

89. Wang, C.; Cheng, L.; Xu, H.; Liu, Z. Towards whole-body imaging at the single cell level using ultra-sensitive stem cell labeling with oligo-arginine modified upconversion nanoparticles.*Biomaterials* **2012**, *33*, 4872–4881.

90. Chari, R.V. Targeted cancer therapy: Conferring specificity to cytotoxic drugs. *Acc. Chem. Res.***2008**, *41*, 98–107.

91. Xiong, L.Q.; Chen, Z.G.; Yu, M.X.; Li, F.Y.; Liu, C.; Huang, C.H. Synthesis, characterization, and *in vivo* targeted imaging of amine-functionalized rare-earth up-converting nanophosphors.*Biomaterials* **2009**, *30*, 5592–5600.

92. Xiong, L.; Chen, Z.; Tian, Q.; Cao, T.; Xu, C.; Li, F. High contrast upconversion luminescence targeted imaging *in vivo* using peptide-labeled nanophosphors. *Anal. Chem.* **2009**, *81*, 8687–8694.

93. Danhier, F.; Le Breton, A.; Préat, V. RGD-based strategies to target alpha(v) beta(3) integrin in cancer therapy and diagnosis. *Mol. Pharm.* **2012**, *9*, 2961–2973.

94. Li, Y.; Jing, C.; Zhang, L.; Long, Y.T. Resonance scattering particles as biological nanosensors *in vitro* and *in vivo*. *Chem. Soc. Rev.* **2012**, *41*, 632–642.

95. Cai, W.; Chen, X. Anti-angiogenic cancer therapy based on integrin alphavbeta3 antagonism.*Anticancer Agents Med. Chem.* **2006**, *6*, 407–428.

96. Meyer, A.; Auernheimer, J.; Modlinger, A.; Kessler, H. Targeting RGD recognizing integrins: Drug development, biomaterial research, tumor imaging and targeting. *Curr. Pharm. Des.* **2006**, *12*, 2723–2747.

97. Yu, X.F.; Sun, Z.; Li, M.; Xiang, Y.; Wang, Q.Q.; Tang, F.; Wu, Y.; Cao, Z.; Li, W. Neurotoxin-conjugated upconversion nanoprobes for direct visualization of tumors under near-infrared irradiation. *Biomaterials* **2010**, *31*, 8724–8731.

98. Shao, Q.; Xing, B. Photoactive molecules for applications in molecular imaging and cell biology.*Chem. Soc. Rev.* **2010**, *39*, 2835–2846.

99. Mayer, G.; Heckel, A. Biologically active molecules with a "light switch". *Angew. Chem. Int. Ed.***2006**, *45*, 4900–4921.

100. Lee, H.; Larson, D.R.; Lawrence, D.S. Illuminating the chemistry of life: Design, synthesis, and applications of "caged" and related photoresponsive compounds. *ACS Chem. Biol.* **2009**, *4*, 409–427.

101. Young, D.D.; Deiters, A. Photochemical control of biological processes. *Org. Biomol. Chem.* **2007**, *5*, 999–1005.

102. Min, Y.; Li, J.; Liu, F.; Yeow, E.K.; Xing, B. Near-infrared light-mediated photoactivation of a platinum antitumor prodrug and simultaneous cellular apoptosis imaging by upconversion-luminescent nanoparticles. *Angew. Chem. Int. Ed.* **2014**, *53*, 1012–1016.

103. Cheng, L.; Yang, K.; Li, Y.; Zeng, X.; Shao, M.; Lee, S.T.; Liu, Z. Multifunctional nanoparticles for upconversion luminescence/MR multimodal imaging and magnetically targeted photothermal therapy. *Biomaterials* **2012**, *33*, 2215–2222.

104. He, M.; Huang, P.; Zhang, C.; Hu, H.; Bao, C.; Gao, G.; He, R.; Cui, D. Dual phase-controlled synthesis of uniform lanthanide-doped $NaGdF_4$ upconversion nanocrystals via an OA/ionic liquid two-phase system for *in vivo* dual-modality imaging. *Adv. Funct. Mater.* **2011**, *21*, 4470–4477.

105. Sun, Y.; Yu, M.; Liang, S.; Zhang, Y.; Li, C.; Mou, T.; Yang, W.; Zhang, X.; Li, B.; Huang, C.; *et al.* Fluorine-18 labeled rare-earth nanoparticles for positron emission tomography (PET) imaging of sentinel lymph node. *Biomaterials* **2011**, *32*, 2999–3007.

106. Yang, Y.; Sun, Y.; Cao, T.; Peng, J.; Liu, Y.; Wu, Y.; Feng, W.; Zhang, Y.; Li, F. Hydrothermal synthesis of $NaLuF_4$:^{153}Sm,Yb,Tm nanoparticles and their application in dual-modality upconversion luminescence and SPECT bioimaging. *Biomaterials* **2013**, *34*, 774–783.

107. Zhou, J.; Yu, M.; Sun, Y.; Zhang, X.; Zhu, X.; Wu, Z.; Wu, D.; Li, F. Fluorine-18-labeled Gd^{3+}/Yb^{3+}/Er^{3+} co-doped $NaYF_4$ nanophosphors for multimodality PET/MR/UCL imaging.*Biomaterials* **2011**, *32*, 1148–1156.

108. Xing, H.; Bu, W.; Zhang, S.; Zheng, X.; Li, M.; Chen, F.; He, Q.; Zhou, L.; Peng, W.; Hua, Y.; *et al.* Multifunctional nanoprobes for upconversion fluorescence, MR and CT trimodal imaging.*Biomaterials* **2012**, *33*, 1079–1089.

109. Chen, G.; Ohulchanskyy, T.Y.; Kachynski, A.; Agren, H.; Prasad, P.N. Intense visible and near-infrared upconversion photoluminescence in colloidal $LiYF_4$:Er^{3+} nanocrystals under excitation at 1490 nm. *ACS Nano* **2011**, *5*, 4981–4986.

110. Idris, N.M.; Gnanasammandhan, M.K.; Zhang, J.; Ho, P.C.; Mahendran, R.; Zhang, Y. *In vivo*photodynamic therapy using upconversion nanoparticles as remote-controlled nanotransducers.*Nat. Med.* **2012**, *18*, 1580–1585.

111. Wang, C.; Tao, H.Q.; Cheng, L.; Liu, Z. Near-infrared light induced *in vivo* photodynamic therapy of cancer based on upconversion nanoparticles. *Biomaterials* **2011**, *32*, 6145–6154.

112. Cui, S.; Yin, D.; Chen, Y.; Di, Y.; Chen, H.; Ma, Y.; Achilefu, S.; Gu, Y. *In vivo* targeted deep-tissue photodynamic therapy based on near-infrared light triggered upconversion nanoconstructure.*ACS Nano* **2013**, *7*, 676–688.

113. Zhan, Q.; Qian, J.; Liang, H.; Somesfalean, G.; Wang, D.; He, S.; Zhang, Z.; Andersson-Engels, S. Using 915 nm laser excited $Tm^{3+}/Er^{3+}/Ho^{3+}$ doped $NaYbF_4$ upconversion nanoparticles for *in vitro*and deeper *in vivo* bioimaging without overheating irradiation. *ACS Nano* **2011**, *5*, 3744–3757.

114. Wang, Y.F.; Liu, G.Y.; Sun, L.D.; Xiao, J.W.; Zhou, J.C.; Yan, C.H. Nd^{3+}-sensitized upconversion nanophosphors: Efficient *in vivo* bioimaging probes with minimized heating effect. *ACS Nano***2013**, *7*, 7200–7206.

115. Zou, W.; Visser, C.; Maduro, J.A.; Pshenichnikov, M.S.; Hummelen, J.C. Broadband dye-sensitized upconversion of near-infrared light. *Nat. Photon.* **2012**, *6*, 560–564.

116. Shen, J.; Chen, G.; Vu, A.-M.; Fan, W.; Bilsel, O.S.; Chang, C.-C.; Han, G. Engineering the upconversion nanoparticle excitation wavelength: Cascade sensitization of tri-doped upconversion colloidal nanoparticles at 800 nm. *Adv. Opt. Mater.* **2013**, *1*, 644–650.

117. Xie, X.; Gao, N.; Deng, R.; Sun, Q.; Xu, Q.H.; Liu, X. Mechanistic investigation of photon upconversion in Nd^{3+}-sensitized core-shell nanoparticles. *J. Am. Chem. Soc.* **2013**, *135*, 12608–12611.

Chapter 7

Simultaneous Quasi-one-Dimensional Propagation and Tuning of Upconversion Luminescence Through Waveguide Effect

Dangli Gao, Dongping Tian, Xiangyu Zhang & Wei Gao

College of Materials & Mineral Resources, Xi'an University of Architecture and Technology, Xi'an, Shaanxi 710055, China

College of Science, Xi'an University of Architecture and Technology, Xi'an, Shaanxi 710055, China

ABSTRACT

Luminescence-based waveguide is widely investigated as a promising alternative to conquer the difficulties of efficiently coupling light into a waveguide. But applications have been still limited due to employing blue or ultraviolet light as excitation source with the lower penetration depth leading to a weak guided light. Here, we show a quasi-one-dimensional propagation of luminescence and then resulting in a strong luminescence output from the top end of a single $NaYF_4$:Yb^{3+}/Er^{3+}microtube under near infrared light excitation. The mechanism of upconversion propagation, based on the optical waveguide effect accompanied with energy migration, is proposed. The efficiency of luminescence output is highly dependent on the concentration of dopant ions, excitation power, morphology, and crystallinity of tube as an indirect evidence of the existence of the optical actived waveguide effect. These findings provide the possibility for the construction of upconversion fiber laser.

INTRODUCTION

Lanthanide ions based micro/nanomaterials have a wide range of applications such as displays1, *in vivo* and *in vitro* imaging2′3′4, the imaging of cancer cells and detection of biomarker molecules5′6′7′8, thermometer9 and enhancement for solar cell devices10′11. Despite these applications, it is necessary to solve the problem of remote propagation of the luminescence to further improve and expand their applications. However, the investigation on the remote propagation of upconversion luminescence is extremely lacking.

In recent years, more and more attention has been paid to the investigations of upconversion mechanism in micro/nanosystems12′13′14′15′16, tuning the laser parameters17′18′19′20′21, and designing new and more complex upconversion micro/nanostructures22′23′24′25 to obtain high upconversion efficiency and controlled spectral modulation. For example, core-shell structure successfully devideded space into activator doped region and sensitizer doped region to avoid harmful cross-relaxation10. While overheating effect could be weakened by separating emission region from absorption region on single micro/nanoparticle via a confocal local excitation14. In addition to, the local excitation of single paticles make some hidden physical effects clear. In 2011 years, Liu group designed a donor and acceptor spatially separated core-shell-shell structure10, where the excitation energy absorbed in the shell undergoes a long journey by energy migration between Gd^{3+} ions to reach the activators inside the core. The special energy migration process is belong to "long-range" interaction, and makes the distance of excited energy propagation be longer than 5 nm. However, the propagation distance is so small to some special application such as optical communication that it always is ignored.

It is well-known that, based on total reflection theory, waveguide generally is served as the optical signal propagation channel and optical interconnection at a long distance range26. Given the difficulties to efficiently couple light into a waveguide, various luminescence-based waveguides have emerged to directly produce the guided light under blue or ultraviolet light excitation27′28′29. Due to the unique one-dimensional (1D) structural characteristics and versatile physical/chemical properties, luminescence nanowires have emerged as the building blocks for a variety of fundamental optical components30′31′32′33′34′35′36′37′38. We also know that rare-earth (RE) doped $NaYF_4$ compounds possessing a high refractive index and a low phonon energy have been regarded as currently one of the most excellent upconversion materials2′12. To my knowledge, someone tried to fabricate active waveguide in fluorite-glass composite by laser micro-machining39. However, 1D $NaYF_4$ compounds as upconversion luminescent waveguides have not yet been reported. Compared to blue and ultraviolet excitations, near infrared light as excitation source has the higher penetration depth and less overheating effect and therefore is preferable for applications in optical communication and bioimaging2'3'4'5'6'40.

Here we show the design and realization of quasi-one-dimensional propagation of excited energy along the length direction of a single $NaYF_4:Yb^{3+}/Er^{3+}$ microtube with a length of 55 µm. The propagration mechanism is attributed to the optical waveguide effect of the single microtube accompanied with energy migration. The effects of codopant concentration, excitation power, the size and crystallinity of microtube on the upconversion signal of individual microtube are also investigated in detail. The present investigation provides theoretical and experimental basis for designing special luminescent materials with targeted luminescence and optical waveguide effect, and has also brought a new idea in tuning multicolor upconversion emission.

RESULTS

Structural characterization of $NaYF_4:Yb^{3+}/Er^{3+}$ (20/2%) microtubes

$NaYF_4$ microtubes codoped with Yb^{3+}/Er^{3+} (20/2 mol%) were prepared by using hydrothermal method with the assistance of trisodium citrate41. The products were composed of uniform tubes with diameters of about 800 nm and lengths of about 5500 nm, as shown by the scanning electron microscopy (SEM) photo in Fig. 1a and bright field optical microscope image in Fig. 1c. The selected area electron diffraction pattern shown in Fig. 1b indicates the formation of $NaYF_4$monocrystal. X-ray powder diffraction pattern shows representative reflections for a hexagonal-phase $NaYF_4$ with space group P63/m (Fig. 1d). Such highly crystalline tubes are resulted from the preferential growth along the [0001] direction (c-axis).

Luminescence output from the top end and the centre of a single microtube

Figure 2 top panels are the bright field optical microscope images of an isolated microtube on glass. The diameter and length of a single microtube were around 8 µm and 55 µm, respectively. Figure 2 middle panels are photoluminescence (PL) photographs showing the single tube being excited locally with a laser focused into a spot of about 1.0 µm in diameter at the end and the center of tube, respectively. The PL photographs indicate that

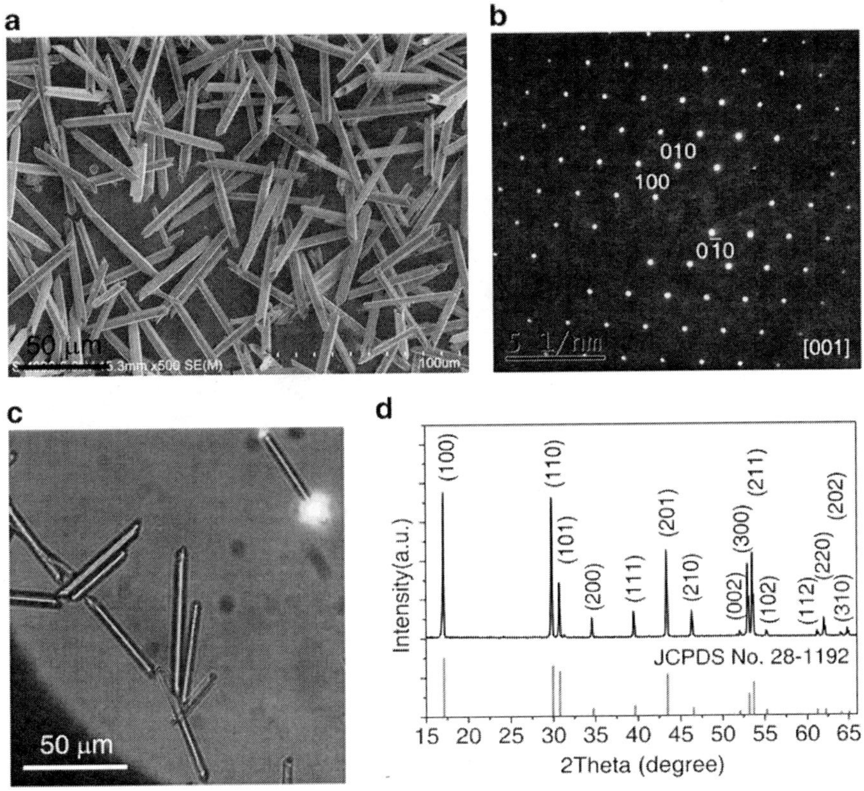

Figure 1: Structural characterization of the as-synthesized $NaYF_4$:Yb^{3+}/Er^{3+} (20/2%) microtubes. (**a**) Typical low-resolution SEM micrograph. (**b**) Selected-area electron diffraction pattern of a single microtube. (**c**) Bright field optical microscope image. (**d**) X-ray powder diffraction pattern of the as-prepared $NaYF_4$: Yb^{3+}/Er^{3+} microtubes showing that all peaks can be well indexed in accordance with hexagonal-phase $NaYF_4$ structure (Joint Committee on Powder Diffraction Standards file number 28–1192).

the excited energy from the excited spot travels along the length direction of tube, which leads to the output of luminescence signals from the top end of a single microtube. Figure 2bottom panels show PL spectra of a single $NaYF_4$ microtube doped with Yb^{3+}/Er^{3+} when the center and end of the microtube are locally excited by a focused laser beam of 980 nm, respectively. All the emission bands are assigned to Er^{3+} ions. Green luminescence corresponding to the $^2H_{11/2}$/$^4S_{3/2} \rightarrow {}^4I_{15/2}$ transition is located in the wavelength region of 500–575 nm, and the $^4F_{9/2} \rightarrow {}^4I_{15/2}$ transition derived from red luminescence ranges from 625 to 700 nm. These upconversion peaks inFig. 2 bottom panels are in agreement with previous reports42'43'44. It is noted that the shapes and positions of peaks are similar when the single tube is excited at various positions by a focused laser beam.

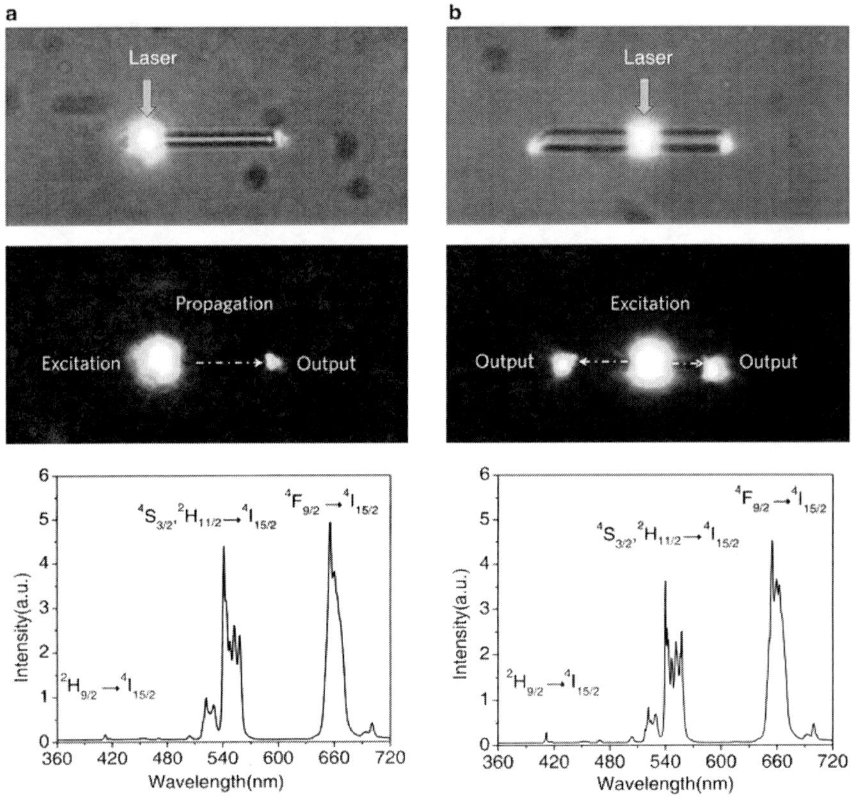

Figure 2: Bright field optical microscope images (top panels), real-color PL photographs (middle panels) and corresponding PL spectra (bottom panels) of a single $NaYF_4$:Yb/Er (20/2 mol%) microtube with local excitation. (**a**) Local excitation at the central part of a single microtube. (**b**) Local excitation at one end position along its length of a single microtube. Scale bars for all images are 20 μm. All the samples were excited with a 980 nm focused laser beam operating at the power density of about 40 mW/cm^2, in which the laser beam is perpendicular to the length direction of the tube.

Self-guided luminescence propagation to the distal tips can be clearly observed when different positions of the tube are excited by focused laser as shown in Fig. 2 top and middle panels. It is well-known that optical waveguide is a kind of device in which the high refractive index medium is surrounded by the low refractive index medium. This kind of novel local luminescence pattern of the single $NaYF_4$ microtube, that is different from the previously reported luminescence pattern of the microtube where microtube is wholly illuminated45, indicates that $NaYF_4$ microtube serving as a kind of new luminescence-based waveguide. Alternatively, we can image that complete reflection of a ray of light in a medium can only take place when the angle of incidence in microtube is greater than the critical angle of total reflection. As a result of total reflection, luminescence output from the tube end could be further promoted by improving the absorption of the excitation energy via increasing the path length of near infrared laser in microtube.

The property of quasi-one-dimensional luminescence propagation

Considering that the property of the luminescence propagation along the length direction is available for 1D $NaYF_4$ rods with average length of 10 µm (Supplementary Fig. 1) that are synthesized through hydrothermal method in the presence of EDTA. Another added benefit is that the operating wavelength of upconversion waveguide in a wide spectral range can be readily adjusted by controlling the dopant composition. For example, blue and pink luminescence output from the top end of microrod can be easily obtained by codoping Yb^{3+}/Tm^{3+} and Yb^{3+}/Ho^{3+}ion pair, respectively (Supplementary Fig. 2). $NaYF_4$ microtube/rod has special optical waveguide effect, which is important to extending their application such as upconversion fiber laser. As we expected that multiple microtubes of varying orientations as waveguide allows one to deliberately propagate light along bent lines as shown in Fig. 3a,b. A similar behavior was observed in CdSSe nanowire29. Noted that the luminescence leakage can be clearly observed in the propagation process due to the presence of the defects and the non-perfect geometries in Fig. 3a–e. So high-quality microtube/rod with smooth surface, uniform diameter and high crystallinity is necessary to the remote propagation of luminescence.

The reason for the quasi-one-dimensional propagation of excited energy could be complex in the single-tube under focused laser excitation. In a sensitizer-activator optical fiber system, the luminescence characteristics and energy propagation efficiency along the length direction of the tube is highly sensitive to the power density of the laser, the concentration of the sensitizer and the activator, and geometrical and physical parameters of tube.

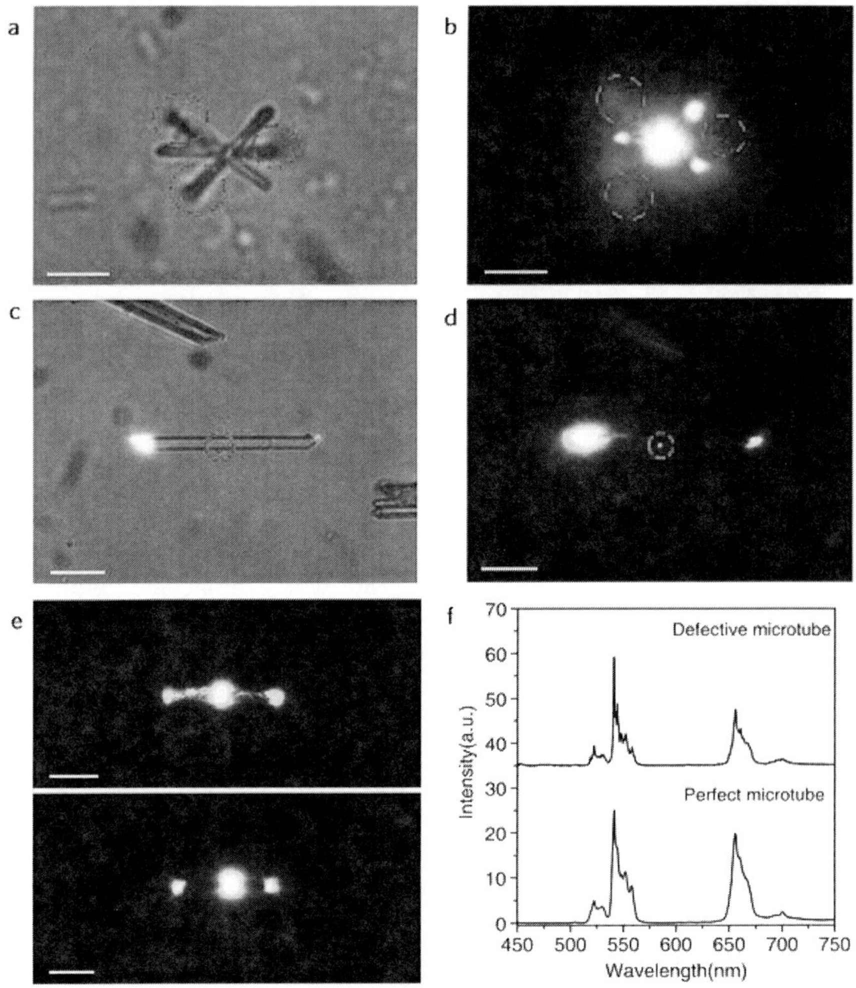

Figure 3: Bright field optical microscope images, real-color PL photographs and PL spectra of doped single one-dimensional upconverting $NaYF_4$ microtube with 980 nm laser local excitation operating at 40 mW/cm^2. (**a**) Bright field image from three crossed $NaYF_4$:Yb^{3+}/Tm^{3+} (20/0.5 mol%) microtubes. (**b**) Real-color PL photograph from three crossed $NaYF_4$:Yb^{3+}/Tm^{3+} microtubes by excitation at the point of intersection. (**c,d**) Bright field optical microscope images and real-color PL photograph from a single $NaYF_4$:Yb^{3+}/Er^{3+} (20/2 mol%) microtube obtained by excitation at one end of the tube. (**e,f**) PL images and their PL spectra from a defective and a perfect $NaYF_4$:Yb^{3+}/Er^{3+} microtubes by excitation at the central part. The dashed red circles indicate the defect positions of one-dimensional microtubes. Scale bars for all images are 20 μm.

The tuning of red-to-green ratio (RGR) of luminescence output

To further get insight of the properties of the quasi-one dimensional propagation of excited energy in a single-tube, we firstly measure the dependence of upconversion luminescence intensity on the concentration of Yb^{3+} ions and pumping laser power density when the laser is focused at local position of microtube. As shown in Fig. 4, the operating wavelength of the upconversion waveguide can be deliberately adjusted from green to yellow spectral region by controlling concentration of Yb^{3+} ions or power density. An increased RGR from 0.9 to 4.2 and from 0.7 to 2.2 can be obtained by elevating Yb^{3+} concentration and pump power, respectively (Supplementary Fig. 3). The tuning mechanisms by elevating Yb^{3+} concentrations can be easily attributed to the elevating of energy migration probability among Yb^{3+} and back-energy-transfer probability from Er^{3+} to Yb^{3+} 12'46'47.

RGR increases by elevating pump power, which is inconsistent with the previous reports in Yb^{3+}/Er^{3+} doped $NaYF_4$ nanocrystals, in which the RGR independent on power density under a diffusion laser excitation is assigned to a saturation effect of the energy transfer processes48. The contradiction may be related to excitation modes and excitation intensities. Theoretically, excitation density is directly related to the initial population of the excited states in an upconversion system, thence affecting the energy transfer process and the upconversion emission properties12. Yet there is no direct evidence to indicate that the energy transfer process depends on excitation power density as excitation density is relatively low. A relative low excitation power density is usually applied to the measurement of massive micro/nanoparticles under a diffusion laser radiation. However, for single micro/nanoparticle measurement under local excitation, high power density excitation is required since it can cause more Yb^{3+} ions in laser facula to be in the excited state until saturation in the micro/nanoparticles. Energy migration from interior to external of laser facula and a high level upconversion process, involving the population of green and red levels, has become an important approach of depopulation when the more Yb^{3+} ions are in the excited state. From this point of view, high excitation power density prefers energy migration and upconversion process, which is equivalent to the effect of higher doping levels of the sensitizer on energy migration, upconversion process and RGR. It is well-known that the length of energy migration is limited by diffusion constant and the lifetime and concentration of Yb^{3+} ions. As shown in Supplementary Fig. 2c,d, RGR tends to be stabilized due to saturation effect and the limted length of energy migration when further elevating pump laser power density up to 120 mW/cm^2.

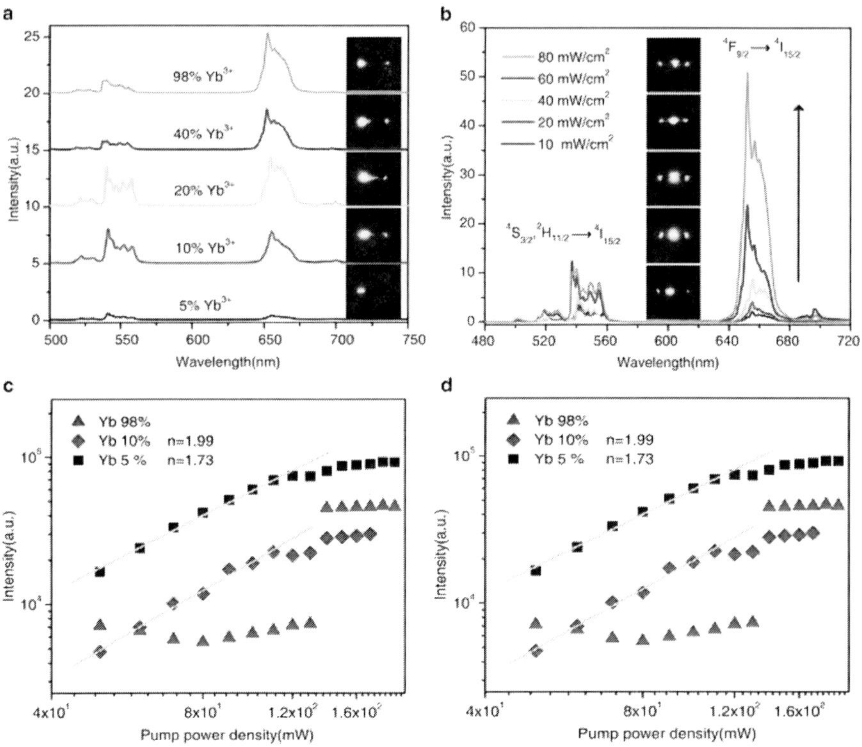

Figure 4: PL spectra and log-log plots of upconversion emission intensities as a function of excitation power densities from a single $NaYF_4:Yb^{3+}/Er^{3+}$ microtube with focused laser excitation. (**a**) PL spectra of a single $NaYF_4:Yb^{3+}/Er^{3+}$ (*x*/2 mol%, *x* = 5, 10, 20, 40, 98) microtube with a 980 nm focused laser excitation from the end part operating at 40 mW/cm^2. (**b**) PL spectra of a single $NaYF_4:Yb^{3+}/Er^3$ (20/2 mol%) microtube excited with a 980 nm laser from the central part at various power densities. (**c**) The log-log plots of green emission intensities as a function of excitation power densities from a single $NaYF_4:Yb^{3+}/Er^{3+}$ (*x*/2 mol%, *x* = 5, 10 and 98) microtube with a 980 nm focused laser excitation at the end and center. (**d**) The log-log plots of red emission intensities as a function of excitation power densities from a single $NaYF_4:Yb^{3+}/Er^{3+}$ microtube. The insets in (**a**,**b**) are real-color PL photographs of a single $NaYF_4:Yb^{3+}/Er^{3+}$ microtube with a 980 nm focused laser excitation from the end part and the central part, respectively.

More interestingly, noted that the bigger size and better crystallinity of a single microtube also results in a higher RGB (Supplementary Fig. 4and Fig. 3e,f), which is in contradiction to the previous result reported on codoped

$Y_2O_3$49. Here the single tube acts as an active optical cavity and the luminescence will travel along the length direction of tube through incessantly repeated re-absorption and re-emitting processes50. During this active propagation, the re-absorption and re-emitting processes will continuously lead to the luminescence intensities ratio change by the cross-relaxation between ions51′52. The energy loss of luminescence mainly come from the non-radiative loss involved in each of the re-absorption and re-emitting process of Er^{3+}activator. While non-perfect geometries and defects lead to the luminescence leakage in propagation process along the length direction of tube, which reduce the path length of luminescence output and cross-relaxation between Er^{3+} ions. Therefore, the ratios of the luminescence output signals depend on microtube size, doping concentration and light path, which indirectly confirmed that a single $NaYF_4$:Yb^{3+}/Er^{3+} microtube belongs to active optic fiber. Note that the change of RGB by the scattering can not be fully excluded because we know the scattering of blue emission will be stronger compared with the red emission. While higher Er^{3+} ions concentration could lead to luminescence quenching in a laser facula and in propagation processes by cross-relaxation (Supplementary Fig. 5)51′53.

DISCUSSION

To well verify the mechanism of the population of $^4S_{3/2}$ and $^4F_{9/2}$ states under NIR irradiation, the upconversion luminescence intensity of green ($^2H_{11/2}$, $^4S_{3/2} \rightarrow {}^4I_{15/2}$) and red ($^4F_{9/2} \rightarrow {}^4I_{15/2}$) emissions as a function of the pump power density was investigated. Figure 4 bottom panels show the log-log plots of the emission intensities as a function of excitation power density for the green and the red emissions, respectively. It can be seen that as the pump power density was lower than 100 mW/cm^2, the emission intensity of upconverson increased with power obeying a rule of $I \propto P^n$ for both the green and the red emissions, where *I* is the emission intensity, *P* is the excitation laser power density, and *n* is the number of photons54′55. In addition to, the slope in the log-log curves varied with the Yb^{3+} concentration. Apparently, the value of *n* firstly increased and then decreased with increasing the Yb^{3+} concentration in the range of 5–98%. It can be also seen that as the power surpassed certain threshold, depending on Yb^{3+}concentration, UCL tend to stabilize. In other words, as the excitation power increases to a certain value such as 120mW/cm^2, the power dependencies do not follow simple power law ($I \propto P^n$). The upconversion luminescence intensity deviates from a linear relationship and the values of *n* decrease with power increase, which can be explained well by the "saturation" effect54′55′56.

The reasons for the *n* value modified by varying Yb^{3+} concentrations could be complex. The complications in the current study could stem from two sources: one is that energy migration channels are introduced as elevating Yb^{3+} concentrations, the other is that the three-photon upconversion process could be involved for populating the red and green luminescence levels.

In present case, when laser was focused at the position of the center or the end of tube, the Yb^{3+} ions that are within the laser facula would absorb photon energy of pumping laser and transfer energy to Er^{3+} ions and Yb^{3+} ions outside the laser facula. The excited energy density wasted by energy migration between Yb^{3+} ions would reduce the slope of *I-P* log-log curve, while high-level upconversion processes increase the slope. The smaller the laser facula and the higher the concentration of Yb^{3+} ions is, the larger the energy density wastes caused by energy migration. If the laser facula was larger than the energy migration length, the energy density waste caused by energy migration would be ignored. In our case, the slope of log-log variation *I-P* curves of two-photon upconversion firstly increase and subsequently violently decrease with Yb^{3+} concentration elevating, which can be mainly attributed to the competition between three-photon upconversion process and energy migration process for wasting the excited energy density within the laser facula.

Generally, the pumping laser power density could be written as $\rho = P_{laser}/S_{laser}$, when pumping laser power is P_{laser} and its facula cross-area is S_{laser}. Thus, the luminescence intensity *I* of two-photon luminescence is about $I \propto \rho^2 S_{laser} = P^2_{laser}/S_{laser}$ 57. If this upconversion mechanism of $NaYF_4:Yb^{3+}/Er^{3+}$ excited by a 980-nm laser is "energy migration among Yb^{3+} ions and then followed by energy transfer between Yb^{3+} and Er^{3+} ions sequentially", the upconversion occurs not only within the laser facula but also in the range of luminescence facula determined by the diffusion length. In such a case, the intensity of two-photon luminescence is about $I \propto \rho^2 S_{lumin} = P^2/S_{limin}$. The $\boldsymbol{S_{lumin}}$ denotes luminescence facula cross-area at the vicinage of excited spot. In such case, $S_{lumin} > S_{laser}$ leads to a significant reduction of the real laser power density. In the log-log plot of the emission intensity as a function of excitation power density, if power density $\rho = P_{laser}/S_{laser}$ is much bigger than the real laser power density $\rho = P_{laser}/S_{lumin}$, a significant reduction of the *n* value can be observed.

The higher excitation density and concentration of Yb^{3+} ions cause more Yb^{3+} ions to be in the excited state within the range of the length of energy migration, and the critical step in upconversion emission is the excited state energy transfer from Yb^{3+} to the activator Er^{3+}. If the number of Er^{3+} ions is not enough, these Er^{3+} ions will get saturated easily in accepting excitation energy via Yb^{3+} ions. From this point of view, the higher the concentration of Yb^{3+} ions is, the lower the power threshold of saturation effect is. Therefore, when the concentration of Yb^{3+} elevating to 98%, even if under 50 mW/cm^2 power density excitation, saturation effect could be observed. However, before reaching saturation of the Er^{3+} ions, more excited state Yb^{3+} ions are conducive to three-photon upconversion process of Er^{3+} ions when elevating Yb^{3+} concentrations from 5% to 20%, which contaminates the red and green luminescence level resulting in an increase of n value. As a result of the competition between three-photon upconversion process and energy migration process, n values firstly increase and then decrease with elevating the Yb^{3+} concentration.

In addition to, color center may also be formed at the femtosecond focused laser focus in microtube, which may further reduce the value of$n$. In fact, we found that the optical output signals from its both ends can be enhanced by nearly two orders of magnitude, when the microtube is locally radiated with 980 nm focused laser with a 120 mW/cm^2 for 10 minutes in turn at various positions along its length (Supplementary Fig. 6 for the corresponding PL spectra). It is still not clear whether the luminescence enhancement is related to color center formed by laser radiation.

Figure 5 displays the schematic illustration of light propagation mechanism and the corresponding upconversion mechanism for a single $NaYF_4:Yb^{3+}/Er^{3+}$ microtube. Also indicated is the most relevant transitions that occur: energy migrations and transfers, cross-relaxation, as well as radiative and nonradiative processes. The upconversion processes are as follows: first, the Yb^{3+} ions within the laser facula absorb pumping photon energy and are excited to $^2F_{5/2}$state from $^2F_{7/2}$ ground state. Then, the Yb^{3+} ions transfer energy to Er^{3+} or another Yb^{3+} ion outside the laser facula. The reciprocal action and energy migration among Yb^{3+} ions is rather strong with the high Yb^{3+} concentration. Not only two adjacent Yb^{3+} ions combine tightly to form a coupling state of cluster, but also there exists some long-distance correlation in the range of the diffusion length and laser elastic scattering region, which leads to bigger luminescence emission area than laser area with about 50 nm in diameter. The luminescence from in/outside the laser spot is propagated along the length direction of the tube due to waveguide effects of microtube, which leads to a strong luminescence output from the distal tips of a single microtube (Fig. 5a). That is to say, the characteristic saturation phenomenon of upconversion luminescence found in this study is mainly due to energy migration under local excitation. The typical saturation caused by population exhaustion has only a quite subordinate action.

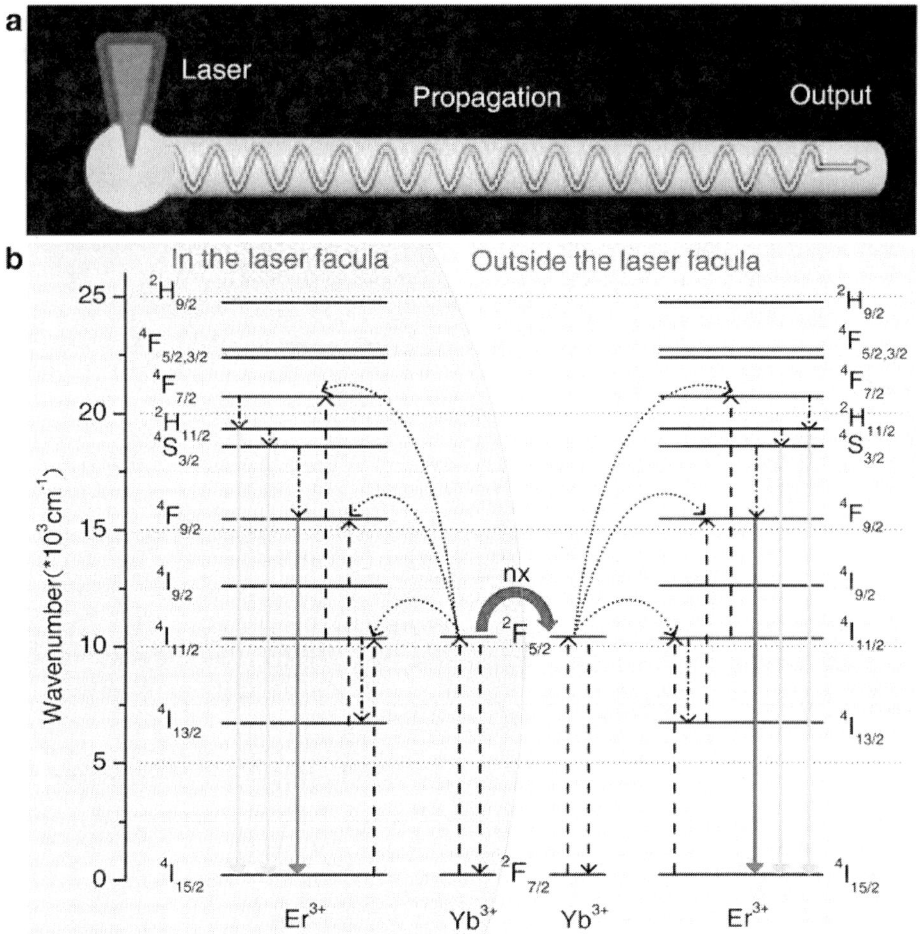

Figure 5: Schematic diagrams of a single $NaYF_4$:Yb^{3+}/Er^{3+}microtube as actived optical waveguide and the corresponding upconversion mechanisms of self-guided luminescence. (a) Schematic illustration of self-guided luminescence propagation, in which green circular region stands for upconversion luminescence facula rather than focused laser facula. (**b**) Proposed mechanism of self-guided luminescence production. Note that only partial energy levels of Yb^{3+} and Er^{3+} are shown for clarity.

Significantly, in this work, we have showed the design and realization of splendid quasi-one-dimensional upconversion luminescence output from the top end and the centre of Yb^{3+}/Er^{3+} doped single $NaYF_4$microtube spin-coated on a glass slide under confocal optical microscope system. This specially designed one-dimensional structure allows the separation of excitation and emission regions of a single microtube. By monitoring the emission of a single microtube that is outside the laser facula, energy migration processes could be readily followed. The mechanism of the quasi-one-dimensional luminescence propagation is explored and is attributed to the optical waveguide effect associated with total internal reflection of the upconverted light at a single microtube in the medium-air interface. The upconversion output signal of an individual microtube is highly dependent on the concentration of dopant ions, excitation power, morphology, and crystallinity of tube.

Interestingly, we have found that the bigger size and better crystallinity of a microtube prefers to the higher ratio of red to green luminescence intensity in a single microtube, which is attributed to the reduce of the luminescence leakage. $NaYF_4$:Yb^3Er^{3+} microtubes as active optical waveguides not only violently absorb the incident excited light without the need for special alignment of the near infrared irradiation, but also propagate luminescence along the length direction of tube. The new physical effect of a single $NaYF_4$ microtube, separated emission from excitation space, produces super bright local luminescence output at the top end and the centre of tube, which may be extending their application in the fields of upconversion waveguide laser, optical communication and biology.

METHODS

Synthesis of microtubes

The $NaYF_4$:$Yb^{3+}Er^{3+}(Tm^{3+})$ microtubes have been fabricated via a facile hydrothermal route assisted by trisodium citrate41. In a typical procedure, 1.5 mL trisodium citrate (0.4 mol L^{-1}) was added to a mixture containing 0.6 mmol RE nitrate (3.0 mL of 0.2 mol L^{-1} $RE(NO_3)_3$, RE = Y, Yb, Er, and Tm) and 20 mL deionized water. The solution was then thoroughly stirred for 30 min to form a chelated RE citrate complex (1:1 molar ratio for Cit^{3-}-RE^{3+}). Then, 6 mL (1.0 mol L^{-1}) NH_4F aqueous solutions were dropped into chelated RE Cit^{3+} complex under thorough stirring. The pH value of the mixture was tuned to 8.5 with ammonia water. Subsequently, the milky colloidal solution was transferred to a 40.0 mL Teflon-lined autoclave, and heated at 200 °C for 24 h. The final products were collected by centrifuging and washed with water and ethanol. The collected microcrystals were dried under 70 °C for 12 h.

Synthesis of microrods

The $NaYF_4:Yb^{3+}/Er^{3+}$ (Tm^{3+}or Ho^{3+}) microrods have been fabricated through hydrothermal method in the presence of the chelating agent EDTA42. Briefly, 0.75 mmol of EDTA (0.5 mol/L) was added to a mixture containing 0.3 mmol of RE nitrate (1.5 mL of 0.2 mol/L $RE(NO_3)_3$, RE = Y, Yb, Er, Ho and Tm) and 15 mL deionized water. The solution was then thoroughly stirred for 30 min to form a chelated RE-EDTA complex. Then, a solution 8 mL (1.0 mol/L) of NH_4F and 4 mL (1.0 mol/L) of NaF aqueous solutions were dropped into chelated RE-EDTA complex under thoroughly stirring. Subsequently, the milky colloidal solution was transferred to a 40.0 mL Teflon-lined autoclave, and heated at 200 °C for 24 h. The as-prepared products were collected by centrifuging and washed with water and ethanol for several times. The collected microcrystals were dried under 70 °C for 12 h.

Structural characterization

The phase compositions of the as-prepared products were examined by XRD with a D/Max2550VB+/PC X-ray diffraction meter with Cu Ka (40 kV, 40 mA) irradiation ($\lambda = 0.15406$ nm). The SEM micrographs were obtained using a Quanta 200 SEM. The selected-area electron diffraction measurements were performed using a JEM 2100 transmission electron microscope (TEM) operating at an acceleration voltage of 200 kV.

Optical characterization

A confocal optical microscope system was employed for microscopic studies of β-$NaYF_4:Yb^{3+}/Er^{3+}$ microtube spin-coated on a glass slide. The schematic diagram of the confocal microscope system is shown in Supplementary Fig. 7. Ti sapphire femtosecond laser (Mira-900) is employed as excitation sources, which passed through a half-wave-plate and a high reflective mirror at 980 nm, and then was focused with a high numerical aperture microscope objective lens (NA = 1.40, 1000×, oil immersion) to a spot diameter of about 50 nm. Emission spectra were collected and sent to the detection part after spatial and spectral filtering, and then detected by a spectrometer (SP2750i) with a spectral resolution of 0.008 nm equipped with a liquid nitrogen cooled CCD detector. All of the spectroscopic measurements are carried out at room temperature.

Additional Information

How to cite this article: Gao, D. *et al.* Simultaneous quasi-one-dimensional propagation and tuning of upconversion luminescence through waveguide effect. *Sci. Rep.* **6**, 22433; doi: 10.1038/srep22433 (2016).

ACKNOWLEDGEMENTS

The authors acknowledge financial support from the China Postdoctoral Science Foundation (2015M570816), the Plan Project of Youth Science and Technology New Star of Shaanxi Province (2015KJXX-33), the Natural Science Foundation of Shaanxi Province of China (2014JQ1008), the Special Foundation of Shaanxi Educational Commission (11JK0520), and the Project of Basic Research Foundation of Xi'an University of Architecture and Technology (JC1518).

References

1. Hinklin, T. R., Rand, S. C. & Laine, R. M. Transparent, polycrystalline upconverting nanoceramics: towards 3-D displays.*Adv. Mater.* **20**, 1270–1273 (2008).

2. Sun, L. D., Wang, Y. F. & Yan, C. H. Paradigms and challenges for bioapplication of rare earth upconversion luminescent nanoparticles: small size and tunable emission/excitation spectra. *Acc. Chem. Res.* **47**, 1001–1009 (2014).

3. Chen, G. *et al.* (α-$NaYbF_4$:Tm^{3+})/CaF_2 core/shell nanoparticles with efficient near-infrared to near-infrared upconversion for high-contrast deep tissue bioimaging. *ACS Nano* **6**, 8280–8287 (2012).

4. Kostiv, U. *et al.* Silica-modified monodisperse hexagonal lanthanide nanocrystals: synthesis and biological properties.*Nanoscale* **7**, 18096–18104 (2015).

5. Mi, C. *et al.* Novel microwave-assisted solvothermal synthesis of $NaYF_4$:Yb, Er upconversion nanoparticles and their application in cancer cell imaging. *Langmuir* **27**, 14632–14637 (2011).

6. Zhou, L. *et al.* Single-band upconversion nanoprobes for multiplexed simultaneous *in situ* molecular mapping of cancer biomarkers. *Nat. Commun.* **6**, 6938 (2015).

7. Zhang, J. *et al.* Synthesis of $NaYF_4$:Yb/Er/Gd up-conversion luminescent nanoparticles and luminescence resonance energy transfer-based protein detection. *Anal. Biochem.* **421**, 673–679 (2012).

8. Wang, C. *et al.* Simultaneous isolation and detection of circulating tumor cells with a microfluidic silicon-nanowire-array integrated

with magnetic upconversion nanoprobes. *Biomaterials* **54**, 55–62 (2015).

9. Hemmer, E., Quintanilla, M., Légaré, F. & Vetrone, F.Temperature-induced energy transfer in dye-conjugated upconverting nanoparticles: a new candidate for nanothermometry. *Chem. Mater.* **27**, 235–244 (2015).

10. Shalav, A., Richards, B. S., Trupke, T., Krämer, K. W. & Güdel, H. U.Application of $NaYF_4$: Er^{3+} up-converting phosphors for enhanced near-infrared silicon solar cell response. *Appl. Phys. Lett.* **86**, 013505 (2005).

11. Martín-Rodríguez, R. *et al.* Highly efficient IR to NIR upconversion in Gd_2O_2S: Er^{3+} for photovoltaic applications. *Chem. Mater.* **25**, 1912–1921 (2013).

12. Tu, L., Liu, X., Wu, F. & Zhang, H. Excitation energy migration dynamics in upconversion nanomaterials. *Chem. Soc. Rev.* **44**, 1331–1345 (2015).

13. Xie, X. *et al.* Mechanistic investigation of photon upconversion in Nd^{3+}-sensitized core–shell nanoparticles. *J. Am. Chem. Soc.* **135**, 12608–12611 (2013).

14. Wang, F. *et al.* Tuning upconversion through energy migration in core–shell nanoparticles. *Nat. Mater.* **10**, 968–973 (2011).

15. Wang, J. *et al.* Enhancing multiphoton upconversion through energy clustering at sublattice level. *Nat. Mater.* **13**, 157–162 (2014).

16. Qin, W. *et al.* Multi-ion cooperative processes in Yb^{3+} clusters.*Light Sci. Appl.* **3**, e193 (2014).

17. Gao, D. *et al.* Up/down conversion switching by adjusting the pulse width of red laser beams in LaF_3: Tm^{3+} nanocrystals. *Opt. Lett.* **40**, 3580–3583 (2015).

18. Schietinger, S., Menezes, L. D. S., Lauritzen, B. & Benson, O.Observation of size dependence in multicolor upconversion in single Yb^{3+}, Er^{3+} codoped $NaYF_4$ nanocrystals. *Nano Lett.* **9**, 2477–2481 (2009).

19. Deng, R. *et al.* Temporal full-colour tuning through non-steady-state upconversion. *Nat. Nanotech*. **10**, 237–242 (2015).

20. Yao, Y. *et al.* Laser polarization and phase control of up-conversion fluorescence in rare-earth ions. *Sci. Rep.* **4**, 7295 (2014).

21. Wei, W. *et al.* Cross relaxation induced pure red upconversion in activator-and sensitizer-rich lanthanide nanoparticles. *Chem. Mater.* **26**, 5183–5186 (2014).

22. Wang, J. *et al.* Photon energy upconversion through thermal radiation with the power efficiency reaching 16%. *Nat. Commun.* **5**, 5669 (2014).

23. Liu, Z. *et al.* Near-infrared light manipulated chemoselective reductions enabled by an upconversional supersandwich nanostructure. *ACS Appl. Mater. Interfaces* **7**, 19416–19423 (2015).

24. Liu, Y. *et al.* A strategy to achieve efficient dual-mode luminescence of Eu^{3+} in lanthanides doped multifunctional $NaGdF_4$ nanocrystals. *Adv. Mater.* **22**, 3266–3271 (2010).

25. Mi, C. C., Tian, Z. H., Han, B. F., Mao, C. B. & Xu, S. K. Microwave-assisted one-pot synthesis of water-soluble rare-earth doped fluoride luminescent nanoparticles with tunable colors. *J. Alloy. Compd.* **525**, 154–158 (2012).

26. Law, M. *et al.* Nanoribbon waveguides for subwavelength photonics integration. *Science* **305**, 1269–1273 (2004).

27. Bulgarini, G. *et al.* Avalanche amplification of a single exciton in a semiconductor nanowire. *Nat. Photon* **6**, 455–458 (2012).

28. Reimer, M. E. *et al.* Bright single-photon sources in bottom-up tailored nanowires. *Nat. Commun.* **3**, 737 (2012).

29. Xu, J. *et al.* Asymmetric light propagation in composition-graded semiconductor nanowires. *Sci. Rep.* **2**, 820 (2012).

30. Yan, Y., Zhang, C., Zheng, J. Y., Yao, J. & Zhao, Y. S. Optical modulation based on direct photon-plasmon coupling in organic/metal nanowire heterojunctions. *Adv. Mater.* **24**, 5681–5686 (2012).

31. Zhao, Y. S., Zhan, P., Kim, J., Sun, C. & Huang, J. Patterned growth of vertically aligned organic nanowire waveguide arrays. *ACS Nano* **4**, 1630–1636 (2010).

32. O'Carroll, D., Lieberwirth, I. & Redmond, G. Microcavity effects and optically pumped lasing in single conjugated polymer nanowires. *Nat. Nanotech*. **2**, 180–184 (2007).

33. Agarwal, R., Barrelet, C. J. & Lieber, C. M. Lasing in single cadmium sulfide nanowire optical cavities. *Nano Lett.* **5**, 917–920 (2005).

34. Duan, X. F., Huang, Y., Agarwal, R. & Lieber, C. M. Single-nanowire electrically driven lasers. *Nature* **421**, 241–245 (2003).

35. Chu, S. *et al.* Electrically pumped waveguide lasing from ZnO nanowires. *Nat. Nanotech*. **6**, 506–510 (2011).

36. Huang, M. H. *et al.* Room-temperature ultraviolet nanowire nanolasers. *Science* **292**, 1897–1899 (2001).

37. Yan, R., Gargas, D. & Yang, P. Nanowire photonics. *Nat. Photon* **3**, 569–576 (2009).

38. Yu, R. *et al.* A new blue-emitting phosphor of Ce^{3+}-activated $CaLaGa_3S_6O$ for white-light-emitting diodes. *Chem. Phys. Lett.***453**, 197–201 (2008).

39. Zhou, S. *et al.* Simultaneous tailoring of phase evolution and dopant distribution in the glassy phase for controllable luminescence. *J. Am. Chem. Soc.* **132**, 17945–17952 (2010).

40. Zhao, Y. S. *et al.* Optical waveguide based on crystalline organic microtubes and microrods. *Angew. Chem. Int. Ed.* **47**, 7301–7305 (2008).

41. Gao, D., Gao, W., Shi, P. & Li, L. pH- and surfactant-mediated tunable morphology and upconversion of rare-earth doped fluoride microcrystals. *RSC Adv*. **3**, 14757–14765 (2013).

42. Gao, D., Zhang, X., Zheng, H., Gao, W. & He, E. Yb^{3+}/Er^{3+} codoped β-$NaYF_4$ microrods: Synthesis and tuning of multicolor upconversion. *J. Alloy. Compd.* **554**, 395–399 (2013).

43. Li, Z., Park, W., Zorzetto, G., Lemaire, J. S. & Summers, C. J.Synthesis protocols for δ-doped $NaYF_4$: Yb, Er. *Chem. Mater.* **26**, 1770–1778 (2014).

44. Gao, D., Zhang, X. & Gao, W. Formation of bundle-shaped β-$NaYF_4$upconversion microtubes via Ostwald ripening. *ACS Appl. Mater. Interfaces* **5**, 9732–9739 (2013).

45. Zhang, Y. *et al.* Multicolor barcoding in a single upconversion crystal. *J. Am. Chem. Soc.* **136**, 4893–4896 (2014).

46. Vetrone, F., Boyer, J. C., Capobianco, J. A., Speghini A. & Bettinelli, M. Significance of Yb^{3+} concentration on the upconversion mechanisms in codoped Y_2O_3: Er^{3+}, Yb^{3+} nanocrystals. *J. Appl. Phys.* **96**, 661–667 (2004).

47. Zhang, J. *et al.* Observation of efficient population of the red-emitting state from the green state by non-multiphonon relaxation in the Er^{3+}-Yb^{3+} system. *Light Sci. Appl.* **4**, e239 (2015).

48. Suyver, J. *et al.* Novel materials doped with trivalent lanthanides and transition metal ions showing near-infrared to visible photon upconversion. *Opt. Mater.* **27**, 1111–1130 (2005).

49. Song, H. *et al.* Three-photon upconversion luminescence phenomenon for the green levels in Er^{3+}/Yb^{3+} codoped cubic nanocrystalline yttria. *Solid State Commun.* **132**, 409–413 (2004).

50. Xu, J. *et al.* Wavelength-converted/selective waveguiding based on composition-graded semiconductor nanowires. *Nano Lett.* **12**, 5003–5007 (2012).

51. Vetrone, F., Boyer, J. C., Capobianco, J. A., Speghini, A. & Bettinelli, M. Concentration-dependent near-infrared to visible upconversion in nanocrystalline and bulk Y_2O_3: Er^{3+}. *Chem. Mater.*.**15**, 2737–2743 (2003).

52. Vetrone, F., Boyer, J. C., Capobianco, J. A., Speghini, A. & Bettinelli, M. Effect of Yb^{3+} codoping on the upconversion emission in nanocrystalline Y_2O_3: Er^{3+} *J. Phys. Chem. B* **107**, 1107–1112 (2003).

53. Martín-Rodríguez, R., Rabouw, F. T., Trevisani, M., Bettinelli, M. &Meijerink, A. Upconversion dynamics in Er^{3+}-doped Gd_2O_2S: influence of excitation power, Er^{3+} concentration, and defects.*Adv. Opt. Mater.* **3**, 558–567 (2015).

54. Pollnau, M., Gamelin, D. R., Lüthi, S. R., Gudel, H. U. & Hehlen, M. P. Power dependence of upconversion luminescence in lanthanide and transition-metal-ion systems. *Phys. Rev. B* **61**, 3337 (2000).

55. Bai, X. *et al.* Size-dependent upconversion luminescence in Er^{3+}/Yb^{3+}-codoped nanocrystalline yttria: Saturation and thermal effects. *J. Phys. Chem. C* **111**, 13611–13617 (2007).

56. Lei, Y. *et al.* Upconversion luminescence, intensity saturation effect, and thermal effect in Gd_2O_3: Er^{3+}, Yb^{3+} nanowires. *J. Chem. Phys.* **123**, 174710–174710 (2005).

57. Chen, X. B., Nie, Y. X., Du, W. M. & Sawanobori, N. The comparison investigation of direct upconversion sensitization luminescence between ErYb: oxyfluoride glass and vitroceramics. *Opt. Commun.* **184**, 289–304 (2000).

Chapter 8

Single-Band Upconversion Nanoprobes for Multiplexed Simultaneous *in Situ*molecular Mapping of Cancer Biomarkers

Lei Zhou, Rui Wang, Chi Yao, Xiaomin Li, Chengli Wang, Xiaoyan Zhang, Congjian Xu, Aijun Zeng,Dongyuan Zhao & Fan Zhang

Department of Chemistry, iChEm (Collaborative Innovation Center of Chemistry for Energy Materials), State Key Laboratory of Molecular Engineering of Polymers, Fudan University, Shanghai 200433, China

ABSTRACT

The identification of potential diagnostic markers and target molecules among the plethora of tumour oncoproteins for cancer diagnosis requires facile technology that is capable of quantitatively analysing multiple biomarkers in tumour cells and tissues. Diagnostic and prognostic classifications of human tumours are currently based on the western blotting and single-colour immunohistochemical methods that are not suitable for multiplexed detection. Herein, we report a general and novel method to prepare single-band upconversion nanoparticles with different colours. The expression levels of three biomarkers in breast cancer cells were determined using single-band upconversion nanoparticles, western blotting and immunohistochemical technologies with excellent correlation. Significantly, the application of antibody-conjugated single-band upconversion nanoparticle molecular profiling technology can achieve the multiplexed simultaneous *in situ*biodetection of biomarkers in breast cancer cells and tissue specimens and produce more accurate results for the simultaneous quantification of proteins present at low levels compared with classical immunohistochemical technology.

INTRODUCTION

Multiplexed, sensitive and specific molecular detection is highly desirable for gene and protein profiling, drug screening, clinical diagnostics and environmental analysis[1,2,3]. For cancer diagnosis, the identification of potential diagnostic biomarkers and target molecules among the plethora of tumour oncoproteins requires facile technology capable of quantitatively analysing multiple biomarkers in tumour cells and tissues[4,5,6,7,8]. Diagnostic and prognostic classifications of human tumours are currently based on immunoenzyme-based immunohistochemical (IHC) methods with single-wavelength detection[9], which are somewhat outdated and focus on approved antibodies for the IHC detection of cancerous markers. Furthermore, IHC remains semi-quantitative and subjective, resulting in considerable inter-observer and intra-observer variations in results[10]. Moreover, a major difficulty in molecular profiling is that most tumours (particularly breast and prostate) are highly heterogeneous and contain a mixture of benign cells, malignant cells, fibroblasts, stromal cells, vascular cells and infiltrating inflammatory cells (such as macrophages and lymphocytes)[11]. Current technologies such as reverse transcription–PCR, gene chips, protein chips, two-dimensional gel electrophoresis and mass spectrometry (for example, matrix-assisted laser desorption/ionization mass spectrometry (MALDI-MS), electrospray mass spectrometry (ES-MS) and surface-enhanced laser desorption/ionization mass spectroscopy (SELDI-MS)) are not suitable for analysing this type of heterogeneous sample[12,13,14,15]. A further limitation of these technologies is that they require destructive preparation of cells or tissue specimens as a homogeneous solution, leading to the loss of valuable three-dimensional cellular and tissue morphological information associated with the original tumour.

Compared with adsorption-based IHC methods, fluorescence imaging with optical microscopy provides a number of advantages, such as higher detection sensitivity, wider signal dynamic ranges and more linear relationships for biomarker quantification and prediction of therapeutic response. Indeed, recent advances have demonstrated that the fluorescence-based analysis of protein biomarkers with both fluorescent dye[16] and quantum dots (QDs)[17,18] is strongly correlated with clinical outcome. However, the use of organic dyes for multicolour fluorescence measurement is often limited by photobleaching and the need for multiple light sources to excite different fluorophores. QDs offer significant advantages over conventional fluorescent dyes because they are more photostable and can be excited in the same spectral range. However, optical interference of QDs with the autofluorescence background of biological tissue with the same excitations in the blue or ultraviolet region remains a major problem. To minimize this spectral interference, QD-stained specimens must be illuminated over an extended period of time to reduce the autofluorescence background before recording images and spectra (typically ranging from seconds to minutes depending on the intensity of the light source)19. Furthermore, because of overlap in the emission spectra of QDs and organic dyes, multiplexed simultaneous *in situ* detection requires image-processing techniques, such as spectral deconvolution19, which limit the screening rate and sensitivity. Thus, the development of appropriate multiplexed molecular profiling technologies remains a challenging task that is crucial for the development of cancer diagnosis technology, biosystem spatiotemporal analysis and the medical sciences.

An attractive alternative approach to this problem involves the use of lanthanide UCNPs, which can be excited in the tissue-transparent infrared region instead of at ultraviolet or visible frequency to produce photoluminescent emissions in the visible range (anti-Stokes process)20′21′22′23′24′25′26′27′28′29′30′31′32′33′34. Triply ionized lanthanide ions in hosts typically have emission line widths of 10–20 nm (full width at half maximum) in the visible portion of the spectrum, which is approximately half the line width observed for QDs (25–40 nm) and much narrower than the line width observed for organic dyes (30–50 nm)17′18′32. This feature increases the number of distinguishable emission bands within a specific spectral bandwidth, enabling a large number of multiplexed detections.

For *in situ* multiplexed molecular mapping, accurate molecular profiling can only be achieved with single-band emission nanoprobes by excluding crosstalk between different labelling signals. However, each lanthanide ion has a unique set of energy levels and generally exhibits a set of sharp emission peaks with distinguishable spectroscopic fingerprints32′35. Although some groups have reported that single-band upconversion emission can be realized by co-doping transition metal ions, only red (650–670 nm)36 and near infrared (NIR; ~800 nm)37 single-band upconversion emissions have been realized. A general method to achieve more single-band upconversion emissions remains a major challenge for applications such as multiplexed intracellular detection and IHC staining without the constraints associated with conventional multi-peak UCNPs.

In this work, we report a general and simple method of achieving single-band upconversion emission with different colours by coating the upconversion nanocrystals with a screen layer containing an organic dye with a high molar absorption coefficient as nanofilters to remove the unwanted emission bands. As a result of the efficient reabsorption of the organic dye, remarkably pure single-band upconversion emissions can be generated in the blue, green and red regions. We use single-band UCNPs (sb-UCNPs) for the quantitative and simultaneous *in situ* profiling of multiple biomarkers in intact breast cancer cells and tissues. The biomarker expression levels in breast cancer cell specimens determined using sb-UCNPs, standard western blotting (WB) and IHC technology exhibit excellent correlation among these three methods; however, the WB and IHC methods can probe only one biomarker at a time. Significantly, the application of antibody-conjugated sb-UCNP molecular profiling technology can achieve the multiplexed simultaneous *in situ* biodetection of biomarkers in breast cancer cells and tissue specimens and produces more accurate results for the simultaneous quantification of proteins present at low levels compared with IHC.

RESULTS

Fabrication of sb-UCNPs

The upconversion nanocrystals were prepared by the successive layer-by-layer strategy developed previously by our group (Supplementary Table 1)38. To obtain green emission sb-UCNPs, β-$NaGdF_4$:20% Yb, 2% Er@$NaGdF_4$ nanocrystals emitting both in the green (540 nm) and red (650 nm) were first prepared (Figs 1 and 2a,c). Then, the nanocrystals were coated with a pure silica spacer layer and amino-reactive organic dyes doped silica layer to form a selective nanofilter around the nanocrystals sequentially using a microemulsion method (Fig. 1; Fig. 2b). The organic dyes were selected based on the criteria of overlapping absorption spectra with only one of the dual-emission bands of the nanocrystals with a high molar absorption coefficient on the scale of 10^5 M^{-1} cm^{-1}. The pure silica spacer layer was used to prevent fluorescence resonance energy transfer (FRET) between the filtered upconversion emission band and the fluorescent dye-doped screen layer. To obtain green single-band emission, nickel (II) phthalocyanine-tetrasulfonic acid tetrasodium salt (NPTAT) organic dyes with a maximum absorption wavelength (λ_{max}) of 657 nm were added with tetraethyl orthosilicate (TEOS) to form NPTAT-doped silica nanofilters on the β-$NaGdF_4$:20% Yb, 2% Er@$NaGdF_4$@SiO_2 nanoparticles to filter the red emission band efficiently. With the β-$NaGdF_4$:20% Yb, 2% Er@$NaGdF_4$@SiO_2@NPTAT-doped SiO_2 nanostructure, only the narrow green emission centred at 540 nm was observed (Fig. 2d; Supplementary Fig. 1), in stark contrast to the dual-upconversion emission bands of β-$NaGdF_4$:20% Yb, 2% Er@$NaGdF_4$ (Fig. 2c). To obtain the blue and red emission sb-UCNPs, β-$NaGdF_4$:20% Yb, 0.2% Tm@$NaGdF_4$ and α-$NaYbF_4$:10% Er@$NaYF_4$ nanocrystals with strong blue to red (Fig. 2e,g) and red to green (Fig. 2i,k) upconversion emission ratios were first prepared. After coating the pure SiO_2 layers, nanofilters doped with NPTAT and rhodamine B isothiocyanate (Fig. 2f,j) were used to filter the red and green emissions to obtain the final blue and red sb-UCNPs, respectively (Fig. 2h,l; Supplementary Figs 2–4). This general approach permits not only the removal of minor emission peaks away from the main peaks using appropriate nanofilters but also the alternative removal of the main peaks to leave the minor peaks for single-band upconversion emission. For example, green emission sb-UCNPs can be obtained by coating the α-$NaYbF_4$:10% Er@$NaYF_4$ nanocrystals with NPTAT-doped nanofilters (Supplementary Fig. 5).

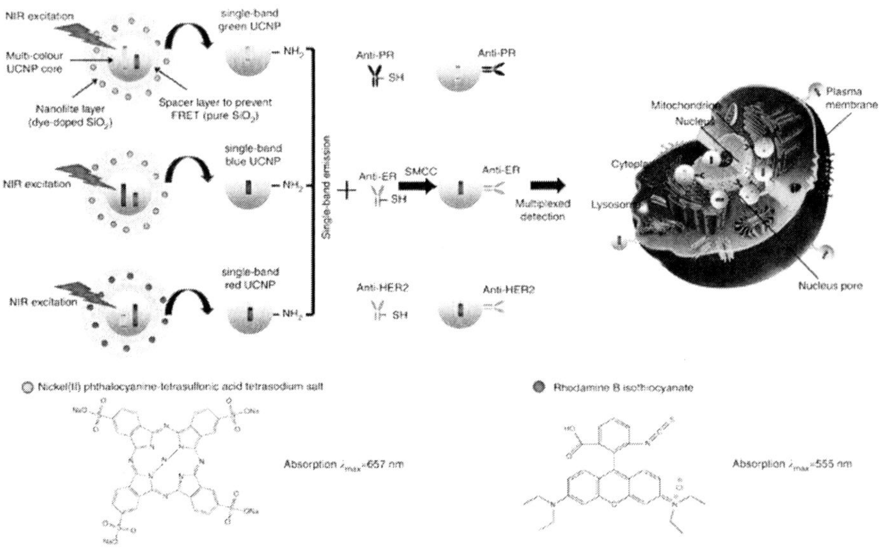

Figure 1: Schematic diagram of sb-UCNP fabrication for multiplexed detection. Surface amino modifications of the multi-layer structure of green, blue and red sb-UCNPs and conjugates with antibodies to the breast cancer biomarkers PR, ER and HER2, respectively, for multiplexed *in situ* molecular mapping of breast cancer biomarkers.

In accordance with the Lambert–Beer law, the filtered intensity can be increased by increasing the dye concentration in the screen layer at a given wavelength39. Therefore,

the intensity of a specific upconversion emission band can be reduced gradually by increasing the dye concentrations. For example, to obtain blue, green and red emission sb-UCNPs under 980-nm laser excitation (10 W cm^{-2}), the molar ratio of doped dyes to lanthanide luminescent centres (Tm or Er) should be adjusted to 9:5, 1:5 and 11:100, respectively (Supplementary Table 2;Supplementary Figs 6–9). In addition, we observed

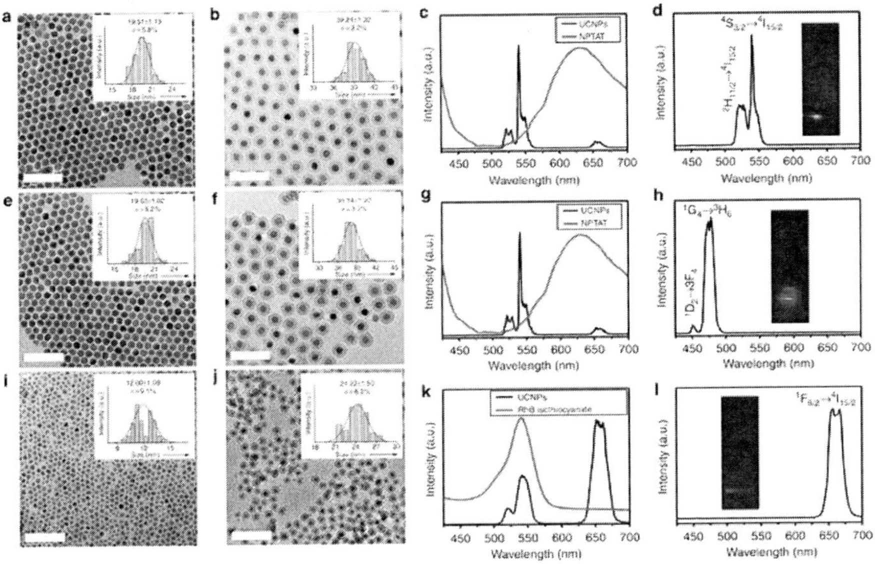

Figure 2: Fabrication and properties of single-band upconversion nanoprobes. TEM images and size distributions of (**a**) β-$NaGdF_4$:20% Yb, 2% Er@$NaGdF_4$ nanocrystals, (**b**) green emission sb-UCNPs (β-$NaGdF_4$:20% Yb, 2% Er@$NaGdF_4$@SiO_2@NPTAT-doped SiO_2), (**e**) β-$NaGdF_4$:20% Yb, 0.2% Tm@$NaGdF_4$ nanocrystals, (**f**) blue emission sb-UCNPs (β-$NaGdF_4$:20% Yb, 0.2% Tm@$NaGdF_4$@SiO_2@NPTAT-doped SiO_2), (**i**) α-$NaYbF_4$:10% Er@$NaYF_4$nanocrystals and (**j**) red emission sb-UCNPs ($NaYbF_4$:10%Er@$NaYF_4$@SiO_2@rhodamine B isothiocyanate-doped SiO_2). Upconversion photoluminescence spectra of the green, blue and red sb-UCNPs (**d**,**h**,**l**) and original upconversion nanocrystals measured in water and cyclohexane, respectively (**c**,**g**,**k**) (insets: corresponding photoluminescent photos of the colloidal solutions under continuous wavelength 980-nm laser excitation). Scale bars, 100µm.

that dye-doped nanofilters can be coated directly on the upconversion nanocrystals to achieve single-band upconversion emission when non-fluorescence dyes are used (Supplementary Fig. 10). However, FRET between the upconversion emission and the dyes with overlapping absorptions cannot be avoided if fluorescent dye-doped nanofilters are used. To obtain sb-UCNPs with high chromatic purities, a silica spacer layer must be inserted before coating the fluorescent dye-doped nanofilter layer (Supplementary Fig. 11a,b). Because the FRET effect is dependent on the distance between the donor and acceptor40, the thickness of the silica spacer layer should be adjusted accordingly. In this work, we determined that the FRET effect can be prevented completely by a 5-nm silica spacer layer for sb-UCNPs with different colours.

Optical properties of sb-UCNPs

In the upconversion photoluminescence processes, the visible output intensity is proportional to the *n*th power of the NIR excitation intensity, that is, $I_{up} \propto (I_{exc})^n$, where I_{up} and I_{exc} represent the upconversion intensity and the NIR excitation intensity, respectively, and n is the number of photons required to produce an upconversion photon41. Figure 3 shows the power dependence of the upconversion intensities of the sb-UCNPs and original upconversion nanocrystals. By increasing the pump power density up to 15.3 W cm^{-2}, the green, blue and red single-band emission intensities were increased accordingly, but the single-band feature of the sb-UCNPs was maintained with increasing excitation power density (Fig. 3b,d,f). Significantly, the single-band feature of the sb-UCNPs was even maintained at the saturation point of upconversion emission intensity under high excitation power density (82 W cm^{-2}; Supplementary Fig. 12). By contrast, before nanofilter coating, the original β-$NaGdF_4$:20% Yb, 2% Er@$NaGdF_4$, β-$NaGdF_4$:20% Yb, 0.2% Tm@$NaGdF_4$ and α-$NaYbF_4$:10% Er@$NaYF_4$ nanocrystals exhibited multipeak emissions with relative intensity ratios closely associated with the pump power density (Fig. 3a,c,e).

The slopes of the log–log plots of emission intensity versus NIR excitation power for the sb-UCNPs were 1.93 (green, $^2H_{11/2}$, $^4S_{3/2}$-$^4I_{15/2}$transition), 2.71 (blue, $^1G_4$-$^3H_6$ transition) and 1.54 (red, $^4F_{9/2}$-$^4I_{15/2}$transition; Fig. 4b,d,f), revealing little variation compared with the original upconversion nanocrystals (1.88 (green), 2.60 (blue) and 1.46 (red); Fig. 4a,c,e). This result indicates that the nanofilter coating did not affect the electronic transition states of lanthanide and that multi-photon features were maintained for the sb-UCNPs. Furthermore, the lifetimes of the three single-band emissions were 1066±17 μs (blue), 89±1 μs (green) and 59±1 μs (red), revealing little variation compared with the pure silica-coated UCNPs (1045±14 μs (blue), 88±3 μs (green) and 55±3 μs (red); Supplementary Fig. 13), which indicates that the phosphorescence properties of the sb-UCNPs are maintained after coating with the nanofilters.

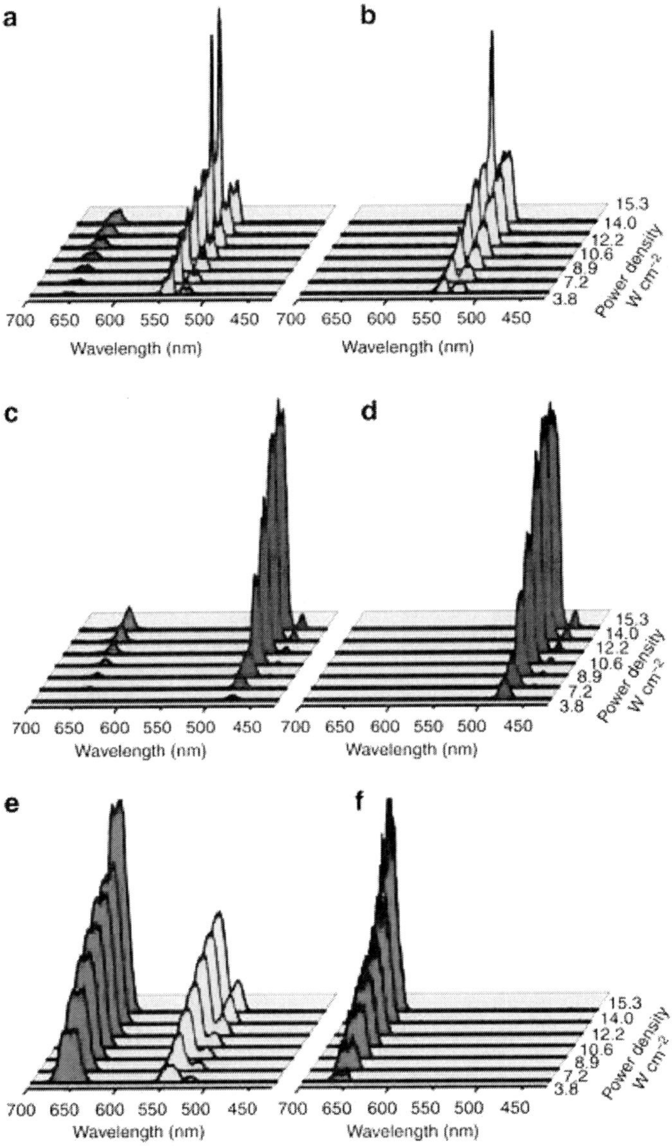

Figure 3: Optical properties of single-band upconversion nanoprobes. Pump-power-dependent upconversion photoluminescent emission spectra of the green, blue and red sb-UCNPs (**b**,**d**,**f**) and original upconversion nanocrystals (**a**,**c**,**e**) measured in water and cyclohexane, respectively. Note that the single-band features of the sb-UCNPs with the nanofilters are independent of the excitation laser power.

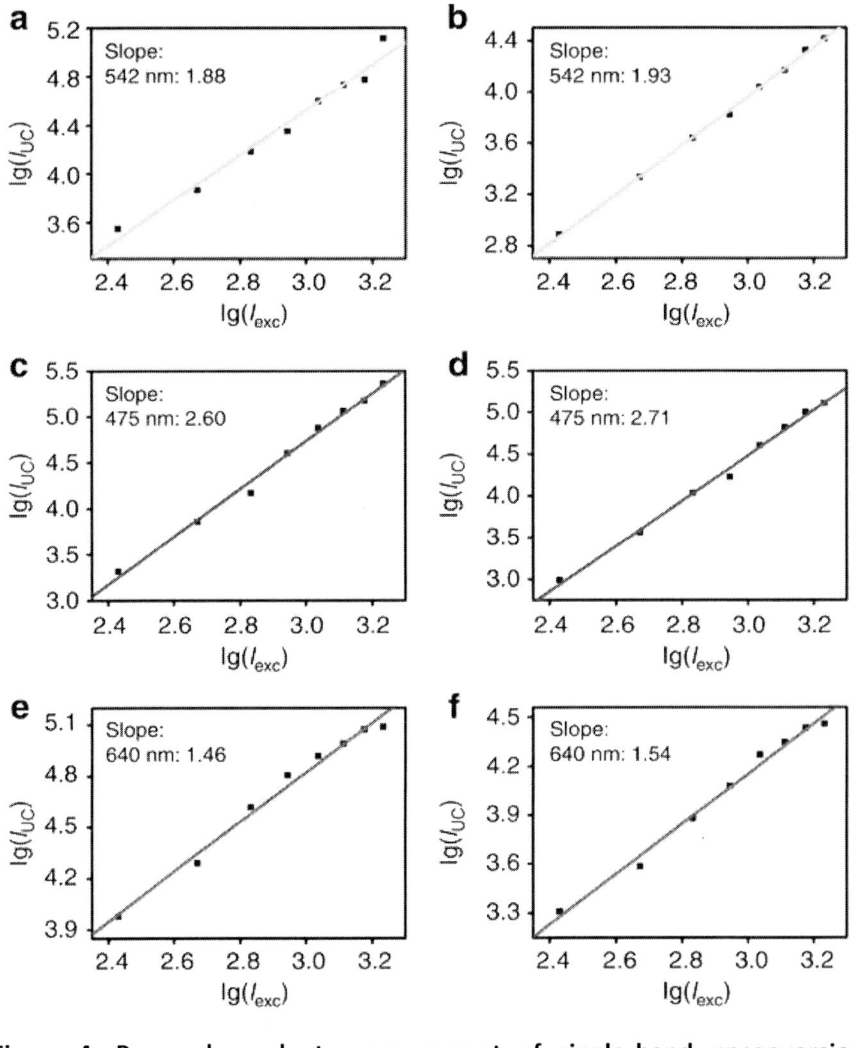

Figure 4: Power-dependent measurement of single-band upconversion nanoprobes. Power density dependence of upconverted emission of original upconversion nanocrystals (**a**,**c**,**e**) and the three single-band upconversion nanoprobes (**b**,**d**,**f**) at 475, 542 and 640 nm.

To determine the photostability of the sb-UCNPs, dilute samples of sb-UCNPs with different emission colours were dispersed on the coverslips and imaged by atomic force microscopy (AFM)-coupled optical microscopy with excitation with a tightly focused 980-nm continuous wave (CW) laser (Fig.

5a)[31,42]. Upconverted photoluminescent images of the samples revealed homogeneous and randomly distributed diffraction-limited spots corresponding to either single particles or aggregates (Fig. 5b,d,f). In addition, as confirmed by the 1:1 correspondence of the optical and AFM images (Fig. 5c,e,g), particles #1 to #3 were unambiguously assigned as single nanoparticles with different colours. The photoluminescence emissions (Fig. 5h) and time traces (Fig. 5i,j) for particle #1 to #3 were studied under continuous illumination with a 980-nm CW laser. Notably, no photoblinking was detected when the bin time for each data point in emission intensity was reduced to 10 ms (Fig. 5i). Furthermore, the sb-UCNPs investigated here did not exhibit photobleaching or photodamage after 15 min of continuous laser irradiation with a 980-nm CW laser (Fig. 5j).

The photostability of the NPTAT and rhodamine B isothiocyanate dyes doped in nanofilters was quantitatively tested in both sb-UCNPs and $NaGdF_4$ or $NaYbF_4$ nanoparticles devoid of emitting lanthanides (Supplementary Fig. 14). The optical absorption peaks for both organic dyes were maintained under 980-nm laser excitation for 30 min. Because the upconversion quantum yield of UCNPs is low (<1%)[20,40], it is not adequate to evaluate the photostability of the organic dyes doped with nanofilters under upconversion visible emission. Therefore, a xenon lamp (500 W) with a prism spectroscope was used to generate higher-power visible irradiation overlapping with the absorption wavelength of the NPTAT and rhodamine B isothiocyanate dyes (Supplementary Fig. 14). The optical absorption peaks for both organic dyes were maintained under xenon lamp excitation for 30 min. The above results indicate that the homogeneous doping of organic dyes in the silica layer did not alter the photostability of the organic dyes under the 980-nm NIR laser or visible light irradiation.

Multiplexed detection for cancer cells and tissue specimens

To demonstrate the feasibility of the multiplexed labelling, multicolour sb-UCNPs emitting at 480, 540 and 650 nm were directly conjugated with primary antibodies (Abs) against the three most important breast cancer biomarkers (oestrogen receptors (ERs), progesterone receptors (PRs) and human epithelial growth fator receptor-2 (HER2)), which are routinely analysed in surgical pathology laboratories and on which therapeutic decisions are based (Fig. 1). MCF-7 and MDA-MB-231 breast cancer cells, which have different expression levels of the three protein markers, were stained with sb-UCNP-Abs bioconjugates (sb-UCNPs-Anti-PR (green), sb-

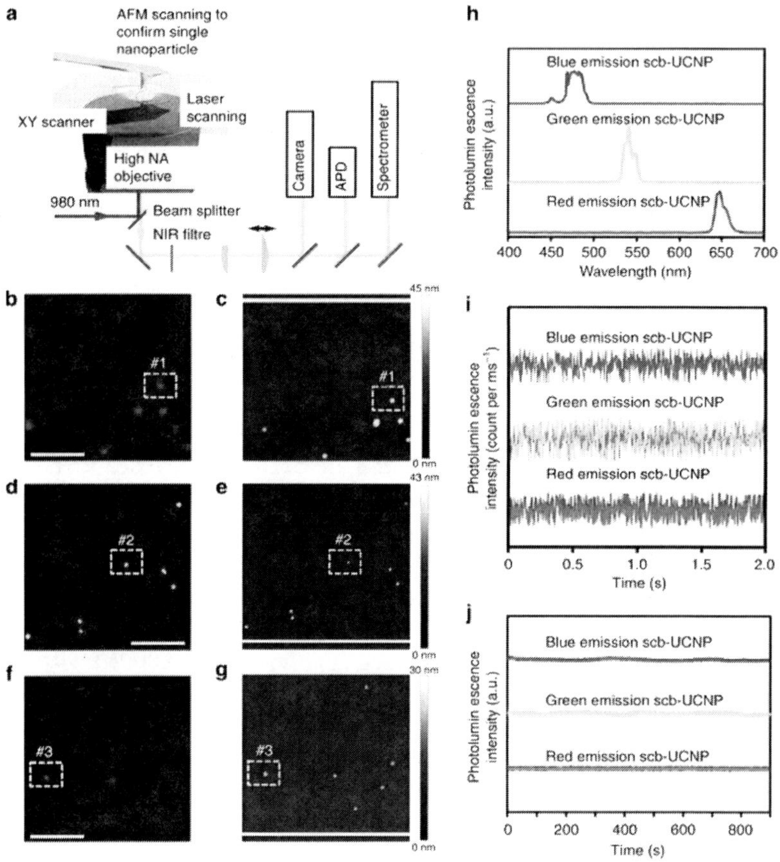

Figure 5: Single-particle optical property measurements of sb-UCNPs. (**a**) Experimental setup. On top of an inverted microscope, an AFM allows direct size determination combined with optical characterization. Upconversion photoluminescence images (**b**,**d**,**f**) and AFM images (**c**,**e**,**g**) of blue (**b**,**c**), green (**d**,**e**) and red (**f**,**g**) emission sb-UCNPs dispersed on a poly(L-lysine)-coated coverslip. Both photoluminescence and AFM images were obtained for the same sample region, as indicated by the 1:1 correspondence of the particle locations. (**h**) Typical spectra of single sb-UCNPs #1, #2# and #3. Emission bands corresponding to the transitions $^1G_4 \rightarrow ^3H_6$ (~480 nm), $^4S_{3/2} \rightarrow ^4I_{15/2}$ (~550 nm), and $^4F_{9/2} \rightarrow ^4I_{15/2}$ (~650 nm) can be observed. (**i**) The photoluminescence time traces for single nanoparticles #1, #2 and #3 acquired with 10-ms time bins. (**j**) Long-lifetime photoluminescence intensity traces for single nanoparticles #1, #2 and #3 acquired with 500-ms time bins. All optical data shown were obtained with continuous wavelength 980-nm laser excitation. Scale bar, 5 μm.

UCNPs-Anti-ER (blue) and sb-UCNPs-Anti-HER2 (red)). Using multispectral confocal microscopy, spectrally separated upconversion fluorescence was clearly visible in both cell lines (Fig. 6a;Supplementary Fig. 15; Supplementary Fig. 16). The remarkable photostability of the sb-UCNPs with low biological background fluorescence under 980-nm excitation allows extended exposure to excitation lights and consequently allows the reconstruction of multicolour projection. As shown in Fig. 6a, HER2 was located on the cell membrane, and both ER and PR were detected predominantly in the cell nuclei. These results demonstrate that sb-UCNP-Abs bioconjugates can be used to detect proteins regardless of their cellular location. To further explore whether fluorescent signals from labelled proteins can be simultaneously quantified, we performed single-cell spectroscopy using a wavelength-resolved spectrometer coupled with a confocal microscope. Individual spectra of each labelled protein with peaks in the region of the sb-UCNP emission maxima (480, 540 and 650 nm) were observed, representing the photoluminescent emission of the sb-UCNP-Abs (Fig. 6b). Due to the narrow upconversion emission bands, there was no spectral overlap for the multiplexed simultaneous detection of the three protein markers with sb-UCNPs, which greatly overcomes the limitations of QDs and organic dyes with clear multicolour emission spectral overlap43′44.

Without adjusting the photoluminescence intensities of individual colours, the relative photoluminescence intensities of a biomarker across different cell lines are more meaningful than the comparison of the photoluminescence intensities of different biomarkers in one type of cells. This distinction is the result of the differential optical and structural properties of sb-UCNPs of different colours. The use of different photoluminescence detectors in different labs can also lead to discrepancies in experiment results. We observed that the 480- and 540-nm sb-UCNPs were 1.1-fold and 1.5-fold, respectively, brighter than the 650-nm sb-UCNPs at the same concentration (Supplementary Fig. 17). When the individual colours were resolved using a previously reported method19′45 and normalized against this 'brightness index', the peaks at 650 and 480 nm were increased 1.5-fold and 1.3-fold with respect to their original values, respectively, whereas the peak at 540 nm was unchanged. The results obtained by quantitative spectrometry (Fig. 6b) correlated well with the biomarker expression patterns obtained using the traditional WB technique (Fig. 6d,e). For statistical significance, the photoluminescence spectra of 100 cells in each cell line were measured, and the mean levels of HER2, ER and PR expression were evaluated (Fig. 6c). The protein expression results obtained using sb-UCNPs-Abs molecular profiling technology and WB were then compared after the absolute measurements of the triplet (HER2, ER

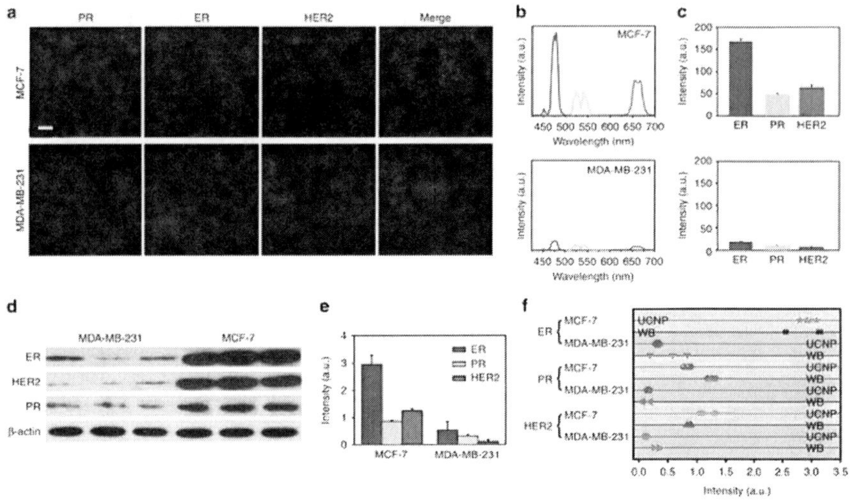

Figure 6: Multiplexed detection in breast cancer cell lines using sb-UCNPs. (**a**) Expression of PR, ER and HER2 signals in MCF-7 and MDA-MB-231 breast cancer cells detected by fluorescence microscopy (pseudo-colour). (**b**) Representative spectra of the protein biomarkers obtained by single-cell spectroscopy measurements. (**c**) Average expression levels of ER, PR and HER2 determined by spectral analysis of 100 single cells. Data points represent mean+s.d. (n=100). (**d**,**e**) ER, PR, and HER2 expression in MCF-7 and MDA-MB-231 cells as measured by WB. Data points represent mean+s.d. (n=3). (**f**) Comparative statistical analysis of sb-UCNPs-Abs bioconjugate profiling data and WB results. Scale bar, 10 μm.

and PR) were transformed to a relative scale in percent (Fig. 6f). Note that biomarker expression levels determined using sb-UCNP molecular profiling technology and standard WB for a given cell line tend to aggregate, revealing excellent correlation between the two methods.

Furthermore, because IHC is also one of the widely used methods for analysing cancer-related proteins, we compared the sb-UCNP molecular profiling technology with a traditional IHC method using the same set of breast cancer cell lines (Fig. 7a,b). The slides of cells were photographed after immunostaining in accordance with standard pathological protocols for ER, PR and HER2 (Fig. 7a; Supplementary Table 3). The IHC results were analysed by three independent observers and scored with a standard scale from 0 to 3+. Cells with IHC scores of 3+ had 100% relative expression of the proteins by sb-UCNP multiplexed detection (the highest expression level for each biomarker was set as unity). ER and PR expression in MDA-MB-231 cells were both classified as 1+ by IHC, which corresponded to 11 and 5%

estimated by the new sb-UCNP molecular profiling technology (Fig. 7b;Supplementary Table 3), while the 4% HER2 expression detected by sb-UCNPs was classified with a 0 IHC score (Fig. 7b). In both cases, the expression of the proteins using sb-UCNPs correlated well with the IHC results. However, the application of conjugated sb-UCNPs and quantitative spectroscopy may be more accurate than IHC for quantifying proteins present at low levels.

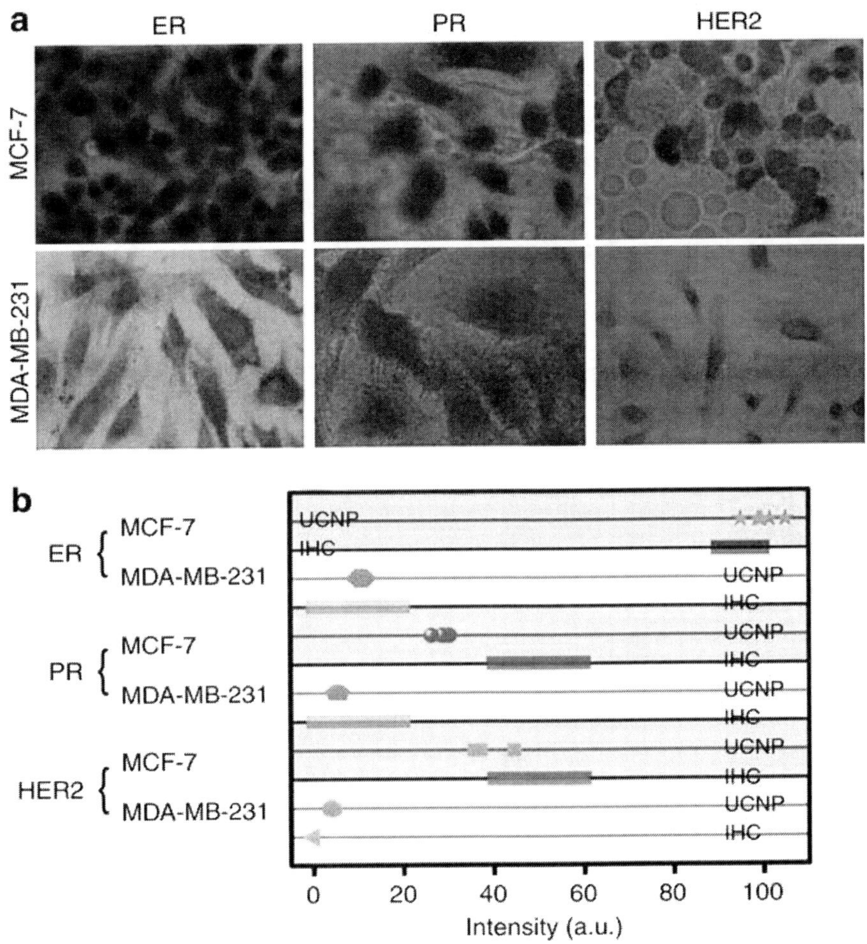

Figure 7: Comparison of multiplexed detection with sb-UCNPs and IHC. (**a**) Expression of ER, PR and HER2 in the breast cancer cell lines as estimated by IHC and (**b**) comparative statistical analysis of the sb-UCNP profiling data and IHC results.

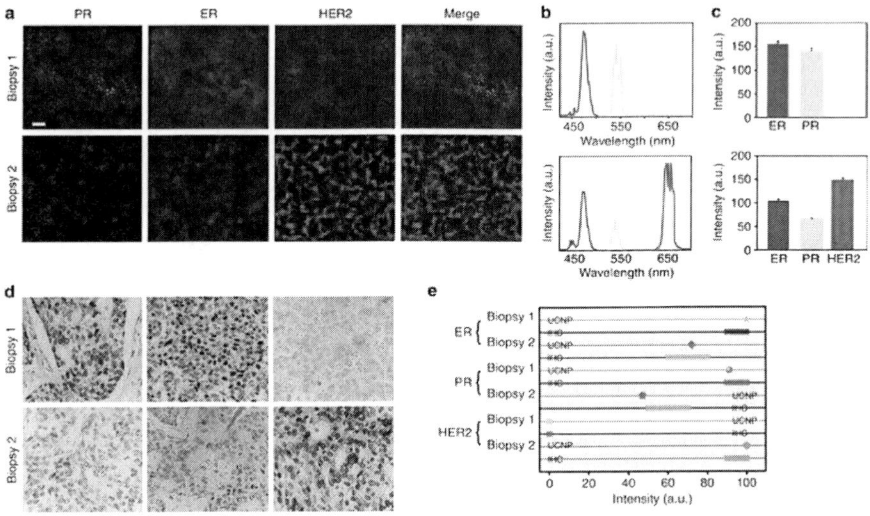

Figure 8: Multiplexed detection in breast cancer tissue specimens using sb-UCNPs. (**a**) Nuclear and cell membrane fluorescent signals were detected by microscopy (pseudo-colour), revealing the expression of ER, PR and HER2 biomarkers as quantified by wavelength-resolved spectroscopy. (**b**) Emission spectra resolved into individual channels and compensated for the differential brightness between different colours. (**c**) Average expression levels of ER, PR and HER2 calculated from the results of spectral analysis obtained by detecting 30 randomly selected spots in the tissue sections. Data points represent mean+s.d. (n=30). (**d**) Expression of ER, PR and HER2 estimated by IHC in the breast cancer tissue specimens and (**e**) comparative statistical analysis of the sb-UCNP profiling data and IHC results. Scale bar, 20μm.

The multiplexed detection results for cell lines indicate that the novel sb-UCNP-based molecular profiling technology is capable of imaging and quantifying multiple tumour biomarkers in intact tumour specimens. We therefore sought to apply this technology to detect PR, ER and HER2 biomarkers in FFPE human breast cancer specimens with Ab-conjugated sb-UCNPs. Furthermore, according to the American Society of Clinical Oncology/College of American Pathologists, the validated techniques to assess these biomarkers are IHC46. In light of this novel sb-UCNP-based molecular profiling technology, our aim was to perform a comparison study of PR, ER and HER2 concordance between IHC and sb-UCNP molecular profiling technology.

In tumour biopsy 1, only PR and ER signals were observed by confocal microscopy (Fig. 8a), corresponding to the two upconversion photoluminescent peaks detected on the spectrogram (Fig. 8b). Bioimaging and spectroscopic detection of biopsy 2 revealed expression of PR, ER and HER2 (Fig. 8a,b). Quantification detection by analysing 30 randomly selected spots in the tissue sections with sb-UCNP molecular profiling technology revealed a definitive difference in HER2 expression between the two tumour specimens. Biopsy 1 did not express HER2, in contrast to the high expression of HER2 (147±10) in biopsy 2 (Fig. 8c;Supplementary Table 4). When compared with the IHC results using relative scales (Fig. 8d; Supplementary Table 4), the average values of biomarker expression measured using sb-UCNPs revealed excellent correlation for high-level protein expression (Fig. 8e; Supplementary Table 4). Similar to the cell line detection, the application of Ab-conjugated sb-UCNPs and quantitative spectroscopy may be more accurate at quantifying proteins than IHC in tumour specimens. For example, ER and PR expression in biopsy 2 were both classified as 2+ by IHC. However, more accurate detection was achieved for ER (72%) and PR (47%) expression using the new sb-UCNP molecular profiling technology.

DISCUSSION

If cancer could be reliably identified at the earliest possible stage of cellular transition, then novel therapeutics could be designed to arrest the process of malignant transformation. Aggressive behaviours or phenotypes may be understood and predicted by a defining set of biomarkers. By critically defining the interrelationships among these biomarkers, cancer could be diagnosed and prognosis predicted based on a patient's molecular profile, leading to personalized and predictive oncology. That is, a unique molecular profile could be used to predict the tumour's invasive and metastatic potential, its ability to survive and grow under androgen-deprived, hypoxic and metabolic stress conditions, and the potential of certain cancer cells to evade host immune surveillance. Current diagnostic and prognostic classifications of human tumours are mainly based on the IHC method with single-wavelength detection, which is not compatible with multiplexed molecular profiling47. Recent advances have demonstrated that the fluorescence-based analysis of protein biomarkers is strongly correlated with clinical outcome. Compared with conventional biological labels, such as organic dyes, QDs and regular lanthanide labels with ultraviolet excitation, UCNPs have many advantages, including virtually zero auto-fluorescence background, which improves the signal-to-noise ratio, large anti-Stokes shifts, which facilitate the separation of photoluminescence from the excitation wavelength, narrow emission bandwidths, which facilitate multiplexed imaging, and high resistance to photobleaching for compatibility with long-term repetitive imaging. In addition, UCNPs are non-blinking, exhibit reduced light scattering and permit deep tissue penetration because their excitation lies in the NIR region within the optical transparency window48.

In this report, we have developed a general and simple method to achieve single-band upconversion emissions with different colours and high chromatic purities by coating the upconversion nanocrystals with organic dye-doped nanofilters48. As a result of the efficient reabsorption of the organic dye with a high molar absorption coefficient, remarkably pure single-band upconversion emissions can be generated in the blue (480 nm), green (550 nm) and red (650 nm) regions. Moreover, lanthanide ions are spectroscopically rich species, and most of the rare earth (RE) ions can exhibit upconversion, including Pr, Nd, Sm, Eu, Tb, Dy, Ho, Er and Tm. By adjusting different combinations of doping ions and host materials20'38'40, the upconversion emission wavelength (from ultraviolet–visible to NIR) and relative intensity of emission peaks can be effectively controlled to modulate the emission colours of the lanthanide-doped nanocrystals. Therefore, this method can be extended to produce additional sb-UCNPs in combination with one or more appropriate filter dye types.

Significantly, in this work, we have demonstrated the use of sb-UCNPs for the multiplexed detection of three tumour biomarkers in both cultured human breast cancer cells and paraffin-embedded clinical tissue sections. The simultaneous quantification of ER, PR and HER2 receptor expression levels in the breast cancer cell specimens correlated closely with the results of the traditional WB method. Furthermore, the application of conjugated sb-UCNPs and quantitative spectroscopy may be more accurate than IHC for the simultaneous quantification of proteins present at low levels in cancer cells and tissue specimens. Thus, sb-UCNP-based technology may be well suited to the molecular profiling of tumour biomarkers *in vitro* and represent a clinical translational application of upconversion nanomaterials for cancer prognosis. The ability to detect multiple target proteins in small samples of cancer tissues could enable more effective therapeutic decisions when used in combination with regular IHC methods. The next step is to conduct large-scale clinical studies to establish protocols and practices for sb-UCNP-based molecular pathology.

Moreover, fluorescence colour codes used in a broad range of fields have been among the most important methods for multiplexing, as demonstrated in this work. Although much narrower emission peaks can be achieved with UCNPs, the crowded spectral domain still limits detection numbers. Creating additional distinguishable coding dimensions should enhance multiplexed detectabilities. Recently, Jin *et al.*49'50 demonstrated that tunable luminescent lifetimes in the microsecond region can be exploited to code individual upconversion nanocrystals (τ-dots) that are independent of both colour and intensity and extend the optical multiplexing capability by adding a temporal dimension to photoluminescent signals. In terms of luminescence intensity-based multiplexing, sb-UCNPs are superior, and thus combining single-band upconversion and the temporal dimension of luminescence signals is expected to further extend the optical multiplexing capability for next-generation applications in the life sciences.

METHODS

Synthesis of lanthanide upconversion nanocrystals

The upconversion nanocrystals, including hexagonal phase β-$NaGdF_4$:20% Yb, 0.2% Tm, β-$NaGdF_4$:20% Yb, 0.2% Tm@$NaGdF_4$, β-$NaGdF_4$:20% Yb, 2% Er, β-$NaGdF_4$:20% Yb, 2% Er@$NaGdF_4$, cubic phase α-$NaYbF_4$:10% Er and α-$NaYbF_4$:10% Er@$NaYF_4$ nanocrystals, were prepared by the successive layer-by-layer strategy (see Supplementary Methods).

Synthesis of sb-UCNPs

The sb-UCNPs were synthesized by a reversed-phase microemulsion method. In a typical synthesis of single red-emission sb-UCNPs, a mixing solution of 0.5 ml of surfactant CO-520 (purchased from Sigma Aldrich), 16 ml of cyclohexane (purchased from Sinopharm Chemical Reagents Co. Ltd) and 1 ml of 0.1 Mmol α-$NaYbF_4$:10% Er@$NaYF_4$ solution in cyclohexane was vigorously stirred for 15 min. Next, 2.1 ml of CO-520 and 0.2 ml of $NH_3 \cdot {}_2O$ (30 wt%, purchased from Beijing Chemical Reagents Co. Ltd) were added to the solution, and the container was sealed and sonicated for 15 min until a transparent reversed-phase microemulsion was formed. Then, 0.1 ml of TEOS (purchased from Sigma Aldrich) was added to the solution. After vigorous stirring for 2 h, 0.1 ml of (3-Aminopropyl) triethoxysilane (APTES) (purchased from Sigma Aldrich)-modified rhodamine B isothiocyanate (purchased from Sigma Aldrich) solution (5 mg ml^{-1}) was added. After continuous stirring at room temperature at 750 r.p.m. for 2 days, the products were precipitated by adding ethanol. The collected precipitates were washed twice with ethanol and dispersed in ethanol for further use. The blue and green emission sb-UCNPs were synthesized in a similar manner using nickel (II) phthalocyanine-tetrasulfonic acid tetrasodium (5 mg ml^{-1}, purchased from Sigma Aldrich) doping dyes.

Wide-field luminescence and AFM imaging of single sb-UCNPs

To prepare an UCNP sample for single-particle measurements, a clean coverslip was first coated with poly(L-lysine; 0.1% w/v in H_2O), and the sb-UCNP solution was then dispersed in it for measurement. We imaged the sb-UCNPs using an in-house wide-field fluorescence imaging setup based on an inverted microscope (X71, Olympus), an NIR excitation source (980-nm CW laser), and a highly sensitive ARC-SP 2360 spectrometer equipped with a Pro EMCCD camera (Pro EM: 1,600 × 200 B/I eXcelon EMCCD Camera System, Princeton Instruments, Inc.). The output of a 980-nm laser (SDL-980LM-500T, 500 mW, Shanghai Dream Lasers Technology) was spatially filtered using a 100-mm pinhole and collimated so that the beam diameter was ~10 mm before entering the microscope. The beam was reflected by a short-pass optical filter (750SP from Chroma Corp) inside the microscope and directed to the microscope objective (numerical aperture 1.40, oil immersion, Olympus). After passing through the dichroic beam splitter, the emission was filtered by a short-pass emission filter (750SP from Chroma Corp) with a transmission range of 400–700 nm, which covers the blue, green and red emission bands of sb-UCNPs. The image of the sb-UCNP sample was further magnified using a set of achromatic lenses outside the microscope and finally detected by a Pro EMCCD camera. To determine whether the bright spots in such optical images corresponded to single sb-UCNPs or their aggregates, we imaged the same region by AFM (Catalyst, Bruker) and analysed the morphology of the particles.

Multiplexed labelling of breast cancer cells

Breast cancer cell lines (MCF-7, MDA-MB-231) were obtained from American Type Culture Collection (ATCC, Manassas, VA, USA). For multiplexed staining of sb-UCNPs-Abs, cells (1×10^4) were incubated in confocal dishes (35 × 10 mm) at 37 °C under 5% CO_2 for 24 h. The cells were deparaffinized by incubation with formaldehyde (4%), followed by a 30-min incubation in a humidified chamber at 37 °C with 2% Triton X-100. After blocking with 5% bovine serum diluted in phosphate-buffered saline for 30 min, cells were stained with a mixture of conjugates targeting ER, PR and HER2 (40 nM each) for 1 h and washed five times in PBS.

Quantitative spectrometry with confocal microscopy

Single-cell spectroscopy was accomplished using scanning mapping of a fluorescence confocal microscope. To measure biomarker expression in cultured cells, spatially separated individual cells were manually positioned in the 'hot-spot' defined by the position and size of the pinhole. To determine biomarker expression in tissue specimens, sb-UCNP spectra from 30 randomly selected areas were measured. Data from the spectroscopic measurements were converted to ASCII format for further quantification and statistical analysis.

ACKNOWLEDGEMENTS

This work was supported by NSFC (grant No. 21322508, 21210004), the China National Key Basic Research Program (973 Project) (No. 2013CB934100, 2012CB224805), the Program for New Century Excellent Talents in University (NCET), and the Shu Guang project supported by the Shanghai Municipal Education Commission and Shanghai Education Development Foundation.

REFERENCES

1. Braun, G. B. *et al.* Etchable plasmonic nanoparticle probes to image and quantify cellular internalization. *Nat. Mater.* **13**, 904–911 (2014).

2. Qing, Q. *et al.* Free-standing kinked nanowire transistor probes for targeted intracellular recording in three dimensions. *Nat. Nanotechnol.* **9**, 142–147 (2014).

3. Shen, Y. *et al.* Plasmonic gold mushroom arrays with refractive index sensing figures of merit approaching the theoretical limit.*Nat. Commun.* **4**, 2381 (2013).

4. Tian, B. *et al.* Three-dimensional, flexible nanoscale field-effect transistors as localized bioprobes. *Science* **329**, 830–834 (2010).

5. Hanash, S. Integrated global profiling of cancer. *Nat. Rev. Cancer***4**, 638–644 (2004).

6. Zheng, G. F., Patolsky, F., Cui, Y., Wang, W. U. & Lieber, C. M.Multiplexed electrical detection of cancer markers with nanowire sensor arrays. *Nat. Biotechnol.* **23**, 1294–1301 (2005).

7. Duan, X. *et al.* Intracellular recordings of action potentials by an extracellular nanoscale field-effect transistor. *Nat. Nanotechnol.***7**, 174–179 (2012).

8. Zrazhevskiy, P. & Gao, X. Quantum dot imaging platform for single-cell molecular profiling. *Nat. Commun.* **4**, 1619 (2013).

9. Umemura, S. & Osamura, R. Y. Utility of immunohistochemistry in breast cancer practice. *Breast Cancer* **11**, 334–338 (2004).

10. Yu, C. C., Woods, A. L. & Levison, D. A. The assessment of cellular proliferation by immunohistochemistry: a review of currently available methods and their applications. *Histochem. J.* **24**, 121–131 (1992).

11. Heppner, G. & Miller, B. Tumor heterogeneity: biological implications and therapeutic consequences. *Cancer Metastasis Rev.* **2**, 5–23 (1983).

12. Chechlinska, M., Kowalewska, M. & Nowak, R. Systemic inflammation as a confounding factor in cancer biomarker discovery and validation. *Nat. Rev. Cancer* **10**, 2–U13 (2010).

13. Lockhart, D. J. & Winzeler, E. A. Genomics, gene expression and DNA arrays. *Nature* **405**, 827–836 (2000).

14. Lee, H., Park, J. E. & Nam, J. M. Bio-barcode gel assay for microRNA. *Nat. Commun.* **5**, 3367 (2014).

15. Uetrecht, C. & Heck, A. J. R. Modern biomolecular mass spectrometry and its role in studying virus structure, dynamics, and assembly. *Angew. Chem. Int. Ed. Engl.* **50**, 8248–8262 (2011).

16. Perry, P. & Wolff, S. New giemsa method for differential staining of sister chromatids. *Nature* **251**, 156–158 (1974).

17. Bruchez, M., Moronne, M., Gin, P. & Weiss, S. Alivisatos AP. Semiconductor nanocrystals as fluorescent biological labels.*Science* **281**, 2013–2016 (1998).

18. Han, M. Y., Gao, X. H., Su, J. Z. & Nie, S. Quantum-dot-tagged microbeads for multiplexed optical coding of biomolecules. *Nat. Biotechnol.* **19**, 631–635 (2001).

19. Yezhelyev, M. V. *et al.* In situ molecular profiling of breast cancer biomarkers with multicolor quantum dots. *Adv. Mater.* **19**, 3146–3151 (2007).

20. Li, X. M., Zhang, F. & Zhao, D. Y. Lab on upconversion nanoparticles: optical properties and applications engineering via designed nanostructure. *Chem. Soc. Rev.* **44**, 1346–1378 (2014).

21. Wu, S. *et al.* Non-blinking and photostable upconverted luminescence from single lanthanide-doped nanocrystals. *Proc. Natl Acad. Sci. USA* **106**, 10917–10921 (2009).

22. Wang, F. *et al.* Simultaneous phase and size control of upconversion nanocrystals through lanthanide doping. *Nature* **463**, 1061–1065 (2010).

23. Lee, J. *et al.* Universal process-inert encoding architecture for polymer microparticles. *Nat. Mater.* **13**, 524–529 (2014).

24. Johnson, N. J. J. & van Veggel, F. C. J. M. Sodium lanthanide fluoride core-shell nanocrystals: A general perspective on epitaxial shell growth. *Nano Res.* **6**, 547–561 (2013).

25. Idris, N. M. *et al.* In vivo photodynamic therapy using upconversion nanoparticles as remote-controlled nanotransducers. *Nat. Med.* **18**, 1580–1585 (2012).

26. Wang, J. *et al.* Photon energy upconversion through thermal radiation with the power efficiency reaching 16%. *Nat. Commun.* **5**, 5669 (2014).

27. Zhao, J. B. *et al.* Single-nanocrystal sensitivity achieved by enhanced upconversion luminescence. *Nat. Nanotechnol.* **8**, 729–734 (2013).

28. Boyer, J. C., Vetrone, F., Cuccia, L. A. & Capobianco, J. A. Synthesis of colloidal upconverting $NaYF_4$ nanocrystals doped with Er^{3+}, Yb^{3+} and Tm^{3+}, Yb^{3+} via thermal decomposition of lanthanide trifluoroacetate precursors. *J. Am. Chem. Soc.* **128**, 7444–7445 (2006).

29. Wang, F., Deng, R. & Liu, X. Preparation of core-shell $NaGdF_4$ nanoparticles doped with luminescent lanthanide ions to be used as upconversion-based probes. *Nat. Protoc.* **9**, 1634–1644 (2014).

30. Wang, J. *et al.* Enhancing multiphoton upconversion through energy clustering at sublattice level. *Nat. Mater.* **13**, 157–162 (2014).

31. Zhang, F. *et al.* Fluorescence upconversion microbarcodes for multiplexed biological detection: nucleic acid encoding. *Adv. Mater.* **23**, 3775–3779 (2011).

32. Dejneka, M. J. *et al.* Rare earth-doped glass microbarcodes. *Proc. Natl Acad. Sci. USA* **100**, 389–393 (2003).

33. Zou, W., Visser, C., Maduro, J. A., Pshenichnikov, M. S. &Hummelen, J. C. Broadband dye-sensitized upconversion of near-infrared light. *Nat. Photonics* **6**, 560–564 (2012).

34. Zhang, Y. *et al.* Multicolor barcoding in a single upconversion crystal. *J. Am. Chem. Soc.* **136**, 4893–4896 (2014).

35. Deng, R. *et al.* Temporal full-colour tuning through non-steady-state upconversion. *Nat. Nanotechnol.* **10**, 237–242 (2015).

36. Tian, G. *et al.* Mn^{2+} dopant-controlled synthesis of $NaYF_4$:Yb/Er upconversion nanoparticles for in vivo imaging and drug delivery.*Adv. Mater.* **24**, 1226–1231 (2012).

37. Wang, J., Wang, F., Wang, C., Liu, Z. & Liu, X. G. Single-band upconversion emission in lanthanide-doped $KMnF_3$ nanocrystals.*Angew. Chem. Int. Ed. Engl.* **50**, 10369–10372 (2011).

38. Li, X. M., Wang, R., Zhang, F. & Zhao, D. Y. Engineering homogeneous doping in single nanoparticle to enhance upconversion efficiency. *Nano Lett.* **14**, 3634–3639 (2014).

39. Gorris, H. H., Ali, R., Saleh, S. M. & Wolfbeis, O. S. Tuning the dual emission of photon-upconverting nanoparticles for ratiometric multiplexed encoding. *Adv. Mater.* **23**, 1652–1655 (2011).

40. Wang, F. *et al.* Tuning upconversion through energy migration in core-shell nanoparticles. *Nat. Mater.* **10**, 968–973 (2011).

41. Suyver, J. F., Aebischer, A., García-Revilla, S., Gerner, P. & Güdel, H. U. Anomalous power dependence of sensitized upconversion luminescence. *Phys. Rev. B* **71**, 125123 (2005).

42. Park, Y. I. *et al.* Nonblinking and nonbleaching upconverting nanoparticles as an optical imaging nanoprobe and T1 magnetic resonance imaging contrast agent. *Adv. Mater.* **21**, 4467–4471 (2009).

43. Liu, J., Lau, S. K., Varma, V. A., Kairdolf, B. A. & Nie, S. Multiplexed detection and characterization of rare tumor cells in hodgkin's lymphoma with multicolor quantum dots. *Anal. Chem.* **82**, 6237–6243 (2010).

44. Huang, D. *et al.* Comparison and optimization of multiplexed quantum dot-based immunohistofluorescence. *Nano Res.* **3**, 61–68 (2010).

45. Goldman, E. R. *et al.* Multiplexed toxin analysis using four colours of quantum dot fluororeagents. *Anal. Chem.* **76**, 684–688 (2003).

46. Hammond, M. E. *et al.* American society of clinical oncology/college of American pathologists guideline recommendations for immunohistochemical testing of estrogen and progesterone receptors in breast cancer (unabridged version). *Arch. Pathol. Lab. Med.* **134**, e48–e72 (2010).

47. Sun, L. *et al.* Expressions of ER, PR, HER-2, COX-2, and VEGF in primary and relapsed/metastatic breast cancers. *Cell Biochem. Biophys.* **68**, 511–516 (2014).

48. Weissleder, R. A clearer vision for in vivo imaging. *Nat. Biotechnol.* **19**, 316–317 (2001).

49. Lu, Y. *et al.* Tunable lifetime multiplexing using luminescent nanocrystals. *Nat. Photonics* **8**, 33–37 (2014).

50. Lu, Y. *et al.* On-the-fly decoding luminescence lifetimes in the microsecond region for lanthanide-encoded suspension arrays. *Nat. Commun.* **5**, 3741 (2014).

Chapter 9

Multifunctional Upconversion–Nanoparticles–Trismethylpyridylporphyrin–Fullerene Nanocomposite: A Near-Infrared Light-Triggered Theranostic Platform for Imaging-Guided Photodynamic Therapy

Mirong Guan[1], Hao Dong[2], Jiechao Ge[3], Daiqin Chen[1], Lingdong Sun[2], Shumu Li[1], Chunru Wang[1], Chunhua Yan[2], Pengfei Wang[3] and Chunying Shu[1]

[1]*Key Laboratory of Molecular Nanostructure and Nanotechnology, Institute of Chemistry, Chinese Academy of Sciences, and Beijing National Laboratory for Molecular Sciences, Beijing, China*

[2]*Beijing National Laboratory for Molecular Sciences, State Key Laboratory of Rare Earth Materials Chemistry and Applications, PKU-HKU Joint Laboratory in Rare Earth Materials and Bioinorganic Chemistry, College of Chemistry and Molecular Engineering, Peking University, Beijing, China*

[3]*Key Laboratory of Photochemical Conversion and Optoelectronic Materials, Technical Institute of Physics and Chemistry (TIPC), Chinese Academy of Sciences, Beijing, China*

ABSTRACT

Photodynamic therapy (PDT) is an excellent therapeutic modality for various malignant and nonmalignant cancers. This approach utilizes reactive oxygen species generated through the reaction between photosensitizer and oxygen in tissues upon light irradiation to achieve effective treatment. However, limited penetration depth and oxygen-deficient microenvironment hinder the efficiency of PDT. In this work, we design a multifunctional near-infrared (NIR)-triggered theranostic agent based on upconversion–nanoparticles-Polyoxyethylene bis (amine)–trismethylpyridylporphyrin–fullerene nanocomposite (UCNP–PEG–FA/PC_{70}) for imaging (fluorescence/upconversion luminescence/magnetic resonance imaging)-guided photodynamic therapy. In this multimodal nanocompsite, UCNPs are employed as light transducers to convert NIR light into ultraviolet–visible light to activate PC_{70} to generate singlet oxygen ($^1O_2$) even under low-oxygen conditions. Meanwhile, the upconversion emission, magnetic resonance imaging and fluorescence signal coming from UCNPs and PC_{70} nanocomposite enable UCNP–PEG–FA/PC_{70} to act as a multimodal imaging diagnostic agent, which facilitates the imaging-guided PDT. Furthermore, folate-mediated active targeting would enhance the accumulation of multifunctional hybrid in tumor. *In vitro* as well as *in vivo*results suggest that this smart nanocomposite is promising as an NIR light-triggered and -targeted theranostic platform for imaging-guided PDT of cancer, which may provide a solution to the bottleneck problems of PDT, namely, limited penetration depth and oxygen-deficient microenvironment.

INTRODUCTION

Theranostic nanoparticles, which combine both therapeutic and diagnostic capabilities in one nanoplatform, have potential to propel the biomedical field toward personalized medicine.[1, 2, 3] Specifically, phototriggered theranostics, combining phototherapies (for example, photodynamic,[4, 5] photothermal,[6, 7] and phototriggered chemo[8, 9]/gene therapy[10]) with real-time photodiagnostics (for example, bioluminescence,[11] fluorescence,[12, 13] and photoacoustic imaging[14,15]) have been actively pursued due to the advantages of spatiotemporal selectivity, non-ionizing radiation and specificity for disease destruction. Among them, near-infrared (NIR) light-triggered photodynamic theranostics, in which nanoparticles are used as photodynamic therapy (PDT) agents, have recently been receiving much attention due to deep penetration and simultaneous generation of reactive oxygen species (ROS) upon the NIR light irradiation, leading to ablation of cancer cells as a result. Up to date, several nanocomposites including gold nanocomposites,[16, 17, 18, 19] carbon nanostructures,[20, 21] $W_{18}O_{49}$,[22] WS_2,[23] porphyrin-functionalized porous nanoparticles,[24] and upconversion nanoparticles (UCNPs)[25] as NIR light-triggered photodynamic theranostic nanoplatforms have shown efficacy *in vitro* and *in vivo*studies. Wherein, NIR light-triggered photodynamic theranostics based on lanthanide-doped UCNPs has been well investigated for biomedical uses due to their intrinsic anti-Stokes upconversion luminescent properties.[26, 27, 28]Compared with traditional optical biomarkers, such as quantum dots and organic dyes, UCNPs exhibit excellent photostability, low background autofluorescence, remarkable penetration depth and high conversion efficiency for NIR photons into visible and ultraviolet (UV) ones, which lead to its unique superiority for bioimaging and bioassay studies.[29, 30] In particular, $NaGdF_4$-based UCNPs have been developed as multimodal imaging probes for upconversion luminescence (UCL) imaging and magnetic resonance imaging (MRI) modality.[31] Such multifunctional nanomaterials offer synergetic advantages of UCL imaging, that is, deep penetration and high sensitivity, along with the anatomic imaging capability of MRI in high spatial resolution.[31] Benefiting from the UV and visible emissions on specific NIR excitation, UCNPs have also been employed as light transducers to activate photosensitizers. UCNP-based PDT nanocomposites are considered promising platforms for cancer therapy due to the non-invasive manner and deep tissue penetration of NIR light, which breaks the dilemma of using UV or visible lights as irradiation source. Since Zhang *et al.*[32] first demonstrated the application of UCNP-based nanocomposite for PDT of bladder cancer cells, recent advances in the integration of UCNPs with PSs such as chlorin e6, Rose Bengal, and IR825 (an NIR-absorbing dye) for their PDT applications *in vitro* and *in vivo* have created many excitements.[33, 34, 35]However, the phototoxic effect of $^1O_2$ on PDT is restricted because of a hypoxic condition in tumor tissue.[36] Therefore, it is important to develop NIR light-triggered and -targeted theranostic platform for imaging-guided PDT of cancer under low-oxygen conditions.

Porphyrins as most common used PSs have been approved by Food and Drug Administration for treatment of various non-malignancies and malignancies. They have excellent photophysical properties, such as stronger absorption bands and high singlet oxygen quantum yield because of their large π-conjugated aromatic domains. Strong fluorescence of porphyrin also can be used as imaging probe to monitor the pharmacokinetics.[37] Very recently, trismethylpyridylporphyrin-C_{70}(PC_{70}) dyad has been reported to act as a promising photosensitizer for its reasonable water dispersibility, high photostability and favorable biocompatibility.[38] Significantly, the present PC_{70} exhibits extraordinary PDT efficiency *in vitro* even under low-oxygen conditions due to its extremely long lifetime triplet state (211.3 μs).[38] Unfortunately, the UV-to-blue excitation of PC_{70}hinders the enhancement of PDT efficiency *in vivo* due to the limited penetration depth of UV-to-blue irradiation. The introduction of UCNPs may pave the way for the application of PC_{70} as a NIR light-triggered PDT agent *in vivo* since UCNPs can serve as phototransducers by absorbing NIR light to activate PC_{70} to generate $^1O_2$. In addition, the involved UCNPs may reduce the effect by energy transfer from UCNPs to porphyrin,[38] and recover part of the quenched fluorescence of porphyrin.[37]

Herein, we design a multifunctional theranostic nanoplatform based on UCNPs/PC_{70} nanocomposite for NIR-triggered fluorescence/UCL/MR trimodal imaging and simultaneous PDT both *in vitro* and *in vivo* (Figure 1). To match the UV-to-blue absorption of PC_{70} for high efficiency of PDT, β-$NaGdF_4$:Yb,Tm@$NaGdF_4$ core/shell nanoparticles were rationally designed. Subsequently, PC_{70} was decorated on the surface of UCNPs through electrostatic interaction. Finally, the UCNPs/PC_{70} nanocomposite was coated with folic acid-modified PEG for active targeting and prolonged circulation time. With such a UCNP–PEG–FA/PC_{70} nanocomposite, trimodal imaging of FL/UCL/MR can be used to guide the PDT treatment of cancer cells efficiently *in vitro* and *in vivo* via the activation of PC_{70} by UV and blue emissions from UCNPs. Besides, folate as a target element can enhance accumulation of nanocomposite in tumor to improve the efficiency of PDT. Most importantly, the synthesized UCNP–PEG–FA/PC_{70} may provide solutions to the bottleneck problems of PDT, namely, limited penetration depth and oxygen-deficient microenvironment.

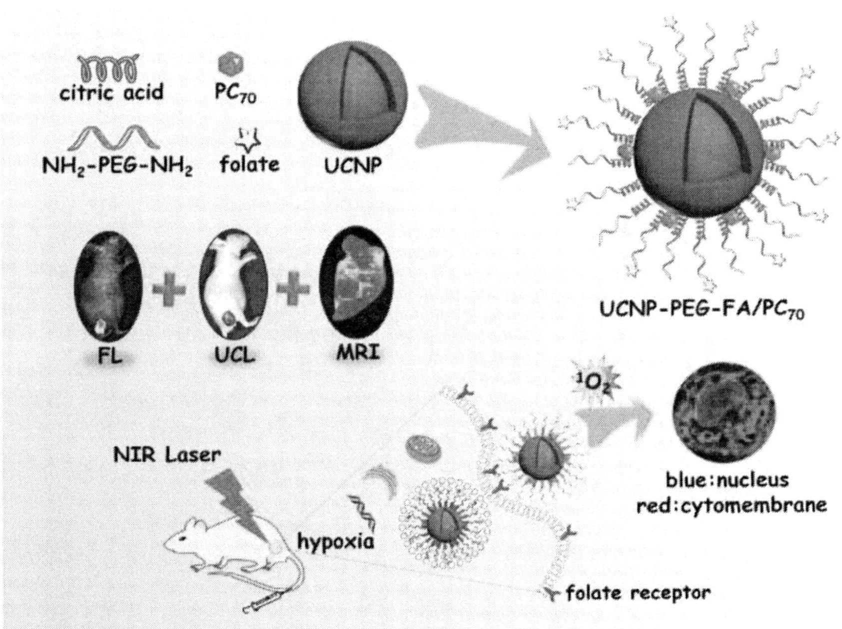

Figure 1. Schematic strategies of trimodal imaging guided UCNP–PEG–FA/PC_{70}(upconversion–nanoparticles–trismethylpyridylporphyrin–fullerene nanocomposite) for photodynamic therapy.

Materials and methods

Instruments

Samples for transmission electron microscopy (TEM) analysis were prepared by drying a colloidal solution containing nanoparticles on amorphous carbon-coated copper grids. Low-resolution TEM was operated on a JEOL-JEM 2100 transmission electron microscope at 200 kV. High-resolution TEM was carried out on a JEOL-JEM 2100F field emission transmission electron microscope at 200 kV. Powder X-ray diffraction patterns were recorded on a Rigaku D/MAX-2000 diffractometer (Tokyo, Japan) with a slit of 1/2° at a scanning rate of 2° min/min, using Cu Kαradiation (λ=0.5406 Å). Upconversion emission spectra were recorded on a Hitachi F-4500 spectrometer, with the Photomultiplier tubes (PMT) voltage of 700 V. The particle size and zeta potential were determined by dynamic light scattering using Malvern Nano-ZS90 apparatus (Malvern Instruments, Worcestershire, UK). Absorbance measurements were carried out using UV2550 spectrophotometer (SHIMADZU, Shimane, Japan). Fluorescence spectra were acquired though Fluorolog spectrophotometer. IR spectra were measured using Thermo instrument (NICOLET In10 MX, Shanghai, China).

Materials

NH_2-PEG3000-NH_2 and folate were purchased from Innochem (Beijing, China). 1,10-Diamino-4,7-dioxadecane, 1-ethyl-3-(3-dimethylaminopropyl)carbodiimide hydrochloride (EDC) and *N*-hydroxysuccinimide (NHS) were purchased from Acros (Shanghai, China). The culture medium Dulbecco's modified Eagle's medium (Invitrogen, Shanghai, China) supplemented with 10 % fetal bovine serum (Hyclone Company, South Logan, UT, USA), penicillin (100 μg ml^{-1}) and streptomycin (100 μg ml^{-1}) (Gibco, Grand Island, NY, USA). Dil, MitoRed, Lyso, PI and Hoechst 33258 were purchased from PuJingKangLi (Beijing, China). All other chemicals used were analytical reagents.

Synthesis of β-$NaGdF_4$:Yb,Tm@$NaGdF_4$

The β-$NaGdF_4$:Yb,Tm@$NaGdF_4$ core/shell nanoparticles were synthesized with a modified thermolysis method. Three steps are included as following.

Synthesis of α-$NaGdF_4$:Yb,Tm nanoparticles (α-core)

A given amount of $Ln(CF_3COO)_3$ (1 mmol), 0.29 mmol $Gd(CF_3COO)_3$, 0.70 mmol $Yb(CF_3COO)_3$, and 0.01 mmol $Tm(CF_3COO)_3$, and CF_3COONa (1 mmol) were added to a mixture containing oleic acid (OA), oleylamine (OM), and octadecene (ODE) (40 mmol, molar ratio: 1:1:2) into a three-necked flask (100 ml) at room temperature. The slurry was heated to 110 °C to remove low-boiling water, and oxygen with vigorous stirring under vacuum. Then, the solution was heated to 310 °C at a heating rate of 15 °C min^{-1} and kept for 15 min under N_2 atmosphere. After cooling to room temperature, the nanoparticles were collected with centrifugation (7800 r.p.m., 10 min) after precipitation by adding excess amount of ethanol. The final product was dispersed in 10 ml of cyclohexane.

Synthesis of β-$NaGdF_4$:Yb,Tm nanoparticles (β-core)

Five milliliter of as-prepared α-$NaGdF_4$:Yb,Tm nanoparticles were redispersed in a mixture of OA and ODE (40 mmol, molar ratio: 1:1). Then, $Ln(CF_3COO)_3$ (0.5 mmol, 0.145 mmol $Gd(CF_3COO)_3$, 0.35 mmol $Yb(CF_3COO)_3$, and 0.005 mmol $Tm(CF_3COO)_3$), and CF_3COONa (0.5 mmol) were added. After removing low boiling cyclohexane, water and oxygen with vigorous stirring under vacuum, the mixture was then heated to 325 °C at a heating rate of 15 °C min^{-1} and kept for 60 min under N_2atmosphere. Then, the reaction was cooled to room temperature. Excess amount of ethanol was added to precipitate the nanoparticles, which was collected by further centrifugation (7800 r.p.m., 10 min) and redispersed in 10 ml of cyclohexane.

Synthesis of β-$NaGdF_4$:Yb,Tm@$NaGdF_4$ nanoparticles (β-core@shell)

Five milliliter of as-prepared β-$NaGdF_4$:Yb,Tm nanoparticles were redispersed in a mixture of OA and ODE (40 mmol, molar ratio: 1:1). Then, $Gd(CF_3COO)_3$(1 mmol) and CF_3COONa (1 mmol) were added. The slurry was heated to 110 °C to remove low boiling cyclohexane, water, and oxygen with vigorous stirring under vacuum. Then, the solution was heated to 320 °C at a heating rate of 15 °C min^{-1} and kept for 40 min under N_2 atmosphere. After cooling to room temperature, the nanoparticles were collected with centrifugation (7800 r.p.m., 10 min) after precipitation by adding excess amount of ethanol. The final product was dispersed in 5 ml of cyclohexane.

Preparation of citric acid modified nanoparticles (citric-UCNP)

One milliliter cyclohexane dispersions containing hydrophobic nanoparticles were precipitated with ethanol by centrifugation and redispersed in 5 ml of chloroform. Citric acid (200 mg) was dissolved in 5 ml of deionized water. The pH value of the citric acid solution was tuned ~7 by ammonia. Subsequently, the two solutions were mixed under vigorous stirring at room temperature overnight. After that, the chloroform phase was deserted, whereas the water phase was transferred to a centrifuge tube. Citric acid-modified UCNPs were collected after high-speed centrifugation at 18 000 r.p.m. for 15 min and then washed with water for several times. Finally, the products were redispersed in water.

Synthesis of folate-NH-PEG3000-NH_2

Folate–NH–PEG3000–NH_2 conjugate was synthesized according to the previous method. Briefly, folate was activated with EDC and NHS at a 1:2:2 molar ratio in dimethyl sulfoxide for 2 h. Under nitrogen atmosphere, folate was filtered with a syringe filter (pore size: 0.2 μm). A hetero-bifunctional poly (ethylene glycol) derivate (NH_2–PEG3000–NH_2 Mw 3000) was added to the above solution at a molar ratio of 15:1 (folate:PEG). After reaction for 4 h under nitrogen atmosphere, the reactants were precipitated with acetone and centrifuged to remove the folate residue. Folate-NH–PEG3000–NH_2 conjugation in the supernatant was dialyzed against deionized water.

Synthesis of UCNP-ca/PC_{70} (UCNP/PC_{70})

Citric-UCNP (37.6 mg) was dispersed in water (10 ml). Then 18.8 mg of PC_{70} was added into the above solution and stirred overnight to obtain UCNP/PC_{70}.

Synthesis of UCNP–NH–PEG3000–NH–folate/PC_{70} (UCNP–PEG–FA/PC_{70})

Fifteen milligram of EDC and 8.5 mg of NHS were added into the above UCNP/PC_{70} solution and stirred for 2 h. Then, folate–NH–PEG3000–NH_2 (500 mg) was added into the above solution, followed by vigorous stirring for 24 h. The nanoparticles were dialyzed by deionized water for 1 week.

Synthesis of PC_{70}

In brief, terephthalaldehydic acid methyl ester, pyridine-4-aldehyde and pyrrole were reacted in propionic acid for 1.5 h, and the resultant mixture was separated through a silica gel. Then $LiAlH_4$ was used to reduce the ester group to primary hydroxyl group at room temperature. Subsequently a Swern oxidation reaction was applied to obtain aldehyde group, the final product was purified by silica gel column chromatography. After that, Prato reaction was carried out to obtain crude product which was subjected to silica gel column chromatography. The obtained final product was mixed with methyl tosylate and refluxed in argon atmosphere for 1 h and then passed through an anion-exchange resin repeatedly to yield PC_{70} as a chloride salt.

Loading capacities and release behaviors PC_{70}

A gradient concentration (4, 8, 12, 16, 20 and 25%) of PC_{70} in UCNPs solutions was centrifuged at 5000 r.p.m. min^{-1} for 30 min. PC_{70} loading content was calculated according to the following formula: PC_{70} loading content=weight of PC_{70}/(weight of PC_{70}+weight of UCNPs). The release behavior of PC_{70} from the nanocomposites was evaluated in 10% fetal bovine serum. The resulting solution was filtrated through ultrafiltration ($M.W._{cut\ off}$=10 000) and the filtrate was detected by ultraviolet absorbance spectra.

Detection of singlet oxygen from UCNP–PEG–FA/PC_{70}

Water soluble disodium 9,10-anthracendipropionic acid (Na_2–ADPA) was used as the $^1O_2$ trapping agent. In the experiments, 100 μl of Na_2–ADPA solution (1 mg ml^{-1}) was added into 1.5 ml of UCNP–PEG–FA/PC_{70} solution and then exposed to 980 laser irradiation at a power density of 1 W cm^{-2}. To eliminate the inner-filter effect, the maxima absorption of UCNP–PEG–FA/PC_{70} was adjusted to about 0.2 OD. The absorption of Na_2–ADPA from 380 to 420 nm was recorded at different irradiation times.

***In vitro* PDT**

The photo-cytotoxicity of UCNP–PEG–FA/PC_{70} *in vitro* was carried out against luciferase-expression cervical cancer (HeLa-*luc*). Hela-*luc* cells were seeded in a 96-well plate with density of ca. 5×10^4 per well and cultured in 5% CO_2 at 37 °C for 24 h. Hela-*luc* cells were then treated with UCNP–PEG–FA/PC_{70} in different concentration gradients for 3 h in the dark, and then irradiated with a 980 nm laser for 10 min (0.8 W cm^{-2}, 1 min break after 1 min irradiation) under air condition and nitrogen condition, respectively. Afterward, the aged medium was replaced with the fresh medium, and cells were cultivated for 24 h in the dark at 37 °C. Cytotoxicity was evaluated by a WST-8 assay with a Cell Counting Kit-8 (CCK-8; DOJINDO, Kumamoto, Japan). A 96-well plate reader (iMarkmicroplate reader, Bio-RAD, Hercules, CA, USA) was used to determine cell viability. Hela-*luc* cells treated only with UCNP–PEG–FA/PC_{70} in the dark for 24 h were used as control. As for other controls, three kinds of cells including Hela-*luc* cells, A549 and HBE cells were treated with either UCNP–PEG/PC_{70} (156 μg ml^{-1}) or PC_{70} (40 μg ml^{-1}, 23.8 μM) and UCNP (80 μg ml^{-1}). Cell viability was then detected. To further confirm the targeting property of UCNP–PEG–FA/PC_{70}, folate blocking experiment was done by pre-incubated Hela-*luc*, A549 and HBE cells with folate (40 μg ml^{-1}) for 0.5 h. The cells were further incubated with UCNP–PEG–FA/PC_{70}(200 μg ml^{-1}) for 3 h. Afterward, the cell viability was detected by CCK-8. Corresponding results are expressed as mean±s.d.

Cell imagings

Hela-*luc* cells were incubated with UCNP–PEG–FA/PC_{70} (200 μg ml^{-1}) at 37 °C, 5%CO_2 for 3 h. The aged cell culture was replaced with the fresh Dulbecco's modified Eagle's medium (without phenol red). Subsequently, the cells were illuminated by an external 980 nm (0.8 W cm^{-2}, 1 min break after 1 min irradiation) laser for 10 min. Either Hoechst 33258 (1 mg ml^{-1}, 5 μl, incubated with cells for 20 min), PI (1 mg ml^{-1}, 5 μl, incubated with cells for 15 min), Dil (1 mg ml^{-1}, 5 μl, incubated with cells for 20 min), MitoRed (1 mg ml^{-1}, 5 μl, incubated with cells for 30 min), or Lyso (1 mg ml^{-1}, 5 μl, incubated with cells for 45 min) was then added according to the actual conditions. The cells were washed with icy phosphate-buffered saline buffer for three times before observation. The images were obtained by FV 1000-IX81 confocal laser scanning microscope (Olympus, Miyazaki, Japan). The cells were dyed without laser irradiation as a control.

In vivo PDT

Female Balb/c nude mice (16–20 g) were purchased from Center for Experimental Animals, Institute of Process Engineering, Chinese Academy of Science (Beijing, China). *In vivo* study conforms to the guidelines of The National Regulation of China for Care and Use of Laboratory Animals. The tumor model was established by subcutaneous injection of Hela/*luc* cells in the right buttock of each mouse. The tumors were allowed to grow for around 5 days to reach the size of ca. 30–60 mm^3.

The tumor-bearing mice were randomized into six groups (n=3, each group). Group 1 was treated by intravenous injection with UCNP–PEG–FA/PC_{70} at a dosage of 200 μl (2 mg ml^{-1}). After 24 h post-injection, the tumor sites were irradiated with 980 nm NIR light for 3 min (0.7 W cm^{-2}, 30 s break after 30 s irradiation). The other three groups were intravenously injected with UCNP–PEG–FA (1.6 mg ml^{-1}, 200 μl), PC_{70} (0.4 mg ml^{-1}, 200 μl), and saline (200 μl), respectively, and followed the same procedure as controls. The fifth and sixth groups were treated with UCNP–PEG–FA/PC_{70} (2 mg ml^{-1}, 200 μl) and saline (200 μl) without laser irradiation as another control. All the groups were treated with only one injection. The body weights and tumor volumes were monitored every 2 days after treatment. The tumor volume was calculated using the following equation: tumor volume (V)=length × $width^2/2$. Relative tumor volume was calculated as V/V_0 (V_0 was the corresponding tumor volume when the treatment was initiated).

Trimodal FL/UCL/MR imaging *in vivo*

Fluorescence imaging was performed on an *in vivo* imaging system (CRI Maestro 2) with a 600 nm laser as the excited source. The fluorescence images were corrected at 560–800 nm and analyzed with Maestro2 software. The Balb/c nude mice were anesthetized with 1.5% pentobarbital sodium by intraperitoneal injection at dosage of 100 μl. UCNP–PEG–FA/PC_{70} nanoparticles (200 μl, 2 mg ml^{-1}) were injected by intravenous injection, and fluorescence imaging was performed at timed intervals (0, 2, 4, 8, 24 and 72 h).

In vivo UCL imaging experiments were performed with a CALIPER Lumina II *in vivo* imaging system with an adjustable 980 nm laser as the excited source for *in vivo* imaging. Before imaging, the Balb/c nude mice were anesthetized with 1.5%pentobarbital sodium by intraperitoneal injection at dosage of 100 μl. UCNP–PEG–FA/PC_{70} nanocomposites (200 μl, 2 mg ml^{-1}) were injected by intravenous injection, and UCL imaging was performed after 24 h post-injection with exposure time of 0.3 s.

The *in vitro* MR imaging experiments were performed in a 0.5 T MRI magnet (NM120-Analyst, Shanghai Niumag Corporation). UCNP–PEG–FA/PC_{70}nanocomposites were dispersed in water at various Gd concentrations (by ICP measurement). T_1 was acquired using an inversion recovery sequence. Finally, the r_1 value was determined through the curve fitting of $1/T_1$ relaxation time (s^{-1}) versus the Gd concentration (mM).

In vivo MR imaging experiments were performed on a 7 T MRI scanner (Micro MRI, Varian 7T). The tumor-bearing mouse was anesthetized with isoflurane. The mouse was scanned before and after 24 h intravenous injection of UCNP–PEG–FA/PC_{70} nanoparticles (200 µl, 4 mg ml^{-1}).

Histopathological examination

The tissues (heart, liver, spleen, kidneys and lung) were harvested and fixed in a 4% formalin solution. The histopathological tests were performed according to standard laboratory procedures. Tissue samples were numbered and given blind to the pathologist for conventional processing and analysis. Briefly, the tissue samples were embedded in paraffin blocks, sectioned into 5 µm slices, and mounted onto glass slides. The sections were observed after hematoxylin and eosin staining, and photographs were taken using an optical microscope.

Results and Discussion

Synthesis and characterization of UCNP–PEG–FA/PC_{70}

PC_{70} dyad was synthesized according to our previously reported paper.[25](Supplementary Scheme S1, step 1). To obtain prominent UV–visible emission, $NaGdF_4$:Yb,Tm@$NaGdF_4$ core/shell nanoparticles were prepared as the upconversion energy donor. Moreover, Gd^{3+} ions-based host lattice was employed to support the MRI capability by shortening the relaxtion time of T_1. The synthesis of the core/shell UCNPs was enabled with a modified thermolysis method (Supplementary Scheme S1, step 2).[39, 40]

As given in Supplementary Figure S1a, the core and core/shell UCNPs were sphere-like, with high uniformity in size and morphology. After epitaxial growth of the nanoshells (Figure 2a), the mean particle size increased by ~20 nm. Furthermore, the high-resolution TEM image of core/cell nanoparticles (Figure 2b) exhibited legible lattice fringes of 0.52 nm, indicating the $(10\bar{1}0)$ plane of β-$NaGdF_4$, whereas the successful formation of β-phase UCNPs was also confirmed by the X-ray diffraction patterns (Figure 2d). The upconversion

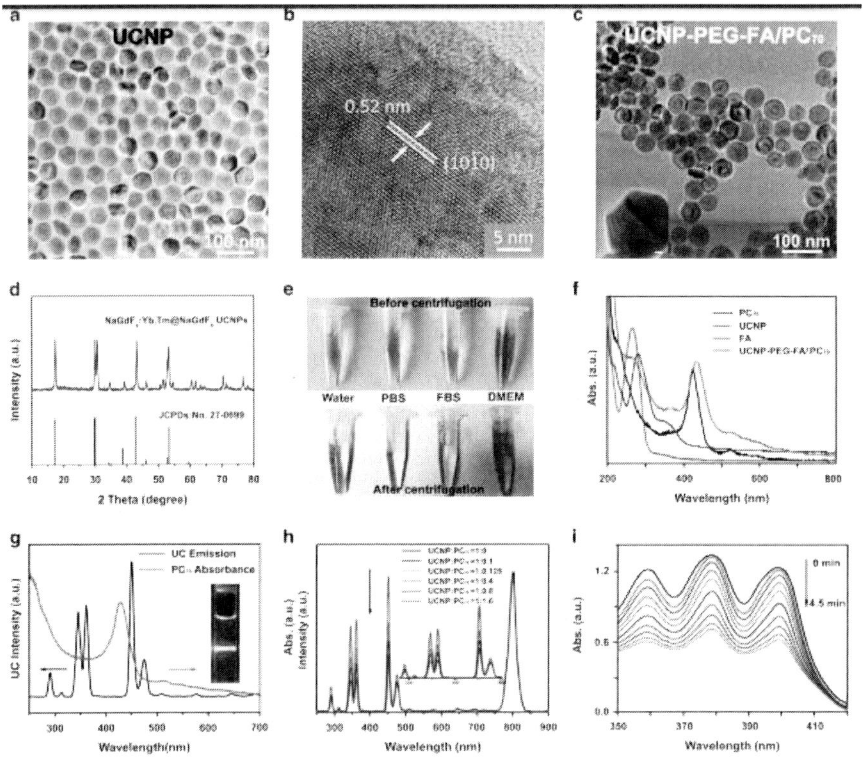

Figure 2. (a) Transmission electron microscope (TEM) image of β–$NaGdF_4$:Yb,Tm@$NaGdF_4$ nanoparticles. (b) High-resolution TEM image of β–$NaGdF_4$:Yb,Tm@$NaGdF_4$ nanoparticle. (c) TEM images of UCNP–PEG–FA/PC_{70} (upconversion–nanoparticles–trismethylpyridylporphyrin–fullerene nanocomposite). Inset is enlarged image and the scale bar is 5 nm. (d) X-ray diffraction pattern of as-prepared β-$NaGdF_4$:Yb,Tm@$NaGdF_4$ nanoparticles, consistent with the standard pattern of β-$NaGdF_4$ (JCPDS 06-0342). (e) Optical photographs of UCNP–PEG–FA/PC_{70} in water, phosphate-buffered saline, fetal bovine serum and Dulbecco's modified Eagle's medium respectively: before centrifugation (top) and after centrifugation (bottom). (f) Absorption spectra of UCNP–PEG–FA/PC_{70}, UCNPs, FA and PC_{70}. (g) Upconversion emission spectrum of UCNPs and absorption spectrum of PC_{70}. Inset is the digital luminescence photograph of UCNPs. (h) Upconversion emission spectra of UCNP–PEG–FA solutions treated with PC_{70} at different concentrations upon 980 nm irradiation (excitation power was 4 W cm^{-2}). The concentrations of UCNPs in samples are identical. Inset is a local amplification profile of spectra. (i) Photodegradation of Na_2-ADPA treated with UCNP–PEG–FA/PC_{70} at different time points due to generation of singlet oxygen upon 980 nm irradiation.

spectra inSupplementary Figure S2 show that the emission intensity of $NaGdF_4$:Yb,Tm@$NaGdF_4$ core/shell UCNPs enhanced 13 times compared with the $NaGdF_4$:Yb,Tm counterpart, indicating the higher upconversion efficiency of the core/shell UCNPs. Afterward, a ligand exchange strategy was employed to transfer the hydrophobic oleate-capped UCNPs to aqueous solutions, where the oleate ligands were replaced by citric acid molecules. As shown inSupplementary Figure S3, no obvious change in size and morphology of the UCNPs occurred after the ligand exchange process. Meanwhile, the emission intensity of Tm^{3+} decreased due to the quenching effect induced by the vibration stretching of hydroxyl groups (Supplementary Figure S4).

Subsequently, water-dispersible and carboxyl-functionalized UCNPs were modified by PC_{70} via electrostatic interactions and then encapsulated by folate-modified polyoxyethylenebis (amine). TEM image shows little change in size of the UCNPs after modification by PEG and PC_{70} (Figure 2c), which may be caused by light atoms with lower constrast compared with that of heavy atoms. Whereas the dynamic light scattering distribution indicates that the hydrated ionic radius is ca. 186 nm, which may ascribe to the free stretching of PEG (Supplementary Figure S5). Further investigation illustrates that the as-prepared UCNP–PEG–FA/PC_{70} nanocomposites are reasonably stable in phosphate-buffered saline, fetal bovine serum and Dulbecco's modified Eagle's medium (Figure 2e), which should mainly attribute to the negative charged property (−15 mV) of UCNP–PEG–FA/PC_{70}. The moderate dispersibility facilitates further biomedical applications. The IR spectrum indirectly verifies the formation of amido bond (Supplementary Figure S6). Loading efficiency of PC_{70} was measured as shown in Supplementary Figure S7a, and the result suggested that the loading efficiency is ca. 20%. Release behavior of PC_{70} was also evaluated in 10% fetal bovine serum (Supplementary Figure S7b). PC_{70} could release from the nanocomposite with an amount of ca. 5–7% indicating that the multifunctional composite is reasonably stable under physiological conditions.

Figure 2g reveals that the absorption spectrum of PC_{70} overlaps with the UV emission of UCNPs at 290 nm ($^1I_6$–$^3H_6$), 345 nm ($^1I_6$–$^3F_4$), 361 nm ($^1D_2$–$^3H_6$), and the visible emission at 451 nm ($^1D_2$–$^3F_4$), 475 nm ($^1G_4$–$^3H_6$), indicating that the emissions of UCNPs matched well with the absorption of PC_{70}. Hence, PC_{70} can be activated by the energy transfer from UCNPs upon NIR irradiation. To further verify the efficient energy transfer from the UCNPs to PC_{70}, the emission spectra of UCNP–PEG–FA nanoparticles incubated with different concentration of PC_{70}were monitored upon 980 nm irradiation. As depicted in Figure 2h, the UV–visible emissions of Tm^{3+} were progressively quenched with elevated addition of PC_{70}, whereas the emission of Tm^{3+} at 800 nm ($^3H_4$–$^3H_6$) was almost constant, indicating that the energy transfer indeed occurred between UCNPs and PC_{70} due to the overlap of emission and absorption spectra in UV–visible range. As a result, a slight red-shift of the UV–visible absorption at 428 nm (Figure 2f) and an obvious red-shift of the typical FL spectrum of PC_{70} (Supplementary Figure S8) were observed for UCNP–PEG–FA/PC_{70} due to the interaction between UCNPs and PC_{70}. These results indirectly indicate that fluorescence resonance energy transfer may exist between UCNPs and PC_{70}. Subsequently, the feasibility of UCNP–PEG–FA/PC_{70} for producing $^1O_2$ was investigated by using chemical trapping method (Figure 2i). The result indicates that the photodegradation of Na_2–ADPA is time-dependent once treated with UCNP–PEG–FA/PC_{70} and 980 nm irradiation. All of these results indicate that the as-synthesized UCNP–PEG–FA/PC_{70} nanocomposites not only exhibit reasonable stability under physiological conditions but also efficiently produce singlet oxygen upon 980 nm irradiation.

The photodynamic activity *in vitro*

To investigate the NIR light-triggered PDT effect of UCNP–PEG–FA/PC_{70}, the *in vitro* viability assay of Hela-*luc* cells was performed under normal conditions and N_2-puring conditions (simulate a low oxygen condition) as a model. As given inFigure 3a, the cell viabilities decrease gradually with the increase of nanoparticles from 0 to 200 μg ml^{-1} upon 980 nm laser irradiation at a power density of 0.8 W cm^{-2} (480 J cm^{-2}, lower than that of previously reported: 4.5 W cm^{-2}, 10 800 J cm^{-2}),[8] and 90% inhibitory effect on the cell viability was detected at 0.15 mg ml^{-1} of UCNP–PEG–FA/PC_{70}. Meanwhile, negligible cytotoxicity against the Hela-*luc* cell was observed in the dark, indicating the favorable biocompatibility of UCNP–PEG–FA/PC_{70} nanocomposites. The same experiments were done under the N_2-puring conditions. As shown in Figure 3b, the cell viabilities decrease gradually with the increase of UCNP–PEG–FA/$PC_{70}$from 0 to 200 μg ml^{-1} under N_2-puring condition, and 50% inhibitory effect on the cell viability was detected at 0.2 mg ml^{-1}. The results indicate that UCNP–PEG–FA/PC_{70} can produce $^1O_2$ under low-oxygen conditions, which may be attributable to the formation of $(^3PC_{70})^*$ with a long triplet lifetime state and thus have enough time to interact with free oxygen.[38]

Several subcellular organelles were stained by related probes and observed under confocal microscope to confirm that the cells were destructed after incubation with UCNP–PEG–FA/PC_{70} and subsequent irradiation with 980-nm NIR light. As shown in Figure 3c, the images in the first row were stained by Dil (cell membrane dye) and Hoechst 33258 (nucleus dye), from which we can see clearly that the cell membrane was bubbling and the nucleus was stained in blue after 980-nm irradiation, indicating that the cell membrane was destroyed and the cells died. The second row was stained by MitoRed tracker

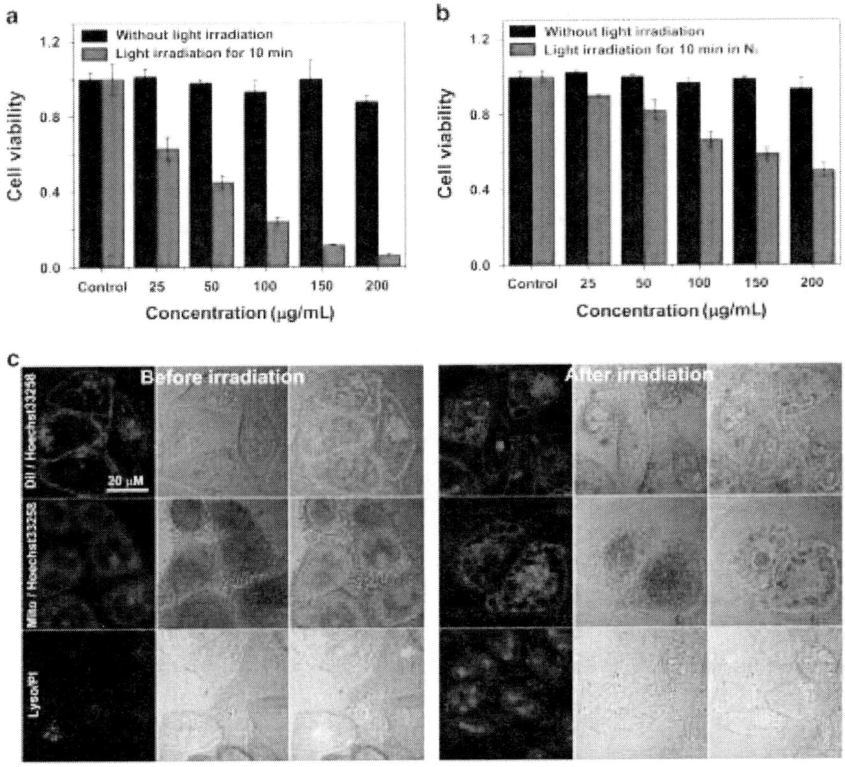

Figure 3. (a) *In vitro* relative viabilities of Hela-*luc* cells incubated with UCNP–PEG–FA/PC_{70} (upconversion–nanoparticles–trismethylpyridylporphyrin–fullerene nanocomposite) at different concentrations, with (red) and without (black) 980 nm NIR light irradiation. (b) Relative viabilities of Hela-*luc* cells under nitrogen-purging conditions incubated with UCNP–PEG–FA/PC_{70} at different concentrations, with (red) and without (black) 980 nm NIR light irradiation. (c) Confocal images before (left) and after (right) NIR light irradiation. First line was stained with DiI and Hoechst 33 258, second line was stained with MitoRed and Hoechst 33258, third line was stained with PI and Lyso. Each group with three columns represent fluorescence image, optical image and merged image, respectively, from left to right.

(mitochondrial dye) and Hoechst 33258, and the third row was stained by Lyso tracker (lysosome dye) and PI (nucleus dye). The compact mitochondria and lysosomes become incomplete after irradiation, indicating that the subcellular organelles were destroyed by $^1O_2$ generated from UCNP–PEG–FA/PC_{70} after 980 nm irradiation. The lightened subcellular organelles indirectly illustrated that the nanocomposites were internalized into cells and produced intracellular ROS to destroy the subcellular organelles when exposed to NIR light irradiation.

Improved photodynamic activity by conjugation with folate

To reduce the photodynamic damage of synthesized nanoparticles to normal cells and enhance the PDT efficiency toward tumor cells, folate-modified UCNP–PEG–FA/PC_{70} was designed. As shown in Figure 4a, UCNP–PEG–FA/PC_{70} hybrid exhibits a dramatic decrease in cell viability even at low concentration of 0.2 mg ml^{-1}, whereas UCNPs or PC_{70} alone showed a negligible effect under the same conditions. Moreover, A549 and HBE cells with negative expression of folate receptor were used as controls to evaluate the PDT efficiency. The results indicate that all the above materials including UCNP–PEG–FA/PC_{70} show negligible cytotoxicity toward A549 and HBE cells, suggesting that the superior PDT effect should benefit from the higher binding affinity of the folate-modified nanoprobe to folate receptor overexpressed Hela-*luc* cells.

Competitive inhibition experiment was carried out to further confirm the targeting specificity of UCNP–PEG–FA/PC_{70}. UCNP–PEG/PC_{70} (without folate) and UCNP–PEG–FA/PC_{70} blocking with an excess of folate during the incubation were used in PDT treatment as controls. As shown in Figure 4a, UCNP–PEG/PC_{70}exhibited negligible effect on the above three kinds of cells, suggesting that UCNP–PEG/PC_{70} fails to effectively attach to the cell surface at absence of folate. Moreover, the blocking group of UCNP–PEG–FA/PC_{70} reveals that the existence of free folate could inhibit the affinity of UCNP–PEG–FA/PC_{70} to Hela-*luc* cells and therefore effectively preserve the cells from PDT damage (Figure 4b). In addition, confocal imaging experiment was explored to further confirm the above results. Strong red fluorescence could be observed when the Hela-*luc* cells were treated with UCNP–PEG–FA/PC_{70} and subsequently exposed to 980-nm irradiation and stained by PI. Meanwhile, controls groups (UCNP–PEG–FA/PC_{70}-treated A549 cells and UCNP–PEG/PC_{70}-, UCNPs- and PC_{70}-treated A549 cells, as well as HBE cells and Hela-*luc* cells) exhibited negligible fluorescence under the same conditions (Figure 4c). These results highlight the targeting capability and specificity of UCNP–PEG–FA/PC_{70} nanocomposites toward Hela-*luc* cells via the recognition between folate and folate receptor on the Hela-*luc* cell surface.

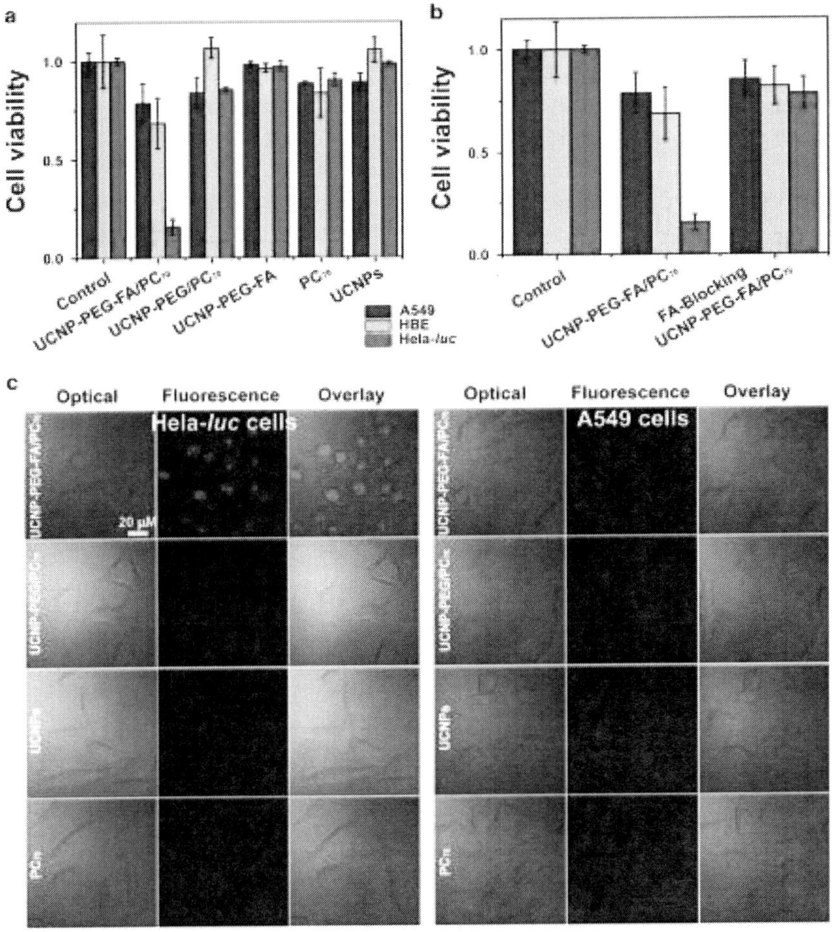

Figure 4. (a) The viabilities of three kinds of cells (Hela-*luc*, A549 and HBE cells) incubated with UCNP–PEG–FA/PC_{70} (upconversion-nanoparticles-trismethylpyridylporphyrin-fullerene nanocomposite), UCNP–PEG/PC_{70}, UCNP–PEG–FA, PC_{70} and UCNPs respectively. (b) The viabilities of three kinds of cells (Hela-*luc*, A549 and HBE cells) incubated with either UCNP–PEG–FA/PC_{70} or the mixture of UCNP–PEG–FA/PC_{70} and an excess of folate. (c) Confocal images of PI stained Hela-*luc* cells (left) and A549 cells (right) after photodynamic therapy treatment with UCNP–PEG–FA/PC_{70}, UCNP–PEG/PC_{70}, UCNP and PC_{70} from top to bottom, respectively.

Fluorescence/UCL/MR/trimodal imaging

Despite the promising results obtained *in vitro*, one of the most demanding challenges regarding multifunctional platforms still lies in good sensitivity *in vivo*and real-time monitoring for optimized treatment. Herein, the fluorescence intensity *in vivo* was detected to evaluate the real-time biodistribution of UCNP–PEG–FA/PC_{70} nanoparticles. For *in vivo* optical imaging studies, tumor-bearing nude mice were injected with UCNP–PEG–FA/PC_{70} through the tail vein (20 mg kg^{-1}). The resulting fluorescence avoids the autofluorescence from biological tissues and could be used to image the specifics sensitively *in vivo*. From the chronological observation of the images in Figure 5a, we detected the appearance of a rapid tropism associated with the biodistribution of UCNP–PEG–FA/PC_{70} nanoparticles. The fluorescence intensity of UCNP–PEG–FA/PC_{70} in tumor increases to the maximum after 24 h post injection and then decreases with the prolonging time. At the time point of 24 h, the tumor and other organs of the nude mouse injected with UCNP–PEG–FA/PC_{70} were harvested for *ex vivo* imaging to quantify tissue/organ distribution (Figure 5b). The result revealed that distinct fluorescence was presented in tumor site and also in the liver and kidney. By contrast, as given in Supplementary Figure S9, the mouse treated with UCNP–PEG/PC_{70} showed negligible fluorescence after 4 h post-injection. The *ex vivo*imaging obtained after 24 h post injection (Supplementary Figure S10) indicated the maximal accumulation in the liver and kidney and only a weak fringe of fluorescence located in tumor sites. These results demonstrate that the active targeting of folic and enhanced permeability and retention effect mediated passive targeting play a synergistic role for the accumulation of UCNP–PEG–FA/PC_{70} nanocomposites in tumor. Most interestingly, the real-time fluorescence images not only confirm the feasibility of fluorescence-guided PDT but also contribute to the enhanced efficiency of PDT.

In addition, the emission of UCNPs at 800 nm could be used as the output signal to realize the NIR-to-NIR UCL imaging. The images from left to right in Figure 5crepresent the optical image of mouse, UCL image, and their merged images at 24 h post injection, which depict the tumor clearly. The *ex vivo* UCL imaging (Figure 5d) further indicated that UCNP–PEG–FA/PC_{70} mainly accumulated in the liver, spleen and tumor, which were similar to the result from fluorescence image.

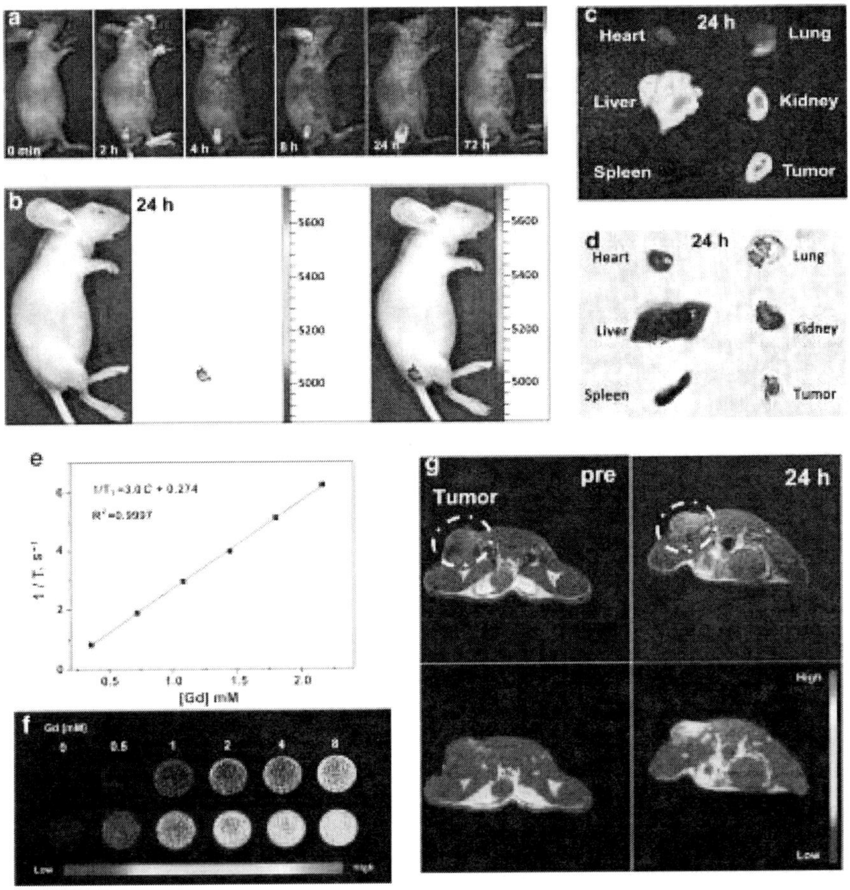

Figure 5. (a) *In vivo* fluorescence images of a tumor-bearing mouse at different time points after treatment with UCNP–PEG–FA/PC_{70} (upconversion–nanoparticles–trismethylpyridylporphyrin–fullerene nanocomposite). (b) UCL images at the time point of 24 h, and images from left to right represent optical image, UCL image, and merged image. *Ex vivo* images of tumor and other organs after 24 h post-injection: (c) fluorescence image and (d) UCL image. (e) Relaxation rate R_1 versus different molar concentrations of UCNP–PEG–FA/PC_{70} nanoparticles. (f) *In vitro* T_1-weighted MR images of UCNP–PEG–FA/PC_{70} at different Gd concentrations. (g) *In vivo* T_1-weighted MR image of a tumor-bearing nude mouse: preinjection (left) and after 24-h post injection (right). The color images are treated by pseudo-color.

To demonstrate the MRI capability of UCNP–PEG–FA/PC_{70}, T_1 relaxation time andT_1-weighted MR images at different Gd concentrations were obtained with a 0.5 T MRI scanner. The measured r_1 value was ca. 3.0 $mM^{-1} s^{-1}$ for UCNP–PEG–FA/PC_{70}nanocomposites (Figure 5e), which was comparable to that of clinically used Gd-DTPA (3–4 $mM^{-1} s^{-1}$) under the same conditions. T_1-weighted MR images (Figure 5f) enhanced with the increase of Gd concentration from 0 to 8 mM. For better visual effect, pseudocolor image was used to enhance the sensitivity. *In vivo* MR imaging was performed for a tumor-bearing mouse with a 7 T MR scanner (Figure 5g). The tumor site was clearly depicted after 24-h post injection, indicating that UCNP–PEG–FA/PC_{70} can be used as a potential T_1 MRI contrast agent. The results from MR image are reasonably consistent with those from fluorescence and UCL images, which also indicate the better tumor targeting performance of UCNP–PEG–FA/PC_{70} as an entity. In a word, the obtained results indicate that UCNP–PEG–FA/PC_{70} could be employed as an efficient and trimodal FL/UCL/MR imaging agent *in vivo* and direct the optimized PDT treatment.

***In vivo* PDT**

The PDT experiments were conducted *in vivo* to evaluate the tumor inhibition efficacy of the UCNP–PEG–FA/PC_{70} with 980 nm NIR light irradiation. The luciferase-expressing cervical cancer cell line Hela-*luc* was selected as the xenograft model, and tumor-bearing nude mice were randomly divided into six groups (n=3). The first group was irradiated with 980 nm NIR light (0.7 W cm^{-2}, 126 J cm^{-2}, 30 s break after 30 s irradiation) in the tumor site for 3 min (lower than that of previously reported: 2.5 W cm^{-2}, 30 min, 4500 J cm^{-2})[8] after injection with UCNP–PEG–FA/PC_{70} (2 mg ml^{-1}, 200 μl) for 24 h through caudal vein. In addition, short-interval irradiation method was developed to avoid over-heating by the NIR light. The next three groups were injected with UCNP–PEG–FA, PC_{70}, and saline, respectively, and then irradiated under the same conditions as controls. Another two groups injected with either UCNP–PEG–FA/PC_{70} or saline separately without irradiation were used as controls. Figure 6a demonstrates that the group treated with UCNP–PEG–FA/PC_{70}+ 980 nm NIR light irradiation exhibited optimal tumor growth inhibition efficacy. By contrast, the tumors of control groups grew continuously under the same conditions. Meanwhile, the digital photographs of representative mice with neoplasm (Supplementary Figure S11) before and after irradiation for 6 days showed that the tumor sizes of mice treated with UCNP–PEG–FA/PC_{70}+980 nm NIR light irradiation were much smaller than those of control groups, as confirmed by the representative bioluminescence images of mice (Supplementary Figure S12) given at the 14th day (Figure 6c). Bioluminescence from the UCNP–PEG–FA/PC_{70}+980 nm NIR light irradiation-treated group nearly disappeared compared with saline-treated mice, suggesting that the tumor of treated group was effectively destroyed. Hematoxylin and eosin staining images of main organs and tumor are presented in Figure 6d. The clear separation of the

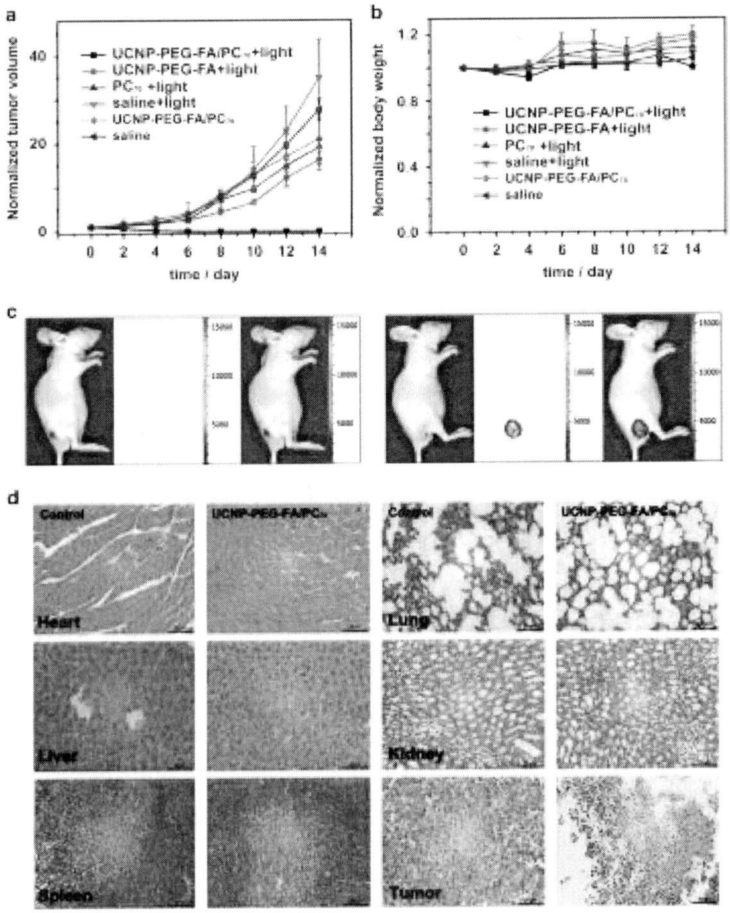

Figure 6. (a) Growth of tumors after treatments. The relative tumor volumes were normalized to their initial sizes. (b) The relative body weights of the nude mice increased continuously after different treatments. Error bars indicate standard deviations; *n*=3. (c) Bioluminescence images of different treatments after 14 days: UCNP–PEG–FA/PC_{70}+980 nm NIR light irradiation (left) and saline+980 NIR light irradiation (right). (d) Hematoxylin and eosin stained images of tumors and other organs under different photodynamic therapy treatments.

nuclear and cytoplasm confirmed that the tumor cells were destroyed after treated with UCNP–PEG–FA/PC_{70}+980 nm NIR light irradiation. By contrast, tumor from the mouse treated with saline exhibited cramped construction of nuclear and cytoplasm. Importantly, there were no obvious lesions were found in other organs, indicating that UCNP–PEG–FA/PC_{70}has no significant toxicity to organisms. In addition, body weight, as an important parameter to evaluate the systemic toxicity of nanomaterials, did not show abnormal change for all groups (Figure 6b), indicating that UCNP–PEG–FA/PC_{70} nanocomposite does not have obvious adverse side effects.

CONCLUSION

In summary, we have successfully prepared UCNP–PEG–FA/PC_{70} nanocomposite as an NIR light-triggered and targeted theranostic platform for trimodal imaging-guided PDT *in vitro* as well as *in vivo*. The developed UCNP–PEG–FA/PC_{70}nanocomposites have the following advantages: (1) Upon NIR irradiation, UCNPs could convert NIR light into UV-to-visible light to effectively activate PC_{70} to produce $^1O_2$ for killing tumor under low oxygen conditions; (2) Three kinds of imaging methods (FL/UCL/MR) were integrated into one system to provide the complementary information for accurate diagnosis and guide treatment of tumor; (3) Folate-mediated active targeting together with the PEG leaded enhanced permeability and retention effect effect (passive targeting) enhances the accumulation of multifunctional nanoparticles in tumor. Importantly, the synthesized UCNP–PEG–FA/PC_{70} may provide solutions to the two bottleneck problems of PDT, namely, limited penetration depth and oxygen-deficient microenvironment. These results highlight the potential of UCNP–PEG–FA/PC_{70} as a multifunctional theranostic agent for imaging-guided PDT of tumor and a promising candidate for overcoming the problem of existing PDT agents in deep cancerous tissue under hypoxic conditions.

ACKNOWLEDGEMENTS

This work is supported by 973 Program (No. 2011CB302100), the Ministry of science and technology support program (No. 2011BAK15B04), the Chinese Academy of Sciences (No. XDA09030302), the National Natural Science Foundation of China (Nos. 51372251, 51472252) and the Key Research Program of the Chinese Academy of Sciences (Grant no. KGZDEW-T02).

AUTHOR CONTRIBUTIONS

CS and JG planned and supervised the project. MG carried out the detailed experiments. MG, JG and CS analyzed the data and wrote the whole paper. HD contributed to the synthesis and characterization of UCNPs and helped the analysis of the data. DC, LS, CY, CW and PW contributed to discussion of the paper.

CONFLICT OF INTEREST

The authors declare no conflict of interest.

REFERENCES

1. He, Q. & Shi, J. MSN anti-cancer nanomedicines: chemotherapy enhancement, overcoming of drug resistance, and metastasis inhibition.*Adv. Mater.* **26**, 391–411 (2014).

2. Chen, X., Gambhir, S. S. & Cheon, J. Theranostic nanomedicine. *Acc. Chem. Res.* **44**, 841 (2011).

3. Wang, J., Hu, Z., Xu, J. & Zhao, Y. Therapeutic applications of low-toxicity spherical nanocarbon materials. *NPG Asia Mater.* **6**, e84 (2014).

4. Tian, B., Wang, C., Zhang, S., Feng, L. & Liu, Z. Photothermally enhanced photodynamic therapy delivered by nano-graphene oxide. *ACS Nano* **5**, 7000–7009 (2011).

5. Ge, J., Lan, M., Zhou, B., Liu, W., Guo, L., Wang, H., Jia, Q., Niu, G., Huang, X., Zhou, H., Meng, X., Wang, P., Lee, C. S., Zhang, W. & Han, X. A graphene quantum dot photodynamic therapy agent with high singlet oxygen generation. *Nat. Commun.* **5**, 4596–4603 (2014).

6. Chen, D., Wang, C., Nie, X., Li, S., Li, R., Guan, M., Liu, Z., Chen, C., Wang, C., Shu, C. & Wan, L. Photoacoustic imaging guided near-infrared photothermal therapy using highly water-dispersible single-walled carbon nanohorns as theranostic agents. *Adv. Funct. Mater.* **24**, 6621–6628 (2014).

7. Liang, C., Diao, S., Wang, C., Gong, H., Liu, T., Hong, G., Shi, X., Dai, H. & Liu, Z. Tumor metastasis inhibition by imaging-guided photothermal therapy with single-walled carbon nanotubes. *Adv. Mater.* **26**, 5646–5652 (2014).

8. Dai, Y., Xiao, H., Liu, J., Yuan, Q., Ma, P., Yang, D., Li, C., Cheng, Z., Hou, Z., Yang, P. & Lin, J. *In vivo* multimodality imaging and cancer therapy by near-infrared light-triggered trans-platinum pro-drug-conjugated

upconverison nanoparticles. *J. Am. Chem. Soc.* **135**, 18920–18929 (2013).

9. Zhu, H., Chen, H., Zeng, X., Wang, Z., Zhang, X., Wu, Y., Gao, Y., Zhang, J., Liu, K., Liu, R., Cai, L., Mei, L. & Feng, S. S. Co-delivery of chemotherapeutic drugs with vitamin E TPGS by porous PLGA nanoparticles for enhanced chemotherapy against multi-drug resistance. *Biomaterials* **35**, 2391–2400 (2014).

10. Shieh, M. J., Peng, C. L., Lou, P. J., Chiu, C. H., Tsai, T. Y., Hsu, C. Y., Yeh, C. Y. & Lai, P. S. Non-toxic phototriggered gene transfection by PAMAM-porphyrin conjugates. *J. Control Release* **129**, 200–206 (2008).

11. Yuan, H., Chong, H., Wang, B., Zhu, C., Liu, L., Yang, Q., Lv, F. & Wang, S. Chemical molecule-induced light-activated system for anticancer and antifungal activities. *J. Am. Chem. Soc.* **134**, 13184–13187 (2012).

12. Li, J., Cai, P., Shalviri, A., Henderson, J. T., He, C., Foltz, W. D., Prasad, P., Brodersen, P. M., Chen, Y., DaCosta, R., Rauth, A. M. & Wu, X. Y. A multifunctional polymeric nanotheranostic system delivers doxorubicin and imaging agents across the blood-brain barrier targeting brain metastases of breast cancer. *ACS Nano* **8**, 9925–9940 (2014).

13. Huang, P., Wang, D., Su, Y., Huang, W., Zhou, Y., Cui, D., Zhu, X. & Yan, D. Combination of small molecule prodrug and nanodrug delivery: amphiphilic drug-drug conjugate for cancer therapy. *J. Am. Chem. Soc.* **136**, 11748–11756 (2014).

14. Zhang, L., Gao, S., Zhang, F., Yang, K., Ma, Q. & Zhu, L. Activatable hyaluronic acid nanoparticle as a theranostic agent for optical/photoacoustic image-guided photothermal therapy. *ACS Nano* **8**, 12250–12258 (2014).

15. Fan, Q., Cheng, K., Yang, Z., Zhang, R., Yang, M., Hu, X., Ma, X., Bu, L., Lu, X., Xiong, X., Huang, W., Zhao, H. & Cheng, Z. Perylene-diimide-based nanoparticles as highly efficient photoacoustic agents for deep brain tumor imaging in living mice. *Adv. Mater.* **27**, 843–847 (2015).

16. Vijayaraghavan, P., Liu, C. H., Vankayala, R., Chiang, C. S. & Hwang, K. C. Designing multi-branched gold nanoechinus for NIR light activated dual modal photodynamic and photothermal therapy in the second biological window. *Adv. Mater.* **26**, 6689–6695 (2014).

17. Yuan, H., Wilson, C. M., Xia, J., Doyle, S. L., Li, S., Fales, A. M., Liu, Y., Ozaki, E., Mulfaul, K., Hanna, G., Palmer, G. M., Wang, L. V. & Vo-Dinh,

T. Plasmonics-enhanced and optically modulated delivery of gold nanostars into brain tumor. *Nanoscale* **6**, 4078–4082 (2014).

18. Huang, P., Rong, P., Lin, J., Li, W., Yan, X., Zhang, M. G., Nie, L., Niu, G., Lu, J., Wang, W. & Chen, X Triphase interface synthesis of plasmonic gold bellflowers as near-infrared light mediated acoustic and thermal theranostics. *J. Am. Chem. Soc.* **136**, 8307–8313 (2014).

19. Wang, B., Wang, J. H., Liu, Q., Huang, H., Chen, M., Li, K., Li, C., Yu, X. F. & Chu, P. K. Rose-bengal-conjugated gold nanorods for *in vivo*photodynamic and photothermal oral cancer therapies. *Biomaterials* **35**, 1954–1966 (2014).

20. Wang, Y., Wang, H., Liu, D., Song, S., Wang, X. & Zhang, H. Graphene oxide covalently grafted upconversion nanoparticles for combined NIR mediated imaging and photothermal/photodynamic cancer therapy.*Biomaterials* **34**, 7715–7724 (2013).

21. Wang, L., Shi, J., Liu, R., Liu, Y., Zhang, J., Yu, X., Gao, J., Zhang, C. & Zhang, Z. Photodynamic effect of functionalized single-walled carbon nanotubes: a potential sensitizer for photodynamic therapy. *Nanoscale* **6**, 4642–4651 (2014).

22. Kalluru, P., Vankayala, R., Chiang, C. & Hwang, K. C. Photosensitization of singlet oxygen and *in vivo* photodynamic therapeutic effects mediated by PEGylated W(18)O(49) nanowires. *Angew. Chem. Int. Ed.* **52**, 12332–12336 (2013).

23. Cheng, L., Liu, J., Gu, X., Gong, H., Shi, X., Liu, T., Wang, C., Wang, X., Liu, G., Xing, H., Bu, W., Sun, B. & Liu, Z. PEGylated WS(2) nanosheets as a multifunctional theranostic agent for *in vivo* dual-modal CT/photoacoustic imaging guided photothermal therapy. *Adv. Mater.* **26**, 1886–1893 (2014).

24. Secret, E., Maynadier, M., Gallud, A., Chaix, A., Bouffard, E., Gary-Bobo, M., Marcotte, N., Mongin, O., Cheikh, K., El, Hugues, V., Auffan, M., Frochot, C., Morere, A., Maillard, P., Blanchard-Desce, M., Sailor, M. J., Garcia, M., Durand, J. O. & Cunin, F. Two-photon excitation of porphyrin-functionalized porous silicon nanoparticles for photodynamic therapy. *Adv. Mater.* **26**, 7643–7648 (2014).

25. Barreto, J. A., O'Malley, W., Kubeil, M., Graham, B., Stephan, H. & Spiccia, L. Nanomaterials: applications in cancer imaging and therapy. *Adv. Mater.***23**, H18–H40 (2011).

26. Maji, S. K., Sreejith, S., Joseph, J., Lin, M., He, T., Tong, Y., Sun, H., Yu, S. W. & Zhao, Y. Upconversion nanoparticles as a contrast agent for

photoacoustic imaging in live mice. *Adv. Mater.* **26**, 5633–5638 (2014).

27. Wang, F., Deng, R., Wang, J., Wang, Q., Han, Y., Zhu, H., Chen, X. & Liu, X. Tuning upconversion through energy migration in core-shell nanoparticles. *Nat. Mater.* **10**, 968–973 (2011).

28. Sun, L. D., Wang, Y. F. & Yan, C. H. Paradigms and challenges for bioapplication of rare earth upconversion luminescent nanoparticles: small size and tunable emission/excitation spectra. *Acc. Chem. Res.* **47**, 1001–1009 (2014).

29. Li, L. L., Zhang, R., Yin, L., Zheng, K., Qin, W., Selvin, P. R. & Lu, Y. Biomimetic surface engineering of lanthanide-doped upconversion nanoparticles as versatile bioprobes. *Angew. Chem. Int. Ed. Engl.* **51**, 6121–6125 (2012).

30. Feng, W., Zhu, X. & Li, F. Recent advances in the optimization and functionalization of upconversion nanomaterials for *in vivo* bioapplications.*NPG Asia Mater.* **5**, e84 (2013).

31. Qiao, X. F., Zhou, J. C., Xiao, J. W., Wang, Y. F., Sun, L. D. & Yan, C. H. Triple-functional core-shell structured upconversion luminescent nanoparticles covalently grafted with photosensitizer for luminescent, magnetic resonance imaging and photodynamic therapy *in vitro*. *Nanoscale***4**, 4611–4623 (2012).

32. Zhang, P., Steelant, W., Kumar, M. & Scholfield, M. Versatile photosensitizers for photodynamic therapy at infrared excitation. *J. Am. Chem. Soc.* **129**, 4526–4527 (2007).

33. Park, Y. I., Kim, H. M., Kim, J. H., Moon, K. C., Yoo, B., Lee, K. T., Lee, N., Choi, Y., Park, W., Ling, D., Na, K., Moon, W. K., Choi, S. H., Park, H. S., Yoon, S. Y., Suh, Y. D., Lee, S. H. & Hyeon, T. Theranostic probe based on lanthanide-doped nanoparticles for simultaneous *in vivo* dual-modal imaging and photodynamic therapy. *Adv. Mater.* **24**, 5755–5761 (2012).

34. Wang, C., Cheng, L., Liu, Y., Wang, X., Ma, X., Deng, Z., Li, Y. & Liu, Z. Imaging-guided pH-sensitive photodynamic therapy using charge reversible upconversion nanoparticles under near-infrared light. *Adv. Funct. Mater.***23**, 3077–3086 (2013).

35. Chen, Q., Wang, C., Cheng, L., He, W., Cheng, Z. & Liu, Z. Protein modified upconversion nanoparticles for imaging-guided combined photothermal and photodynamic therapy. *Biomaterials* **35**, 2915–2923 (2014).

36. Bratasz, A., Kulkarni, A. C. & Kuppusamy, P. A highly sensitive biocompatible spin probe for imaging of oxygen concentration in tissues.*Biophys. J.* **92**, 2918–2925 (2007).

37. Rieffel, J., Chen, F., Kim, J., Chen, G., Shao, W., Shao, S., Chitgupi, U., Hernandez, R., Graves, S. A., Nickles, R. J., Prasad, P. N., Kim, C., Cai, W. & Lovell, J. F. Hexamodal imaging with porphyrin-phospholipid-coated upconversion nanoparticles. *Adv. Mater.* **27**, 1785–1790 (2015).

38. Guan, M., Qin, T., Ge, J., Zhen, M., Xu, W., Chen, D., Li, S., Wang, C., Su, H. & Shu, C. Amphiphilic trismethylpyridylporphyrin-fullerene (C_{70} dyad: an efficient photosensitizer under hypoxia condition. *J. Mater. Chem. B* **3**, 776–783 (2015).

39. Mai, H., Zhang, Y., Si, R., Yan, Z., Sun, L., You, L. & Yan, C. High-quality sodium rare-earth fluoride nanocrystals: controlled synthesis and optical properties. *J. Am. Chem. Soc.* **128**, 6426–6436 (2006).

40. Yin, A., Zhang, Y., Sun, L. & Yan, C. Colloidal synthesis and blue based multicolor upconversion emissions of size and composition controlled monodisperse hexagonal $NaYF_4$:Yb,Tm nanocrystals. *Nanoscale* **2**, 953–959 (2010).

Chapter 10

One-Pot Solvothermal Synthesis of Highly Emissive, Sodium-Codoped, LaF_3 and $BaLaF_5$ Core-Shell Upconverting Nanocrystals

Joshua T. Stecher, Anne B. Rohlfing and Michael J. Therien *

Department of Chemistry, Duke University, French Family Science Center, 124 Science Drive, Durham, NC 27708, USA

ABSTRACT

We report a one-pot solvothermal synthesis of sub-10 nm, dominant ultraviolet (UV) emissive upconverting nanocrystals (UCNCs), based on sodium-codoped LaF_3 and $BaLaF_5$(0.5%Tm; 20%Yb) and their corresponding core@shell derivatives. Elemental analysis shows a Na-codopant in these crystal systems of ~20% the total cation content; X-ray diffraction (XRD) data indicate a shift in unit cell dimensions consistent with these small codopant ions. Similarly, X-ray photoelectron spectroscopic (XPS) analysis reveals primarily substitution of Na^+ for La^{3+} ions (97% of total Na^+ codopant) in the crystal system, and interstitial Na^+ (3% of detected Na^+) and La^{3+} (3% of detected La^{3+}) present in (Na)LaF_3 and only direct substitution of Na^+ for Ba^{2+} in Ba(Na)LaF_5. In each case, XPS analysis of La 3d lines show a decrease in binding energy (0.08–0.25 eV) indicating a reduction in local crystal field symmetry surrounding rare earth (R.E.$^{3+}$) ions, permitting otherwise disallowed R.E. UC transitions to be enhanced. Studies that examine the impact of laser excitation power upon luminescence intensity were conducted over 2.5–100 W/cm^2 range to elucidate UC mechanisms that populate dominant UV emitting states. Low power saturation of

Tm^{3+} $^3F_3$ and $^3H_4$states was observed and noted as a key initial condition for effective population of the $^1D_2$ and $^1I_6$UV emitting states, via Tm-Tm cross-relaxation.

Keywords: near-infrared to ultraviolet upconversion; photoluminescence; upconversion nanocrystals; impurity doping; lanthanide; LaF_3; $BaLaF_5$; core-shell

1. INTRODUCTION

The design of upconverting nanocrystals (UCNCs) with specific, tunable electronic spectroscopic profiles [1,2,3,4], based on multiple dopants and mixed host crystal structures [5,6], has shown great promise for applications that range from biomedical detection [7,8,9] to specifically triggered photoresponsive reactions [10,11]. The advantages of rare earth-based UCNCs capable of near-infrared (NIR) sensitization [excitation wavelength (λ_{ex}) = 980 nm] of a ytterbium dopant have been well documented in the literature, and are exemplified by the capability of activating alternate emitting rare earth (R.E.) ions (e.g., Ho, Er, Tm) through 4f-4f phonon-assisted energy transfer processes; these non-resonant spin and Laporte forbidden processes result in anti-Stokes emission from multiple excited R.E. f-block states [3,12,13]. Due to the long-lived excited states of R.E. dopants (μs–ms), multiple R.E. emitting ions are capable of undergoing complex upconversion and relaxation processes in concert [13,14,15], producing a range of emission profiles via energy transfer reactions that give rise to high energy R.E.-based emissive states [16,17], and through activation of tandem organic chromophores [18].

As R.E.-based upconversion relies on atomic 4f-orbital emitting states, emission wavelengths are essentially fixed; modulation of UCNC composition provides control of both the relative intensity of emissive states and the nature of the multi-colored emission over the UV-vis-NIR spectral regime [1,2,3,4]. However, as energy transfer upconversion (ETU) processes require the assistance of host crystal phonon modes to aid in energetically matching non-resonant transitions, tuning of f-block transitions on the order of several hundred wavenumbers can be achieved with varied crystal systems. Recent work has shown general enhancement of upconversion luminescence through doping of relatively small, non-emitting ions (Li, Sc, Ca, *etc.*) in low concentrations, which serve to either replace ions in the crystal lattice or occupy interstitial sites. Interstitial site occupation by these dopants leads to lattice distortions and a reduction in crystal field symmetry [19,20,21,22], as evidenced by X-ray diffraction (XRD) and X-ray photoelectron spectroscopic (XPS) studies. Diminished symmetry proximal to emitting R.E. ions permits otherwise disallowed ETU transitions to become more favorable. These studies have reported between two- and 50-fold enhancements of luminescence intensity in two-photon upconversion processes extending from the NIR to visible spectral regime [19,20,21,22], but do not address the potential for enhancement of higher energy, UV emitting transitions or their possible upconversion mechanisms.

While excellent progress has been made regarding surface modifications of non-toxic UCNCs [23,24,25] for biological applications [26,27], conjugation of UCNCs to organic chromophores [18,28] and photoactive drug moieties [10,18,23], enhancement of specific high energy, UV R.E. ion emission bands has proven challenging. Though dominant UV emission has been achieved in Tm/Yb codoped upconverter compositions in 80 nm, 12 μm, and bulk [29,30,31] size regimes, these materials capitalize on low surface area to volume ratios and complementary diminished surface quenching characteristic of UCNCs having diameters greater than 50 nm [32,33]. Due to the reliance on small surface area:volume ratios, it has proven challenging to generate small, dominant UV emitting UCNCs; to date, there are no known examples in the literature of sub-10 nm UCNCs with a dominant UV emission profile. To this end, we introduce sub-10 nm UCNCs of sodium-codoped LaF_3 (0.5%Tm, 20%Yb) and $BaLaF_5$ (0.5%Tm, 20%Yb) and their core@shell derivatives, generated via a one-pot solvothermal approach, which exhibit dominant UV emission or broadly tunable UV-vis-NIR emission profiles. High sodium codopant (>20% *versus* rare earth content) upconverting materials are shown to enhance UV emission intensity from $^1I_6 \rightarrow {}^3F_4$ (349 nm) and $^1D_2 \rightarrow {}^3H_6$ (362 nm) Tm^{3+} transitions by 800- and 540-fold in $(Na)LaF_3$ and 65- and 50-fold in $Ba(Na)LaF_5$, respectively, compared to their non-sodium included analogues. Additionally, studies of laser excitation power density *vs.* luminescence intensity, plotted on a ln-ln scale, provide insight into the number of ETU events required to populate known Tm^{3+} emissive states. Similar power dependence studies of upconverted emission intensity conducted on the core@shell materials *vs.* their unshelled counterparts reveal shifts in the upconversion mechanism at low laser densities (<30 W/cm^2), demonstrating improved efficiency of upconversion into high energy states, as evinced by a decreased number of Yb-to-Tm ETU processes required to populate these states, as well as relaxation processes that produce comparatively strong NIR emissions. Both of these properties highlight the suitability of these nanomaterials for applications requiring dominant UV emission with concomitant intense NIR emission.

2. RESULTS AND DISCUSSION

2.1. UCNC Materials and Emission Profiles

Sodium codoped LaF_3 and $BaLaF_5$ (herein denoted as (Na)LaF_3 and Ba(Na)LaF_5) UCNCs were generated via a solvothermal synthesis, similar to that reported by Wang *et al.* (see Section 3.1) [34]. Both (Na)LaF_3 and Ba(Na)LaF_5 materials were doped with 0.5%Tm and 20%Yb, concentrations, *versus* total R.E., optimized to produce the highest emission intensity from the $^1I_6 \rightarrow {}^3F_4$ (349 nm) and $^1D_2 \rightarrow {}^3H_6$(362 nm) Tm^{3+} transitions. LaF_3 and $BaLaF_5$ UCNCs, synthesized in the absence of sodium in the same size regime, were similarly doped with 0.5%Tm and 20%Yb (Figure 1) to minimize any possible surface defect site variability as a cause of luminescence quenching (Figure S1). Shells of (Na)LaF_3 and Ba(Na)LaF_5 host crystals were solvothermally grown following initial preparation of core seed crystals, in the presence of excess fluoride, in the same Teflon®-lined autoclavable bombs through addition of aqueous $La(NO_3)_3$ and $Ba(NO_3)_2$ salts with no intermediate purification required (see Section 3.1). These shells were applied, congruent with earlier studies, to reduce the number of surface defect trap sites capable of quenching surface presenting R.E. ion emission or quenching of core-residing R.E. excited states through R.E.-to-R.E. energy migration to the surface, thereby increasing the overall luminescence output (Figure 2) [8,18,35,36,37].

In each case, compared to 0.5%Tm, 20%Yb doped nanocrystals lacking a sodium codopant, these materials demonstrate significant luminescence enhancement at identical 60 W/cm^2, 980 nm CW laser excitation conditions. As sodium codopant is introduced into the host crystal lattice, a noted shift in relative emission intensity is seen from lower energy Tm^{3+} transitions $^3H_4 \rightarrow {}^3H_6$ (801 nm) in LaF_3 and $^1G_4 \rightarrow {}^3H_6$ (480 nm) in $BaLaF_5$ to the $^1I_6 \rightarrow {}^3F_4$ (349 nm) transition (Figure 2).

Global emission enhancement of two orders of magnitude is observed in the case of (Na)LaF_3compared to LaF_3 (Figure 2a, Blue and Green traces, respectively). More noteworthy is the appearance of dominant UV emission (349 nm; $^1I_6 \rightarrow {}^3F_4$) over that of the NIR (776–801 nm; $^3F_3 \rightarrow {}^3H_6$ and $^3H_4 \rightarrow {}^3H_6$ transitions), and an enhancement of 800-fold in the $^1I_6 \rightarrow {}^3F_4$ emissive transition in the sodium-codoped LaF_3 over that characteristic of the control. Even under modest 20 W/cm^2 980 nm excitation conditions, the $^1I_6 \rightarrow {}^3F_4$, 349 nm transition displays greater intensity than the NIR emissive transitions derived from two-photon ETU processes. The UV manifold of Na-codoped *versus* control $BaLaF_5$(highlighted in Figure 2b) shows a relative emission enhancement of 65-fold in the $^1I_6 \rightarrow {}^3F_4$ transition, with a similar observed dominant UV emissive signature.

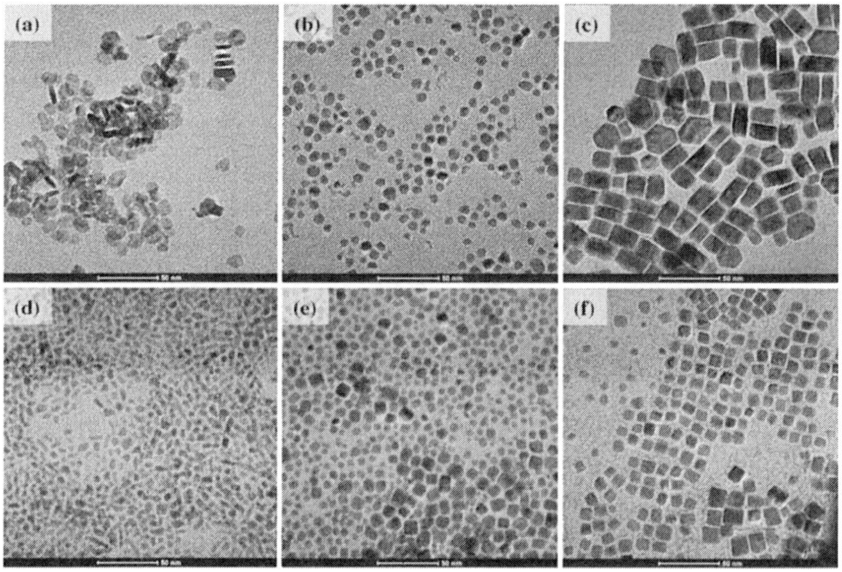

Figure 1. Transmission electron microscopy (TEM) images of 0.5%Tm, 20%Yb codoped Upconverting Nanocrystals (UCNCs) in host lattices: (**a**) LaF_3 (12.6 ± 2.5 nm); (**b**) $(Na)LaF_3$ (7.7 ± 1.9 nm); (**c**) $(Na)LaF_3@(Na)LaF_3$ (22.7 ± 4.5 nm); (**d**) $BaLaF_5$ (6.0 ± 1.4 nm); (**e**) $Ba(Na)LaF_5$ (7.5 ± 2.3 nm); (**f**) $Ba(Na)LaF_5@Ba(Na)LaF_3$ (9.6 ± 2.3 nm): (200 kx magnification; 50 nm scale bars).

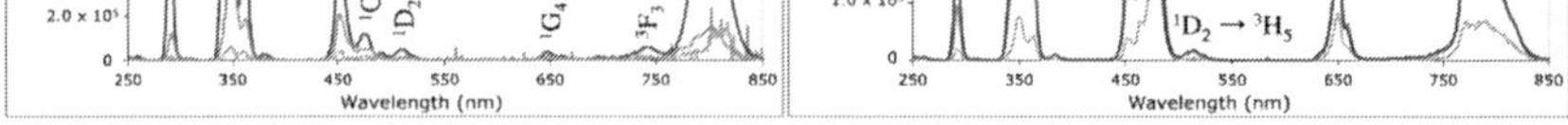

Figure 2. Comparative emission spectra of 0.5%Tm, 20%Yb codoped UCNCs in host lattices. (**a1**) LaF_3 (12.6 ± 2.5 nm) (intensity shown scaled ×1000) (**Green**); (**a2**) $(Na)LaF_3$ (7.7 ± 1.9 nm) (**Blue**); (**a3**) $(Na)LaF_3@(Na)LaF_3$ (22.7 ± 4.5 nm) (**Violet**); (**b1**) $BaLaF_5$ (6.0 ± 1.4 nm) (intensity shown scaled ×10 (**Green**); (**b2**) $Ba(Na)LaF_5$ (7.5 ± 2.3 nm) (**Blue**); (**b3**) $Ba(Na)LaF_5@Ba(Na)LaF_3$ (9.6 ± 2.3 nm) (**Violet**); experimental conditions: 1 mg/mL solutions in toluene; 980 nm CW laser excitation; 60 W/cm^2. Tm^{3+} transitions assigned.

As precedent literature has shown, further enhancement of luminescence can be achieved by shelling UCNCs with a solvothermally grown host crystal, lacking in R.E. emitting ions [8,15,35,37,38,39]. This shelling both diminishes the extent of vibrational relaxation processes involving electronically excited, R.E. dopants and solvent, and augments the emitting ion separation distance from surface crystal defect sites, which can diminish the extent of defect-site luminescence quenching and potentially impact the nature of phonon-assisted energy transfer through the host crystal. Applying a thin shell of inorganic host crystal augments dominant UV emission with nanocrystal sizes below 30 nm, avoiding the need to generate 100 nm-to-μm sized UCs that feature greater concentrations of emitting ions within the UCNC core relative to the surface, where non-radiative relaxation processes dominate. Oftentimes the enhancement of UC emission through inorganic shelling favors, proportionally, the more dominant emissive processes, as is observed for $Ba(Na)LaF_5$@$Ba(Na)LaF_5$. In the case of $(Na)LaF_3$@$(Na)LaF_3$, however, a disproportionate enhancement of the NIR $^3F_3 \rightarrow {}^3H_6$ and $^3H_4 \rightarrow {}^3H_6$ transitions (9.6-fold) over the dominant $^1I_6 \rightarrow {}^3F_4$, UV transition (2.7-fold) is seen.

The UC emission enhancements observed through the introduction of sodium-codopant into the host crystals and solvothermal shelling of UCNC cores derives from shifts in the relative populations of emitting states. The distribution of excited state populations can be affected via several factors, including shifts in unit cell dimensions and crystallographic phases, alteration of crystal field symmetry surrounding the emitting Tm^{3+} ions, and modulation of the excited state lifetimes of either the Yb^{3+}sensitizer or Tm^{3+} activator caused by changes to the local crystal field environment.

2.2. Crystal Structure and Emissive Ion Environment

Unlike previous reports of Ca^{2+} and Ba^{2+} codoping of $R.E.F_3$, wherein the crystal structure is changed from hexagonal to cubic and a resultant shift in both unit cell dimensions and crystal field environment is seen [21], Na-codoping into $(Na)LaF_3$ does not drive a compositional or phase change to $NaLaF_4$, nor does $Ba(Na)LaF_5$ show hexagonal $NaLaF_4$ impurities (Figure 3). Most notably, Na-codoping of LaF_3results in a decrease in 2θ values of all crystallographic peaks (Figure S2) compared to the undoped control, indicating an increase in unit cell dimension most likely due to Na^+ insertion into interstitial sites in the hexagonal unit cell, though direct La^{3+} to Na^+ substitution is also expected [19,22,40,41]. The reverse trend can be seen in $Ba(Na)LaF_5$, where an increase in 2θ values (Table S1) indicates a reduction in unit cell volume, likely through substitution of Ba^{2+} ions for Na^+ [19,20,42,43]. Inductively coupled plasma optical emission spectroscopy

(ICP-OES) elemental analysis of (Na)LaF_3 and Ba(Na)LaF_5 (Table 1) indicates a high level of Na-codopant (>20%) in both UCNC systems, and a marked replacement of Ba^{2+} ions for Na^+ in the Ba(Na)LaF_5 UCNC cores. In contrast to the expected 1:1 ratio of Ba:rare earth (R.E.) elements (La, Tm, and Yb) in $BaLaF_5$, a 0.74:1 (Ba:R.E.) ratio was determined, along with a 0.60:1 Na:R.E. molar ratio. Either by substitution of La^{3+} or Ba^{2+}, insertion into interstitial sites, or a combination of the two, the change in unit cell dimensions would decrease the crystal field symmetry surrounding emitting Tm^{3+} ions and facilitate luminescence enhancement [19,40].

XPS characterization of Na-codoped LaF_3 and $BaLaF_5$ UCNCs, compared to their undoped controls (Figure 4), reveals insights into the surface composition and local crystal field environments of the host matrices. XPS analysis and Gaussian deconvolution of the of Na 1s and La 3d orbital spectra in (Na)LaF_3show a low energy component indicative of interstitial ion insertion into the crystal structure, that is clearly absent in the undoped LaF_3 control (Table 2; Figure S3) [42,44]. The presence of this low energy shoulder in both Na 1s and La 3d orbital XPS spectral regions suggests that Na^+ inclusion contributes to La^{3+} substitution, as evidenced by the presence of interstitial La^{3+} components. The Na 1s line also displays a low binding energy, interstitial component, though less in contribution (2.65%) in total Na^+incorporation *versus* La^{3+} substitution (~97%). Analysis and deconvolution of the Na 1s, La 3d, and Ba 3d orbital lines of Ba(Na)LaF_5 display a low energy spectral shoulder exhibited in the Ba 3d orbital alone, indicating direct substitution of Na^+ for Ba^{2+} and interstitial Ba^{2+} inclusion, with no observed La^{3+}substitution (Table 2).

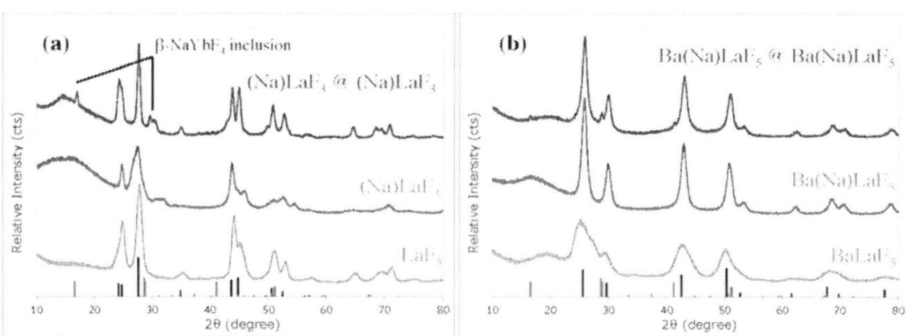

Figure 3. X-ray powder diffraction (XRD) data acquired for: (**a1**) LaF_3 (**Green**); (**a2**) (Na)LaF_3 (**Blue**); (**a3**) (Na)LaF_3@(Na)LaF_3(**Violet**); (**a4**) line spectrum of hexagonal LaF_3, JCPDS: 72-1435 (**Black**); hexagonal $NaYbF_4$ inclusions denoted in (Na)LaF_3@(Na)LaF_3: (**b1**) $BaLaF_5$ (**Green**); (**b2**) Ba(Na)LaF_5 (**Blue**); (**b3**) Ba(Na)LaF_5@Ba(Na)LaF_3 (**Violet**); (**b4**) displays JCPDS: 43-0394 (cubic $BaCeF_5$ comparative, **Black**); and (**a5**,**b5**) line spectrum of hexagonal $NaLaF_4$, JCPDS: 75-1923 (**Red**).

Table 1. Inductively coupled plasma optical emission spectroscopy (ICP-OES) compositional analysis of 0.5%Tm, 20%Yb codoped (Na)LaF_3, and Ba(Na)LaF_5 upconverting nanocrystals (UCNC) and core@shell materials as total ions detected (mol%/L) and ratios of components *vs.* total rare earth (R.E.) ions detected (%). Note: potassium (K^+) content was measured in trace (<0.3 mol%/L) amounts or not detected (ND).

	Conc. Na mol%/L	Conc. Ba mol%/L	Conc. La mol%/L	Conc. Tm mol%/L	Conc. Yb mol%/L	Conc. Ratio (Yb/Tm): La
(Na)LaF_3 Core	22.15	ND	62.11	0.38	15.36	0.26
(Na)LaF_3 Core@Shell	19.02	ND	67.09	0.34	13.55	0.12
Ba(Na)LaF_5 Core	25.68	31.58	33.99	0.21	8.54	0.25
Ba(Na)LaF_5 Core@Shell	12.82	25.02	55.59	0.18	6.38	0.21
	Na:RE (%)	**Ba:RE (%)**	**La:RE (%)**	**Tm:RE (%)**	**Yb:RE (%)**	**Sum mol% RE**
(Na)LaF_3 Core	28.45	ND	79.78	0.49	19.73	77.85
(Na)LaF_3 Core@Shell	23.48	ND	82.85	*0.42*	*16.73*	80.98
Ba(Na)LaF_5 Core	60.08	73.89	79.53	0.49	19.98	42.74
Ba(Na)LaF_5 Core@Shell	20.63	40.26	89.45	*0.29*	*10.27*	62.15

Table 2. X-ray photoelectron spectroscopy (XPS) Na 1s, La 3d, and Ba 3d regions scans binding energies (eV) in (Na)LaF_3, LaF_3, Ba(Na)LaF_5, and $BaLaF_5$. In cases of multicomponent Gaussian fits, percentage values listed correspond to the extent of ion incorporation by substitution (primary peak) and insertion into interstitial sites (lower energy component). ND denotes Not Detectable.

Regions (fitted):	(Na)LaF_3	LaF_3	Ba(Na)LaF_5	$BaLaF_5$
Na 1s	1071.25 eV (97.35%)	ND	1071.25 eV	ND
Na 1s interstitial	1069.2 eV (2.65%)	ND	ND	ND
La $3d_{3/2}$	852.94 eV (38.64%)	853.21 eV	852.87 eV	852.87 eV
La $3d_{3/2}$ interstitial	851.44 eV (1.23%)	ND	ND	ND
La $3d_{5/2}$	836.134 eV (58.71%)	836.38 eV	836.04 eV	836.12 eV
La $3d_{5/2}$ interstitial	834.16 eV (1.43%)	ND	ND	ND
Ba $3d_{3/2}$	ND	ND	794.94 eV (38.94%)	795.07 eV
Ba $3d_{3/2}$ interstitial	ND	ND	793.77 eV (1.55%)	ND
Ba $3d_{5/2}$	ND	ND	779.64 eV (58.75%)	779.77 eV
Ba $3d_{5/2}$ interstitial	ND	ND	778.21 eV (0.77%)	ND

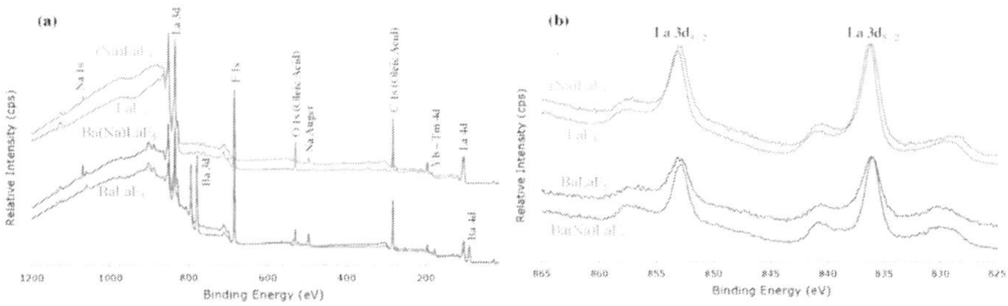

Figure 4. XPS survey scans of (**a1**) $(Na)LaF_3$ (**pale Blue**); (**a2**) LaF_3(**pale Green**); (**a3**) $Ba(Na)LaF_5$ (**Blue**); and (**a4**) $BaLaF_5$ (**Green**) indicating the presence and absence of Na-codopant; normalized XPS La 3d spectral region (**b1**) $(Na)LaF_3$ (**pale Blue**); (**b2**) LaF_3 (**pale Green**); (**b3**) $Ba(Na)LaF_5$ (**Blue**); and (**b4**) $BaLaF_5$ (**Green)** indicating a shift in binding energy in La $3d_{5/2}$ core orbital upon Na^+ inclusion.

The effect of substitution of both Na^+ into LaF_3 and $BaLaF_5$ and charged ions in interstitial sites proximal to R.E.$^{3+}$ ions, is seen in the XPS spectra of $(Na)LaF_3$ and $Ba(Na)LaF_5$ as a decrease in binding energy of the La $3d_{5/2}$ orbital line. This global decrease in La $3d_{5/2}$ binding energy upon Na-codoping denotes perturbation in the local crystal field environment surrounding rare earth emitting ions and an overall decrease in local field symmetry [19,40,42,44]. This reduction in symmetry leads to increasingly allowed R.E.$^{3+}$-R.E.$^{3+}$ energy transfer UC processes and emissive transitions that are formally spin and Laporte forbidden, which in turn result in Tm^{3+} luminescence enhancement [12].

2.3. Core@Shell Compositional Analysis

Shells of $(Na)LaF_3$ and $Ba(Na)LaF_5$, without Tm/Yb dopants, were solvothermally grown, sequentially following UCNC core synthesis in the same Teflon®-lined autoclavable bombs (see Section 3.1) in the interest of further enhancing dominant UV emissions [38]. The conditions utilized resulted in radial growth of 7.5 nm $(Na)LaF_3$ and 1.1 nm $Ba(Na)LaF_5$ shells, respectively, on their same host material cores as measured from TEM images (Figure 1). Conditions for $Ba(Na)LaF_5$ shelling were conducted at a low $Ba(NO_3)_2$ concentration in order to avoid formation of BaF_2 shells; thicknesses of $Ba(Na)LaF_5$ shells were thus significantly reduced relative to those of corresponding $(Na)LaF_3$ shells, and resulted in diminished emission enhancements that derive from reduction in surface defects near emitting ions or augmentation of distance between emitting ions and solvent.

Decreased Yb(Tm):La ratios detected in ICP-OES of dissolved core@shell samples (Table 3), in tandem with consistent particle monodispersity (see TEM data, Figure 1), are consistent with UCNC shelling. Analysis of the Yb(Tm):La ratios of the surfaces of the core@shell nanomaterials is far more indicative of effective shelling. XPS probes the surface compositions of these materials (between 4 and 6 nm) and analysis of Yb(Tm):La ratios on the surface of core@shell materials show a marked reduction in the Yb(Tm):La molar ratio compared to that of their UCNC core counterparts. Congruent with expectations of XPS *versus* ICP-OES compositional analysis, XPS data evince a reduction in the Yb(Tm):La molar ratio relative to that determined via ICP-OES, as XPS interrogates ion composition within 4–6 nm of the surface. These experiments signal discrete shell formation under these solvothermal synthetic conditions, as redissolution of UCNC cores and redistribution of emitting ions in the shell layer would give rise to similar Yb(Tm):La ratios in the ICP-OES and XPS experiments. In the Ba(Na)LaF_5 core@shell *versus* UCNC core material, ICP-OES shows a minimal change in Yb(Tm):La ratio, due to the addition of a thin shell (~1 nm), whereas XPS composition analysis shows a significant reduction in Yb(Tm):La ratio (0.5:1), indicating the presence of a distinct shell composition.

It should also be noted that solvothermal shelling permits secondary, thermodynamically-driven, recrystallization processes to occur in restructuring of the UCNCs to a more stable crystallographic phase [1,32,45,46]. In the (Na)LaF_3 core@shell nanomaterial, XRD indicates that β-$NaYbF_4$ is present (Figure 3) as a minor inclusion and β-$NaLaF_4$ appears in the Ba(Na)LaF_5 core@shell product. Though binding energy shifts were undetected in the XPS spectrum of the Yb 4d region, a noteworthy increase in the NIR $^3H_4 \rightarrow {}^3H_6$ (803 nm) Tm^{3+} transition intensity in (Na)LaF_3 core@shell material is observed relative to (Na)LaF_3 UCNC cores (Figure 2a), consistent with highly efficient Yb-to-Tm ETU processes and dominant NIR emission previously reported for Tm^{3+} doped, $NaYbF_4$ [7]. This topic is further discussed with respect to UC mechanism considerations in Section 2.4, Power Dependence.

Table 3. XPS survey scans of (Na)LaF_3, (Na)LaF_3 core@shell, Ba(Na)LaF_5, and Ba(Na)LaF_5 core@shell surface compositions (% element detected) and Yb/Tm:La ratio. Yb and Tm 4d lines were integrated together.

XPS Survey Scan	Na (%)	Ba (%)	La (%)	Yb/Tm (%)	F (%)	Yb/Tm: La
(Na)LaF_3 Core	3.43	ND	13.9	1.37	81.31	0.10
(Na)LaF_3 Core@Shell	2.14	ND	17.32	0.61	79.92	0.04
Ba(Na)LaF_5 Core	4.94	11.56	6.61	0.65	76.24	0.10
Ba(Na)LaF_5 Core@Shell	0.0	8.56	13.58	0.69	77.17	0.05

2.4. Emission Intensity *versus* Power (IvP) Dependence Studies

Laser power dependence emission studies were conducted on (Na)LaF_3 and Ba(Na)LaF_5 UCNCs and their core@shell counterparts to aid in determining an upconversion mechanism consistent with the observed dominant UV emission at low laser powers (<30 W/cm^2). Emission intensity *versus* laser power density (IvP) studies were conducted using a 980 nm CW laser with power densities varying over 2.5–100 W/cm^2 ensuring a clear transition from low to high power regimes, indicated by saturation of one or more emissive states [12,13,15]. As described by Pollnau *et al.* [47], and elaborated on by Suyver *et al.* [48], the energy transfer upconversion (ETU) mechanism, characterized by sequential energy transfers from a Yb^{3+} sensitizer to activator emitting ion (e.g., Tm^{3+}), can be elucidated on a per excited state basis in the ln-ln plot of emission intensity (arbitrary units) *versus* laser power density (W/cm^2). In the low power regime of the ln-ln IvP plot, the slopes of the linear fits of each emissive transition provide insight into a number of photons or ETU events required to populate the emitting states via a standard Yb-to-Tm sequential ETU process, in the limit where no emitting state achieves saturation, wherein the upconversion process dominates over relaxation via radiative or non-radiative decay [12,47].

Literature precedent demonstrates 2–5 photon ETU processes across Tm^{3+} 4f-4f transitions, often involving a Tm-Tm centered cross-relaxation (Tm-Tm CR) event to by-pass the $^1G_4$ to $^1D_2$ energy level gap [17,30], an energy gap too small to be successfully bridged with a 10,200 cm^{-1} ETU from Yb coupled to a several thousand wavenumber phonon mode, which would otherwise provide for non-radiative decay of lower energy f-block states (Figure 5d) [49]. Even under the low laser power densities utilized, the expected two photon ETU processes required to populate the $^3H_4$ and $^3F_3$ states are linearly fit with a slope of ~1, indicating saturation of these states even at power densities as low as 2.5 W/cm^2. It has been suggested that core lattice defects of small UCNCs may promote CR events to occur under condition of low laser density (<30 W/cm^2) [15], a defect condition intrinsic to Na-codoped UCNCs of sub-10 nm size. The consequences of this initial saturation of NIR emitting states in the ETU mechanism are immediately apparent in global slope reductions in each higher energy transition, shown in the ln-ln IvP plot (Figure 5). The significantly reduced slope, showing only dependence on Yb sensitizer excited state population density [47,48], serves to place Tm^{3+}, when $^3H_4$ and $^3F_3$ states are under saturation, in a co-sensitizer role for activation of higher energy Tm^{3+} excited states (Figure 5b) [48]. From this initial condition, a single-photon, Yb-to-Tm ETU event may populate the $^1G_4$ state, permitting $^1G_4 \rightarrow {}^3F_4$:$^3H_4 \rightarrow {}^1D_2$ Tm-Tm CR followed by another sequential Yb-to-Tm ETU to yield a population in the 292- and 349-nm emitting, $^1I_6$ state. Similar Tm-Tm CR of $^3F_2 \rightarrow {}^3H_6$:$^3F_2 \rightarrow {}^1D_2$ serves to provide a secondary route to high energy UV emitting states with diminished ln-ln IvP slopes of ~2, suggesting 2 photon Yb-to-Tm ETU processes (Figure 5a). From a combination of these Tm-Tm CR events, UV emission increases in intensity as saturation of NIR emitting $^3F_3$ and $^3H_4$ states is achieved, invariably, to a ln-ln IvP slope of 1, congruent with the diminished radiative output of these states due to their roles in CR. Under higher laser densities (30–100 W/cm^2), transitions originating from the $^1I_6$ state exhibit significantly reduced slopes compared to $^1G_4$ and $^1D_2$ derived visible transitions, consistent with relaxation of these states through Tm-Tm CR ($^1G_4 \rightarrow {}^3F_4$:$^3H_4 \rightarrow {}^1D_2$).

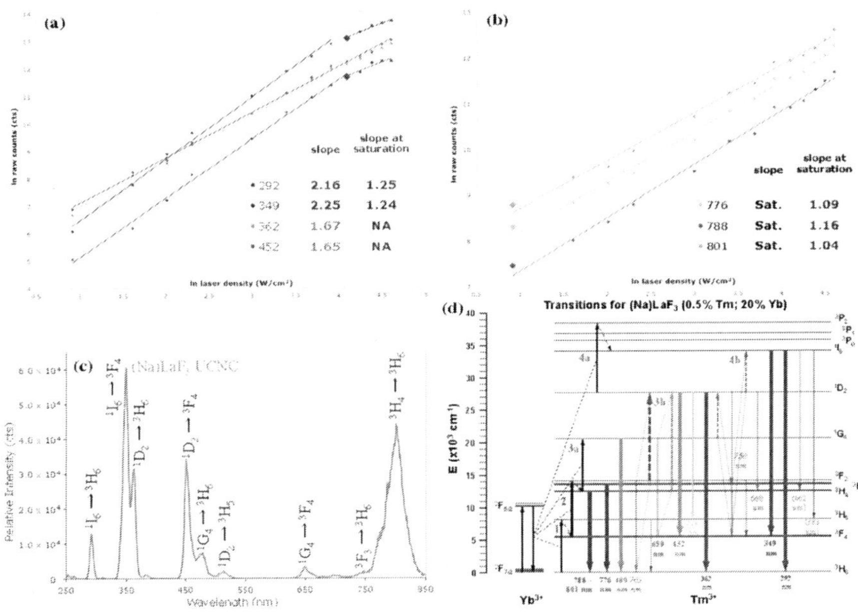

Figure 5. Emission intensity *vs.* 980 nm laser power density (W/cm^2), ln-ln IvP plot of 0.5%Tm, 20%Yb codoped (Na)LaF$_3$ UCNC in (**a**) UV/blue and (**b**) NIR regime, noting low power saturation of Tm^{3+} $^3F_3$ and $^3H_4$ states; (**c**) Emission spectrum at 20 W/cm^2 and (**d**) ETU transition diagram for Yb^{3+}-Tm^{3+}. Experimental conditions: 1 mg/mL solutions in toluene at 23 °C; 980 nm CW laser excitation varying from 2.5 to 100 W/cm^2.

The ln-ln IvP plot of the (Na)LaF$_3$ core@shell system (Figure 6a,b) shows a further departure from the standard, sequential ETU model, including a low power density saturated $^1G_4$ state (479 nm transition) in addition to a saturated $^3H_4$ state. The onset of saturation of these UV and blue emitting states at decreased laser power densities (20–30 W/cm^2) suggests a higher probability of a $^1G_4 \rightarrow {}^3H_6$:$^3F_3 \rightarrow {}^1I_6$Tm-Tm CR. Shelling of UCNCs has also been seen to increase the lifetimes of emitting states, allowing for further complex radiative and CR pathways to occur [18,36,39]. The appearance of $NaYbF_4$ inclusions in the $(Na)LaF_3$ core@shell as indicated in the XRD data (Figure 3a), an UCNC composition established to have a high efficiency UC mechanism into NIR emitting $^3F_3$ and $^3H_4$ states [5,7], serves in part to explain the significant 803 nm emission intensity seen in the ln-ln IvP plot.

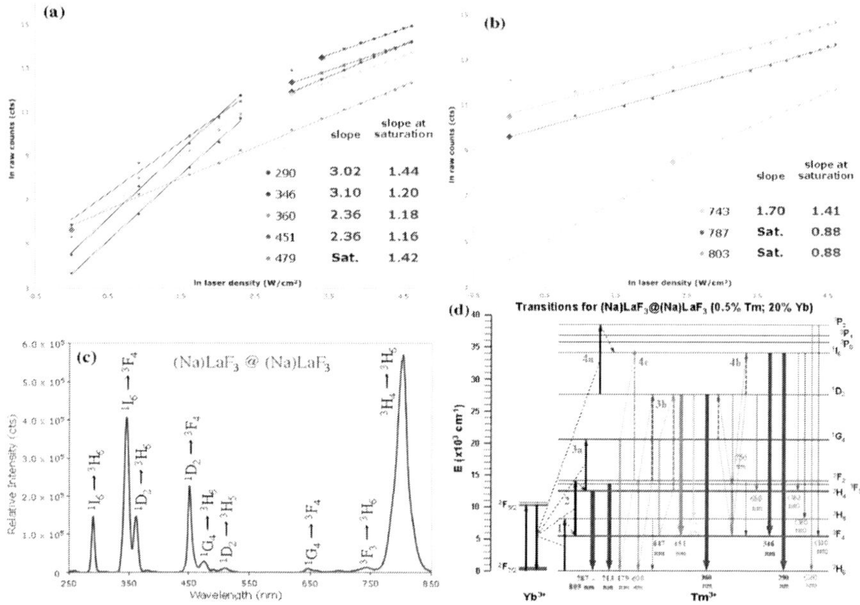

Figure 6. Emission intensity *vs.* 980 nm laser power density (W/cm^2), expressed in ln-ln IvP plots of 0.5% Tm, 20% Yb codoped $(Na)LaF_3$ core@shell in the (**a**) UV/blue and (**b**) NIR regimes, noting low power saturation of Tm^{3+} $^3H_4$ state and onset of saturation (diamond markers); (**c**) Emission spectrum recorded at 20 $W/cm^2$and (**d**) ETU transition diagram of Yb^{3+}-Tm^{3+}. Experimental conditions: 1 mg/mL solutions in toluene at 23 °C; 980 nm CW laser excitation varying from 2.5 to 100 W/cm^2.

$Ba(Na)LaF_5$ UCNC and $Ba(Na)LaF_5$ core@shell materials exhibit similar ln-ln IvP plotted slope reductions per radiative transition in the NIR emitting states in the low laser density regime (<30 W/cm^2) as was observed for $(Na)LaF_3$ UCNCs (Table 4) (see Supplemental Information for detailed ln-ln IvP plots and proposed ETU transition state diagrams). In kind, ETU processes dominate over linear radiative and non-radiative decay of intermediate $^3F_3$ and $^3H_4$ states, with Tm-Tm CR of NIR transitions significantly contributing in the population of higher energy excited states and providing a means of increasing UV emissive output [47]. The lack of a dominant UV emission band in Ba(Na)LaF_5, *in lieu* of a broad band emission profile throughout the UV-vis-NIR regime (Figure 2b), may be due in part to a lack of saturation of the NIR emitting states at low laser power, thus, diminishing the probability of Tm-Tm CR as a viable mechanism for population of higher energy, UV emissive, Tm^{3+} excited states.

Table 4. Emission intensity *vs.* 980 nm laser power density (W/cm^2), ln-ln IvP plotted linear slopes of 0.5%Tm, 20%Yb codoped **(a)** (Na)LaF_3 UCNC; **(b)** (Na)LaF_3 core@shell; **(c)** Ba(Na)LaF_5 UCNC; and **(d)** Ba(Na)LaF_5 core@shell; emission bands and transitions, ln-ln IvP slopes per radiative transition at low laser density, and slopes at saturation are given. Slopes in **bold**, *italics*, and plain text correspond to matched Tm states of origin. NA denotes state saturation Not Achieved.

(a)

(Na)LaF_3 UCNC	Transition	slope	slope sat.
292	$^1I_6 \rightarrow {}^3H_6$	*2.16*	*1.25*
349	$^1I_6 \rightarrow {}^3F_4$	*2.25*	*1.24*
362	$^1D_2 \rightarrow {}^3H_6$	*1.67*	*NA*
452	$^1D_2 \rightarrow {}^3F_4$	*1.65*	*NA*
480	$^1G_4 \rightarrow {}^3H_6$	1.57	NA
514	$^1D_2 \rightarrow {}^3H_5$	*1.83*	*NA*
650	$^1G_4 \rightarrow {}^3F_4$	1.61	NA
776	$^3F_3 \rightarrow {}^3H_6$	Sat.	1.09
788	$^3H_4 \rightarrow {}^3H_6$	Sat.	1.16
801	$^3H_4 \rightarrow {}^3H_6$	Sat.	1.04

(b)

(Na)LaF_3 core@shell	Transition	slope	slope sat.
290	$^1I_6 \rightarrow {}^3H_6$	*3.02*	*1.44*
346	$^1I_6 \rightarrow {}^3F_4$	*3.10*	*1.2*
360	$^1D_2 \rightarrow {}^3H_6$	*2.36*	*1.18*
451	$^1D_2 \rightarrow {}^3F_4$	*2.36*	*1.16*
479	$^1G_4 \rightarrow {}^3H_6$	Sat.	1.42
510	$^1D_2 \rightarrow {}^3H_5$	*2.12*	*1.41*
647	$^1G_4 \rightarrow {}^3F_4$	1.63	1.37
743	$^3F_3 \rightarrow {}^3H_6$	1.70	1.41
787	$^3H_4 \rightarrow {}^3H_6$	Sat.	0.88
803	$^3H_4 \rightarrow {}^3H_6$	Sat.	0.88

(c)

Ba(Na)LaF_5 UCNC	Transition	slope	slope sat.
292	$^1I_6 \rightarrow {}^3H_6$	*2.37*	*1.38*
349	$^1I_6 \rightarrow {}^3F_4$	*2.38*	*1.36*
362	$^1D_2 \rightarrow {}^3H_6$	*1.69*	*NA*
452	$^1D_2 \rightarrow {}^3F_4$	*1.69*	*NA*
463	$^1D_2 \rightarrow {}^3F_4$	*1.63*	*NA*
480	$^1G_4 \rightarrow {}^3H_6$	1.49	NA
515	$^1D_2 \rightarrow {}^3H_5$	*1.70*	*NA*
650	$^1G_4 \rightarrow {}^3F_4$	1.32	NA
772	$^3F_3 \rightarrow {}^3H_6$	1.54	NA
788	$^3H_4 \rightarrow {}^3H_6$	1.24	NA

(d)

Ba(Na)LaF_5 core@shell	Transition	slope	slope sat.
292	$^1I_6 \rightarrow {}^3H_6$	*2.21*	*NA*
349	$^1I_6 \rightarrow {}^3F_4$	*2.28*	*1.38*
362	$^1D_2 \rightarrow {}^3H_6$	*1.71*	*NA*
452	$^1D_2 \rightarrow {}^3F_4$	*1.65*	*NA*
463	$^1D_2 \rightarrow {}^3F_4$	*1.77*	*NA*
480	$^1G_4 \rightarrow {}^3H_6$	1.60	NA
515	$^1D_2 \rightarrow {}^3H_5$	*1.91*	*NA*
650	$^1G_4 \rightarrow {}^3F_4$	1.63	NA
772	$^3F_3 \rightarrow {}^3H_6$	1.50	NA
788	$^3H_4 \rightarrow {}^3H_6$	1.27	NA

3. EXPERIMENTAL SECTION

3.1. Synthesis

3.1.1. Materials

$La(NO_3)_3{\cdot}6H_2O$ (99.999%), $Yb(NO_3)_3{\cdot}5H_2O$ (99.999%), $Tm(NO_3)_3{\cdot}5H_2O$ (99.9%), $Ba(NO_3)_2$ (99.95%), KF (99.99%), NaOH (99.99%), and NH_4F (>99.99%) were purchased from Sigma Aldrich (St. Louis, MO, USA) and used without further purification. Oleic acid (>97%), water (Optima grade), and NH_4OH (Certified ACS Plus) were purchased from Fisher Scientific (Pittsburgh, PA, USA). Ethanol (200 proof, anhydrous, USP) was purchased from KOPTEC (King of Prussia, PA, USA).

Teflon® autoclave sleeves (120 mL) were soaked overnight in a base bath of 2-propanol and potassium hydroxide (saturated), rinsed with distilled water, subjected to aqua regia (3:1 HCl:HNO_3; ACS Grade, Fisher Scientific (Pittsburgh, PA, USA)) for 30 min, rinsed again with distilled water, then treated again under base bath for an additional 20 min. The Teflon® sleeves were subsequently rinsed thoroughly with distilled water and ethanol and left to dry in air prior to use.

3.1.2. Na-Doped LaF_3 Synthesis

Following the solvothermal procedure of Wang *et al.* [34], 2.4 g (0.06 mol) NaOH, 4 mL Optima grade water, 20 mL ethanol, and 38.1 mL (0.12 mol) oleic acid were added in sequential order to a round-bottomed flask under vigorous stirring. The mixture was occasionally warmed with a heat gun to facilitate dissolution. Lanthanide nitrate hydrates (1 mmol) at a molar ratio of 0.795 La:0.20 Yb:0.005 Tm in 2 mL Optima grade water were added in one portion and stirred until fully dissolved, followed by KF (174.3 mg, 3 mmol) in 2 mL Optima grade water. Solutions were stirred for 30 min before being transferred into 120 mL Teflon® sleeves, sealed in stainless steel autoclavable bombs, and placed in an oven at 200 °C for 20 h. Once cooled to room temperature, the particles were precipitated with a mixture of 60 mL ethanol and 40 mL *n*-hexane and subjected to centrifugation at 14,000 rpm for 20 min at 5 °C. The particles were then resuspended in 30 mL n-hexane and 2 mL water with sonication and precipitated with 28 mL ethanol, centrifuged at 14,000 rpm for 20 min at 5 °C. This step was followed by a 50% ethanol/water wash and a final ethanol wash, both run at 14,000 rpm for 10 min at 5 °C.

3.1.3. Na-Doped $BaLaF_5$ Synthesis

Following the solvothermal procedure of Wang *et al.* [34], 2.4 g (0.06 mol) NaOH, 4 mL Optima grade water, 20 mL ethanol, and 38.1 mL (0.12 mol) oleic acid were added in sequential order to a round-bottomed flask under vigorous stirring. The mixture was occasionally warmed with a heat gun to facilitate dissolution. Barium nitrate (1 mmol) and lanthanide nitrate hydrates (1 mmol) at a molar ratio of 0.795 La:0.20 Yb:0.005 Tm in 2 mL Optima grade water were added in one portion and stirred until fully dissolved, followed by KF (290.5 mg, 5 mmol) in 2 mL Optima grade water. Solutions were stirred for 30 min before being transferred into 120 mL Teflon® sleeves, sealed in stainless steel autoclavable bombs, and placed in an oven at 200 °C for 20 h. Once cooled to room temperature, the particles were precipitated with a mixture of 60 mL ethanol and 40 mL *n*-hexane and subjected to centrifugation at 14,000 rpm for 20 min at 5 °C. The particles were then resuspended in 30 mL *n*-hexane and 2 mL water with sonication and precipitated with 28 mL ethanol, centrifuged at 14,000 rpm for 20 min at 5 °C. This step was followed by a 50% ethanol/water wash and a final ethanol wash, both run at 14,000 rpm for 10 min at 5 °C.

3.1.4. (Na)LaF_3@(Na)LaF_3 and Ba(Na)LaF_5@Ba(Na)LaF_5 Core/Shell Syntheses

Na-doped LaF_3 or $BaLaF_5$ cores were made according to the solvothermal method outlined above, and shells of the same materials were grown following a modified procedure from Zhang *et al.* [38]. Excess KF, 232.4 mg (4 mmol) for a (Na)LaF_3 shell synthesis or 387.5 mg (6.67 mmol) for a Ba(Na)LaF_5 shell synthesis, was added to the solvothermal preparations of core nanocrystals. After 20 h in the oven at 200 °C for core synthesis, the solution was allowed to cool to room temperature, then stirred vigorously for 30 min. $La(NO_3)_3{\cdot}6H_2O$ (433.01 mg, 1 mmol) or $La(NO_3)_3{\cdot}6H_2O$ (1 mmol) and 87.03 mg (0.33 mmol) $Ba(NO_3)_2$ were dissolved in a 3 mL solution of Optima grade water which was added dropwise, followed by stirring for an additional 30 min. The autoclaves were resealed and placed in an oven at 200 °C for an additioanl 3 h, cooled to room temperature, and the reaction mixtures purified following the above procedure for $BaLaF_5$ core precipitation.

3.2. Characterization

Transmission electron microscopy (TEM) images were acquired using a FEI Tecnai G2 Twin microscope (FEI, Hillsboro, OR, USA) operated at 200 kV and equipped with a TIA digital camera to obtain data on particle morphology, size, and distribution. The UCNCs were dispersed in 1 mg/mL toluene solutions, drop cast, and dried on copper grids (200 mesh) with carbon/formvar support grids.

X-ray powder diffraction (XRD) patterns and phase analysis of the particles were obtained using a Philips Panalytical X'Pert PRO MRD HR X-ray diffraction system (PANalytical, Westborough, MA, USA) with Cu Kα source at 1.5405 Å and pressed powder samples.

Emission and ln-ln IvP plotted spectra were acquired using an Edinburgh FLS920 system (Edinburgh Instruments, Livingston, UK) at 4 nm slit width, with a 980 nm continuous wave diode laser focused to a 320 μm diameter spot size, measured by the razor-edge method at the 90/10 diameter in 5 μm steps. Spectra were recorded over a 250–850 nm spectral range (1 nm steps; integration time = 0.5 s) and corrected for variations in detector response as a function of wavelength using correction files supplied by the manufacturer. All samples were run at concentrations of 1 mg/mL in toluene at 23 °C.

Inductively coupled plasma optical emission spectrometry (ICP-OES) was performed with a Perkin Elmer Optima 2000 DV system (Perkin Elmer, Waltham, MA, USA) on UCNC samples following treatment with boric acid and dissolution in aqua regia; diluted with water 10–20×.

X-ray photoelectron spectroscopy (XPS) was also performed using a Kratos Analytical Axis Ultra system (Kratos Analytical, Wharfside, UK) using a Mono (Al) emission source and charge neutralizer. Pressed powder samples were affixed to copper tape. Regions scans were collected at 0.1 eV steps with a minimum of 3 sweeps to determine binding energies of individual elements. Survey scans were collected at 1 eV steps with 3 sweeps to determine elemental composition. The resulting spectra were deconvolved, fit, and integrated using CasaXPS.

4. CONCLUSIONS

We report the synthesis and characterization of sub-10 nm, dominant UV emissive, Na-codoped $LaF_3$and $BaLaF_5$ UCNCs and their core@shell derivatives, generated via a one-pot solvothermal approach. For all materials, Na-codopant of approximately 20% (*versus* total dissolved cation concentration) was present and determined to play a significant role in the one- to two-order of magnitude enhancement of Yb-to-Tm ETU *vs.* non-doped 0.5% Tm, 20% Yb samples in the same size regime. XRD analysis of (Na)LaF_3 exhibited 2θ value diminution of all crystallographic peaks, indicating an increase in unit cell volume; XPS spectra of La 3d and Na 1s regions revealed a marked decrease in La 3d binding energy (eV) and low energy peak components, showing relocation of La^{3+} into interstitial sites for charge balancing upon La-to-Na substitution. These modifications in crystallographic dimensions and emitting ion environment are consistent with a reduction in local crystal field symmetry and the observed enhancement of UC luminescence. Na-codoping into the $BaLaF_5$ host matrix showed an increase in all 2θ values and an associated reduction in unit cell volume (XRD). XPS La and Ba 3d regions scans showed only single Gaussian fits in both $BaLaF_5$ and $Ba(Na)LaF_5$, indicating a reduction in La 3d binding energy and no substitution with La^{3+}. Though the relative contributions of change in lattice volumes with respect to shifts in rare earth element binding energies cannot be quantified in relation to emission enhancement, they are known to decrease the local crystal field symmetry around sensitizer-activator pairs and serve to enhance UC processes.

$(Na)LaF_3$ and $Ba(Na)LaF_5$ UCNCs and their core@shell derivatives were subjected to emission intensity *vs.* laser power dependence studies (IvP) to mark deviations from known upconversion mechanisms. In all cases, significant slope reductions in the ln-ln IvP plots were noted, and in the case of $(Na)LaF_3$ and its core@shell derivative, NIR emitting $^3F_3$ and $^3H_4$ states were found populated to saturation even at the lowest laser power densities tested (2.5 W/cm^2). The ln-ln IvP plot linear slope reductions, as an initial condition of $^3H_4$ saturation, are surmised to report an increase Tm-Tm cross-relaxation processes that promote excited state populations in UV emitting $^1I_6$ and $^1D_2$ states, exhibiting depressed slopes and virtual 2–3 photon ETU events. Further analysis of these exemplary, dominant UV emissive nanomaterials is warranted in the pursuit of UCNCs with specific emission band enhancement and unique emission signatures.

ACKNOWLEDGMENTS

Funding for this research was provided by the National Science Foundation (NSEC DMR-0425780). The authors would like to thank Kim Hutchison of the Department of Soil Science at North Carolina State University for ICP-OES analysis, Michelle Gignac and Mark Walters of the Shared Materials instrumentation Facility of Duke University for support in TEM, XRD, and XPS data acquisition and analysis, and Ian N. Stanton and Jennifer Ayres for assistance with spectroscopic studies and synthetic design.

CONFLICTS OF INTEREST

The authors declare no conflict of interest.

REFERENCES

1. Xu, C.F.; Ma, M.; Yang, L.W.; Zeng, S.J.; Yang, Q.B. Lanthanide doping-facilitated growth of ultrasmall monodisperse Ba_2LaF_7 nanocrystals with excellent photoluminescence. *J. Colloid Interface Sci.* **2012**, *368*, 49–55.

2. Wang, F.; Liu, X. Upconversion multicolor fine-tuning: Visible to near-infrared emission from lanthanide-doped $NaYF_4$ nanoparticles. *J. Am. Chem. Soc.* **2008**, *130*, 5642–5643.

3. Wang, F.; Liu, X. Recent advances in the chemistry of lanthanide-doped upconversion nanocrystals. *Chem. Soc. Rev.* **2009**, *38*, 976–989.

4. Yi, G.-S.; Chow, G.-M. Colloidal LaF_3:Yb,Er, LaF_3:Yb,Ho and LaF_3:Yb,Tm nanocrystals with multicolor upconversion fluorescence. *J. Mater. Chem.* **2005**, *15*, 4460–4464.

5. Chen, G.; Shen, J.; Ohulchanskyy, T.Y.; Patel, N.J.; Kutikov, A.; Li, Z.; Song, J.; Pandey, R.K.; Ågren, H.; Prasad, P.N. (α-$NaYbF_4$:Tm^{3+})/CaF_2 core/shell nanoparticles with efficient near-infrared to near-infrared upconversion for high-contrast deep tissue bioimaging. *ACS Nano* **2012**, *6*, 8280–8287.

6. Dong, C.; Pichaandi, J.; Regier, T.; van Veggel, F.C.J.M. Nonstatistical dopant distribution of Ln^{3+}-doped $NaGdF_4$ nanoparticles. *J. Phys. Chem. C* **2011**, *115*, 15950–15958.

7. Chen, G.; Ohulchanskyy, T.Y.; Kumar, R.; Ågren, H.; Prasad, P.N. Ultrasmall monodisperse $NaYF_4$:Yb^{3+}/Tm^{3+} nanocrystals with enhanced near-infrared to near-infrared upconversion photoluminescence. *ACS Nano* **2010**, *4*, 3163–3168.

8. Pichaandi, J.; Boyer, J.-C.; Delaney, K.R.; van Veggel, F.C.J.M. Two-photon upconversion laser (scanning and wide-field) microscopy using Ln^{3+}-doped $NaYF_4$ upconverting nanocrystals: A critical evaluation of their performance and potential in bioimaging. *J. Phys. Chem. C* **2011**, *115*, 19054–19064.

9. Chatterjee, D.K.; Fong, L.S.; Zhang, Y. Nanoparticles in photodynamic therapy: An emerging paradigm. *Adv. Drug Deliv. Rev.* **2008**, *60*, 1627–1637.

10. Yang, Y.; Velmurugan, B.; Liu, X.; Xing, B. NIR photoresponsive crosslinked upconverting nanocarriers toward selective intracellular drug release. *Small* **2013**, *9*, 2937–2944.

11. Yang, Y.; Shao, Q.; Deng, R.; Wang, C.; Teng, X.; Cheng, K.; Cheng, Z.; Huang, L.; Liu, Z.; Liu, X.; *et al. In vitro* and *in vivo* uncaging and bioluminescence imaging by using photocaged upconversion nanoparticles. *Angew. Chem.* **2012**, *51*, 3125–3129.

12. Auzel, F. Upconversion and anti-stokes processes with f and d ions in solids. *Chem. Rev.* **2004**, *104*, 139–173.

13. Suyver, J.F.; Grimm, J.; van Veen, M.K.; Biner, D.; Krämer, K.W.; Güdel, H.U. Upconversion spectroscopy and properties of $NaYF_4$ doped with Er^{3+}, Tm^{3+} and/or Yb^{3+}. *J. Lumin.* **2006**, *117*, 1–12.

14. Huang, T.C.; Hsieh, W.F. Er-Yb codoped ferroelectrics for controlling visible upconversion emissions. *J. Fluoresc.* **2009**, *19*, 511–516.

15. Mai, H.-X.; Zhang, Y.-W.; Sun, L.-D.; Yan, C.-H. Highly efficient multicolor up-conversion emissions and their mechanisms of monodisperse $NaYF_4$:Yb,Er core and core/shell-structured nanocrystals. *J. Phys.Chem. C* **2007**, *111*, 13721–13729.

16. Wang, F.; Deng, R.; Wang, J.; Wang, Q.; Han, Y.; Zhu, H.; Chen, X.; Liu, X. Tuning upconversion through energy migration in core–shell nanoparticles. *Nat. Mater.* **2011**, *10*, 968–973.

17. Zheng, K.; Qin, W.; Wang, G.; Wei, G.; Zhang, D.; Wang, L.; Kim, R.; Liu, N.; Ding, F.; Xue, X. Upconversion luminescence properties of Yb^{3+}, Gd^{3+} and Tm^{3+} co-doped $NaYF_4$ microcrystals synthesized by the hydrothermal method. *J. Nanosci. Nanotechnol.* **2010**, *10*, 1920–1923.

18. Wang, Y.; Liu, K.; Liu, X.; Dohnalová, K.I.; Gregorkiewicz, T.; Kong, X.; Aalders, M.C.G.; Buma, W.J.; Zhang, H. Critical shell thickness of core/shell upconversion luminescence nanoplatform for FRET application. *J. Phys. Chem. Lett.* **2011**, *2*, 2083–2088.

19. Cheng, Q.; Sui, J.; Cai, W. Enhanced upconversion emission in Yb^{3+} and Er^{3+} codoped NaGdF4 nanocrystals by introducing Li^{+} ions. *Nanoscale* **2012**, *4*, 779–784.

20. Huang, Q.; Yu, J.; Ma, E.; Lin, K. Synthesis and characterization of highly efficient near-infrared upconversion $Sc^{3+}/Er^{3+}/Yb^{3+}$ tridoped $NaYF_4$. *J. Phys. Chem. C* **2010**, *114*, 4719–4724.

21. Chen, D.; Yu, Y.; Huang, F.; Wang, Y. Phase transition from hexagonal LnF_3 (Ln = La, Ce, Pr) to cubic $Ln_{0.8}M_{0.2}F_{2.8}$ (M = Ca, Sr, Ba) nanocrystals with enhanced upconversion induced by alkaline-earth doping. *Chem. Commun.* **2011**, *47*, 2601–2603.

22. Chen, G.Y.; Liu, H.C.; Somesfalean, G.; Sheng, Y.Q.; Liang, H.J.; Zhang, Z.G.; Sun, Q.; Wang, F.P. Enhancement of the upconversion radiation in Y_2O_3:Er^{3+} nanocrystals by codoping with Li^+ ions.*Appl. Phys. Lett.* **2008**, *92*, 96–100.

23. Wang, C.; Tao, H.; Cheng, L.; Liu, Z. Near-infrared light induced *in vivo* photodynamic therapy of cancer based on upconversion nanoparticles. *Biomaterials* **2011**, *32*, 6145–6154.

24. Xiong, L.; Yang, T.; Yang, Y.; Xu, C.; Li, F. Long-term *in vivo* biodistribution imaging and toxicity of polyacrylic acid-coated upconversion nanophosphors. *Biomaterials* **2010**, *31*, 7078–7085.

25. Jalil, R.A.; Zhang, Y. Biocompatibility of silica coated $NaYF_4$ upconversion fluorescent nanocrystals. *Biomaterials* **2008**, *29*, 4122–4128.

26. Li, C.; Lin, J. Rare earth fluoride nano-/microcrystals: Synthesis, surface modification and application. *J. Mater. Chem.* **2010**, *20*, 6831–6847.

27. Hu, H.; Xiong, L.; Zhou, J.; Li, F.; Cao, T.; Huang, C. Multimodal-luminescence core-shell nanocomposites for targeted imaging of tumor cells. *Chemistry* **2009**, *15*, 3577–3584.

28. Wang, L.; Yan, R.; Huo, Z.; Wang, L.; Zeng, J.; Bao, J.; Wang, X.; Peng, Q.; Li, Y. Fluorescence resonant energy transfer biosensor based on upconversion-luminescent nanoparticles. *Angew. Chem.* **2005**, *117*, 6208–6211.

29. Mahalingam, V.; Vetrone, F.; Naccache, R.; Speghini, A.; Capobianco, J.A. Colloidal Tm^{3+}/Yb^{3+}-doped $LiYF_4$ nanocrystals: Multiple luminescence spanning the UV to NIR regions via low-energy excitation. *Adv. Mater.* **2009**, *21*, 4025–4028.

30. Wang, G.; Qin, W.; Wang, L.; Wei, G.; Zhu, P.; Kim, R. Intense ultraviolet upconversion luminescence from hexagonal $NaYF_4$:Yb^{3+}/Tm^{3+} microcrystals. *Opt. Express* **2008**, *16*, 11907–11914.

31. Balakrishnaiah, R.; Kim, D.W.; Yi, S.S.; Jang, K.; Lee, H.S.; Ahn, C.W.; Jeong, J.H. Infrared to ultraviolet frequency upconversion studies of Tm^{3+} ions in a $YbLiF_4$ crystal. *J. Korean Phys. Soc.***2008**, *53*, 791–795.

32. Shan, J.; Ju, Y. A single-step synthesis and the kinetic mechanism for monodisperse and hexagonal-phase $NaYF_4$:Yb,Er upconversion nanophosphors. *Nanotechnology* **2009**, *20*, 275603:1–275603:13.

33. Chen, X.Y.; Zhuang, H.Z.; Liu, G.K.; Li, S.; Niedbala, R.S. Confinement on energy transfer between luminescent centers in nanocrystals. *J. Appl. Phys.* **2003**, *94*, 5559–5565.

34. Wang, Y.; Qin Wei, P.; Di, W.-H.; Zhang Ji, S.; Cao, C.-Y. Infrared-to-visible and infrared-to-violet upconversion fluorescence of rare earth doped LaF_3 nanocrystals. *Chin. Phys. B* **2008**, *17*, 3300–3305.

35. Yi, G.-S.; Chow, G.-M. Water-soluble $NaYF_4$:Yb,Er(Tm)/$NaYF_4$/polymer core/shell/shell nanoparticles with significant enhancement of upconversion fluorescence. *Chem. Mater.* **2007**, *19*, 341–343.

36. Liu, Y.; Tu, D.; Zhu, H.; Li, R.; Luo, W.; Chen, X. A strategy to achieve efficient dual-mode luminescence of Eu^{3+} in lanthanides doped multifunctional $NaGdF_4$ nanocrystals. *Adv. Mater.* **2010**, *22*, 3266–3271.

37. Chen, G.; Yang, C.; Prasad, P.N. Nanophotonics and nanochemistry: Controlling the excitation dynamics for frequency up- and down-conversion in lanthanide-doped nanoparticles. *Acc. Chem. Res.* **2013**, *46*, 1474–1486.

38. Zhang, Q.; Zhang, Q.-M. Synthesis and photoluminescent properties of α-$NaYF_4$:Nd/α- $NaYF_4$core/shell nanostructure with enhanced near infrared (NIR) emission. *Mater. Lett.* **2009**, *63*, 376–378.

39. Qu, Y.; Li, M.; Zhang, L.; Zhao, L. Preparation and characterization of upconversion luminescent LaF_3:Yb^{3+}, Er^{3+}/LaF_3 core/shell nanocrystals. *Appl. Surf. Sci.* **2011**, *258*, 34–37.

40. Bai, Y.; Wang, Y.; Yang, K.; Zhang, X.; Song, Y.; Wang, C.H. Enhanced upconverted photoluminescence in Er^{3+} and Yb^{3+} codoped ZnO nanocrystals with and without Li^+ ions. *Opt. Commun.* **2008**, *281*, 5448–5452.

41. Chen, G.Y.; Liu, H.C.; Liang, H.J.; Somesfalean, G.; Zhang, Z.G. Enhanced multiphoton ultraviolet and blue upconversion emissions in Y_2O_3:Er^{3+} nanocrystals by codoping with Li^+ ions. *Solid State Commun.* **2008**, *148*, 96–100.

42. Awan, S.U.; Hasanain, S.K.; Bertino, M.F.; Jaffari, G.H. Ferromagnetism in Li doped ZnO nanoparticles: The role of interstitial Li. *J. Appl. Phys.* **2012**, *112*, 103924:1–103924:9.

43. Cates, E.L.; Wilkinson, A.P.; Kim, J.-H. Delineating mechanisms of upconversion enhancement by Li+ codoping in $Y_2SiO_5:Pr^{3+}$. *J. Phys. Chem. C* **2012**, *116*, 12772–12778.

44. Baikie, T.; Ng, G.M.; Madhavi, S.; Pramana, S.S.; Blake, K.; Elcombe, M.; White, T.J. The crystal chemistry of the alkaline-earth apatites $A_{10}(PO_4)_6Cu_xO_y(H)_z$ (A = Ca, Sr and Ba). *Dalton Trans.* **2009**, *34*, 6722–6726.

45. Haase, M.; Schafer, H. Upconverting nanoparticles. *Angew. Chem.* **2011**, *50*, 5808–5829.

46. Mai, H.-X.; Zhang, Y.-W.; Sun, L.-D.; Yan, C.-H. Size- and phase-controlled synthesis of monodisperse $NaYF_4$:Yb,Er nanocrystals from a unique delayed nucleation pathway monitored with upconversion spectroscopy. *J. Phys. Chem. C* **2007**, *111*, 13730–13739.

47. Pollnau, M.; Gamelin, D.R.; Lüthi, S.R.; Güdel, H.U.; Hehlen, M.P. Power dependence of upconversion luminescence in lanthanide and transition-metal-ion systems. *Phys. Rev. B* **2000**, *61*, 3337–3346.

48. Suyver, J.; Aebischer, A.; García-Revilla, S.; Gerner, P.; Güdel, H. Anomalous power dependence of sensitized upconversion luminescence. *Phys. Rev. B* **2005**, *71*, 125123:1–125123:9.

49. Renero-Lecuna, C.; Martín-Rodríguez, R.; Valiente, R.; González, J.; Rodríguez, F.; Krämer, K.W.; Güdel, H.U. Origin of the high upconversion green luminescence efficiency in β-$NaYF_4$:2% Er^{3+}, 20% Yb^{3+}. *Chem. Mater.* **2011**, *23*, 3442–3448.

Chapter 11

Enhanced Upconversion Luminescence in Yb^{3+}/Tm^{3+}-Codoped Fluoride Active Core/Active Shell/Inert Shell Nanoparticles through Directed Energy Migration

Hailong Qiu [1,2], Chunhui Yang [1], Wei Shao [1,2], Jossana Damasco [2], Xianliang Wang [3], Hans Ågren [4], Paras N. Prasad [2] and Guanying Chen [1,2,*]

[1]*School of Chemical Engineering and Technology, Harbin Institute of Technology, Harbin 150001, China*

[2]*Institute for Lasers, Photonics, and Biophotonics, University at Buffalo, State University of New York, Buffalo, NY 14260, USA*

[3]*Department of Chemical and Biological Engineering, University at Buffalo, State University of New York, Buffalo, NY 14260, USA*

[4]*Department of Theoretical Chemistry, Royal Institute of Technology, S-10691 Stockholm, Sweden*

ABSTRACT

The luminescence efficiency of lanthanide-doped upconversion nanoparticles is of particular importance for their embodiment in biophotonic and photonic applications. Here, we show that the upconversion luminescence of typically used $NaYF_4$:Yb^{3+}30%/Tm^{3+}0.5% nanoparticles can be enhanced by ~240 times through a hierarchical active core/active shell/inert shell ($NaYF_4$:Yb^{3+}30%/Tm^{3+}0.5%) /$NaYbF_4$/$NaYF_4$ design, which involves the use of directed energy migration in the second active shell layer. The resulting active core/active shell/inert shell nanoparticles are determined to be about 11 times brighter than that of well-investigated ($NaYF_4$:Yb^{3+}30%/Tm^{3+}0.5%)/$NaYF_4$ active core/inert shell nanoparticles when excited at ~980 nm. The strategy for enhanced upconversion in Yb^{3+}/Tm^{3+}-codoped $NaYF_4$ nanoparticles through directed energy migration might have implications for other types of lanthanide-doped upconversion nanoparticles.

Keywords: upconversion; efficiency; core/shell; energy migration

1. INTRODUCTION

Lanthanide-doped upconversion nanoparticles (UCNPs) are able to convert near infrared light (NIR) into shorter wavelength NIR, visible, and ultraviolet (UV) emissions, and have important technological applications ranging from bioimaging, displays, photovoltaics, drug delivery, biosensors, to photodynamic therapy [1,2,3,4,5]. Unlike multi-photon absorption, which requires high excitation density (~10^6–10^8 W/cm^2) from ultrashort pulsed lasers, the upconversion (UC) processes generally involve the use of a much lower excitation density (~10^{-1}–10^2 W/cm^2) (provided by a continuous-wave NIR diode laser) due to the use of the real ladder-like energy levels in lanthanides [1,4]. Typically, UCNPs are doped with ytterbium (Yb^{3+}) sensitizer ions and activator ions such as erbium (Er^{3+}), thulium (Tm^{3+}), or holmium (Ho^{3+}) [1,2,3,4,5]. The sensitizer ions are utilized to harvest NIR excitation and non-radiatively transfer their absorbed energy to the activator ions to produce luminescence. Despite recent progresses in accurate control of nanocrystal morphology and size [6,7], crystal phase [6], and emission colors [8,9,10], it remains a great challenge to achieve strong upconversion luminescence in UCNPs.

Attempts to overcome this challenge include variation of concentrations of sensitizer and activator dopants [11,12], selecting host lattice of low crystal phase and phonon energy [4], tailoring the local environment around lanthanide ions [13,14], or using noble metal nanostructures to enhance the energy transfer rate by surface plasmons [15]. However, none of these methods address surface-related quenching mechanisms that are pronounced in lanthanide-doped nanoparticles. Surface-related quenching mechanisms are associated with the fact that a majority of doped lanthanide ions are located on the nanoparticle surface due to the high surface/volume ratio created by the nanoscale dimension, and can be easily quenched by surface deactivations (such as surface impurities, ligands, and solvents). Coating nanoparticles with an inert shell is an effective strategy to minimize surface quenching due to the spatial isolation of the core nanoparticle from the surrounding environment [4]. Moreover, it has been demonstrated that the incorporation of sensitizer Yb^{3+} dopants in the inert $NaGdF_4$ shell can further enhance the UC luminescence from the active core/inert shell ($NaGdF_4$:20%Yb^{3+}/2%Er^{3+})/$NaGdF_4$ UCNPs [16]. This active core/active shell strategy has also been investigated in the $BaGdF_5$:Yb^{3+}/Er^{3+}@$BaGdF_5$:Yb^{3+} UCNPs [17], as well as in ($NaYF_4$:Ce^{3+}/Tb^{3+})/$NaYF_4$:Ce^{3+} downconversion nanoparticles [18]. However, for UCNPs, the Yb^{3+}sensitizer in the active shell generally has to be lower than ~20% to avoid quenching effect produced by random excitation energy migration within the active shell to the surface deactivation sites on the outer surface of the active shell. The random excitation energy migration within the Yb^{3+} sub-lattice of the active shell is due to the unique two-energy-level structure of Yb^{3+} in association with its long-lived excited state [4]. The restriction of limited amount of Yb^{3+} sensitizer has impeded the way of increasing Yb^{3+} sensitizer in the active shell to multiply the efficiency of the active core/active shell UCNPs. Moreover, reports on UC luminescence increase using the active core/active shell strategy are limited to Yb^{3+}/Er^{3+}-doped nanoparticles, and there has been no report on other type of lanthanide-doped UCNPs.

Recently, fluoride nanoparticles doped with lanthanide ions have gained wide interests, as fluoride host lattice has high chemical and thermal stability and possesses low phonon cutoff energy, allowing the generation of relatively high UC radiations through curbing nonradiative losses at the intermediate states of lanthanide ions [6,7,8,9,10,11,12]. Among lanthanide-doped fluoride nanoparticles, Yb^{3+}/Tm^{3+}-doped $NaYF_4$ nanoparticels are of particular interests, not only because $NaYF_4$ has been shown to be the most efficient host material for UCNPs [7,19], but also because they display luminescence in the NIR, blue and UV range, which are useful for bioimaging [20], photoactivation [21], and light-activated therapy [22]. However, until this point, there has been no report on utilizing the active shell to increase the upconversion luminescence of Yb^{3+}/Tm^{3+}-doped $NaYF_4$ UCNPs. In this work, we report on a new design of ($NaYF_4$:Yb^{3+}30%/Tm^{3+}0.5%)/$NaYbF_4$/$NaYF_4$ active core/active shell/inert shell UCNPs. Unlike previously reported active core/active shell UCNPs with limited amount of Yb^{3+} concentration (20%) in the active shell, the active shell here incorporates 100% of Yb^{3+} ions to harvest excitation photon more efficiently. Moreover, as a result of the use of outmost inert $NaYF_4$ shell to block the pathways to surface quenching cites, all energy migration within the $NaYbF_4$ active shell layer have been directed to sensitize lanthanide ions in the core nanoparticles, yielding efficient UC luminescence about ~240 times higher than that of the typically used $NaYF_4$:Yb^{3+}30%/Tm^{3+}0.5% UCNPs, and about 11 times brighter than that of the well-investigated active core/inert shell ($NaYF_4$:Yb^{3+}30%/Tm^{3+}0.5%)/$NaYF_4$ UCNPs.

2. RESULTS AND DISCUSSION

2.1. Characterizations of Morphology and Crystal Structure

Figure 1 shows the transmission electron microscopy (TEM) images of: (a) the core $NaYF_4$: 30%Yb^{3+}, 0.5%Tm^{3+} nanoparticles; (b) the active core/inert shell ($NaYF_4$: 30%Yb^{3+}, 0.5%Tm^{3+})/$NaYF_4$nanoparticles; (c) the active core/active shell ($NaYF_4$: 30%Yb^{3+}, 0.5%Tm^{3+})/$NaYbF_4$ nanoparticles; and (d) the active core/active shell/inert shell ($NaYF_4$: 30%Yb^{3+}, 0.5%Tm^{3+})/$NaYbF_4$/$NaYF_4$ nanoparticles. As one can see, all four types of synthesized nanoparticles appear nearly spherical in shape and monodisperse. The average diameters of the core $NaYF_4$: 30%Yb^{3+}, 0.5%Tm^{3+}, the active core/inert shell ($NaYF_4$: 30%Yb^{3+}, 0.5%Tm^{3+})/$NaYF_4$, the active core/active shell ($NaYF_4$: 30%Yb^{3+}, 0.5%Tm^{3+})/$NaYbF_4$, and the active core/active shell/inert shell ($NaYF_4$: 30%Yb^{3+}, 0.5%Tm^{3+})/$NaYbF_4$/$NaYF_4$ nanoparticles are determined to be about 6.6 ± 0.8, 12.3 ± 1.6, 14.2 ± 2.3, and 20.8 ± 1.4 nm, respectively. Compared to the size of the core $NaYF_4$: 30%Yb^{3+}, 0.5%Tm^{3+} nanoparticle, the size increase in the active core/inert shell and the active core/active shell nanoparticles indicates the successful growth of shell layer. In analogy, after coating an inert shell layer, the size of active core/active shell nanoparticles was increased from 14.2 to 20.8 nm, which suggests the formation of core/shell/shell structure. In addition, compared to the core of spherical shape, the core/shell (or core/shell/shell) nanoparticle has a similar spherical shape but at a larger size. Thus, it can be speculated that the shell layer in the core/shell (or core/shell/shell) nanoparticle is homogenous around the core nanoparticle.

Figure 2 displays the X-ray diffraction (XRD) patterns of the core $NaYF_4$: 30%Yb^{3+}, 0.5%Tm^{3+}nanoparticles, the active core/inert shell ($NaYF_4$: 30%Yb^{3+}, 0.5%Tm^{3+})/$NaYF_4$ nanoparticles, the active core/active shell ($NaYF_4$: 30%Yb^{3+}, 0.5%Tm^{3+})/$NaYbF_4$ nanoparticles, the active core/active shell/inert shell ($NaYF_4$: 30%Yb^{3+}, 0.5%Tm^{3+})/$NaYbF_4$/$NaYF_4$ nanoparticles, as well as the standard JCPDS 6-0342 cubic structure of $NaYF_4$. It is evident from the intensity of the peaks in Figure 2 that the prepared UCNPs are highly crystalline in nature. In addition, the positions of all XRD peaks correspond well to that of the standard JCPDS 6-0342 cubic pattern of $NaYF_4$, and no peaks from other phases or impurities are observed.

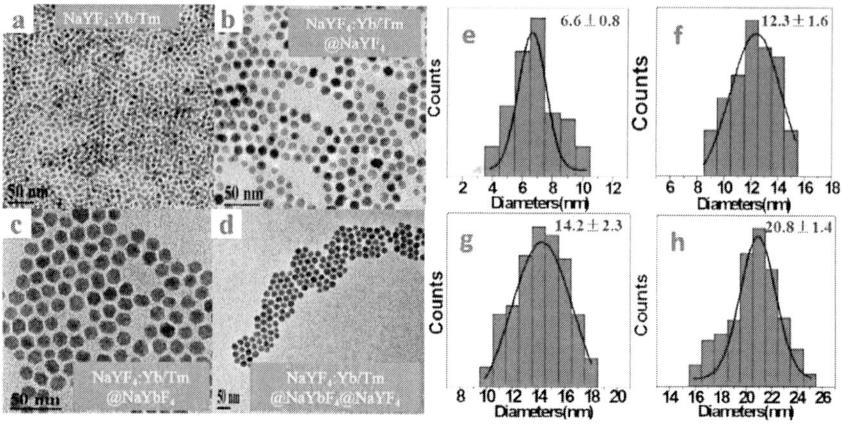

Figure 1. Transmission electron images (TEM) of: (**a**) The core $NaYF_4$: 30%Yb^{3+}, 0.5%Tm^{3+} nanoparticles; (**b**) The active core/inert shell ($NaYF_4$: 30%Yb^{3+}, 0.5%Tm^{3+})/$NaYF_4$ nanoparticles; (**c**) The active core/active shell ($NaYF_4$: 30%Yb^{3+}, 0.5%Tm^{3+})/$NaYbF_4$nanoparticles; and (**d**) The active core/active shell/inert shell ($NaYF_4$: 30%Yb^{3+}, 0.5%Tm^{3+})/$NaYbF_4$/$NaYF_4$ nanoparticles. Histogram of size distribution of (**e**) The core $NaYF_4$: 30%Yb^{3+}, 0.5%Tm^{3+} nanoparticles; (**f**) The active core/inert shell ($NaYF_4$: 30%Yb^{3+}, 0.5%Tm^{3+})/$NaYF_4$ nanoparticles; (**g**) The active core/active shell ($NaYF_4$: 30%Yb^{3+}, 0.5%Tm^{3+})/$NaYbF_4$ nanoparticles; and (**h**) The active core/active shell/inert shell ($NaYF_4$: 30%Yb^{3+}, 0.5%Tm^{3+})/$NaYbF_4$/$NaYF_4$ nanoparticles. The size was evaluated according to TEM images of corresponding nanoparticles dispersed in hexane at a concentration of 0.1 wt.%.

2.2. Upconversion Luminescence

Although $NaYF_4$ host lattice of hexagonal phase are more efficient than its cubic phase [11], lanthanide-doped cubic $NaYF_4$ nanocrystals are selected in the work as the resulting particle can be more easily produced for the size range of 5–20 nm that are essential for a broad spectrum of bio-related applications [3,4]. The UC luminescence spectra (in the spectroscopic range of 300–850 nm) of colloidal nanoparticles of the core $NaYF_4$: 30%Yb^{3+}, 0.5%Tm^{3+}, the active core/inert shell ($NaYF_4$: 30%Yb^{3+}, 0.5%Tm^{3+})/$NaYF_4$, and the active core/active shell ($NaYF_4$: 30%Yb^{3+}, 0.5%Tm^{3+})/$NaYbF_4$ (hexane dispersion) are shown in Figure 3a. Six apparent UC emission bands are resolved, centered at 345, 360, 450, 475, 650 and 800 nm, which correspond

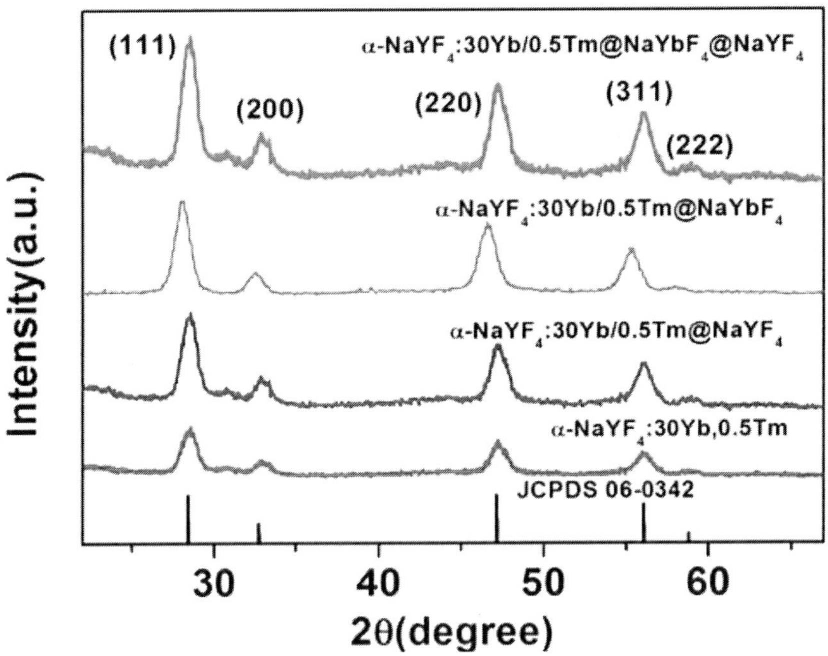

Figure 2. X-ray diffraction patterns of the core $NaYF_4$: 30%Yb^{3+}, 0.5%Tm^{3+} nanoparticles, the active core/inert shell ($NaYF_4$: 30%Yb^{3+}, 0.5%Tm^{3+})/$NaYF_4$ nanoparticles, the active core/active shell ($NaYF_4$: 30%Yb^{3+}, 0.5%Tm^{3+})/$NaYbF_4$ nanoparticles, and the active core/active shell/inert shell ($NaYF_4$: 30%Yb^{3+}, 0.5%Tm^{3+})/$NaYbF_4$/$NaYF_4$ nanoparticles, contrasted with the standard cubic $NaYF_4$ structure of JCPDS 06-0342.

to the $^1I_6 \rightarrow ^3F_4$, $^1D_2 \rightarrow ^3H_6$, $^1D_2 \rightarrow ^3F_4$, $^1G_4 \rightarrow ^3H_6$, $^1G_4 \rightarrow ^3F_4$, and $^3H_4 \rightarrow ^3H_6$ transitions of Tm^{3+} ions, respectively [7,11]. The weak emission at 511 nm arises from the transition of $^1D_2 \rightarrow ^3H_5$ of Tm^{3+} ions. The dependence of the intensities of these emission bands on the excitation power density indicates that five-, four-, three-, and two-photon processes are involved to populate the $^1I_6$, $^1D_2$, $^1G_4$, and $^3H_4$ state, respectively (Consult Figure 4a). The peaks at ~690 and ~720 nm marked by asterisk correspond to the second order of the peaks at 345 and 360 nm, respectively. We would like to highlight that the same batch of core nanoparticles is utilized to grow the active core/inert shell, the active core/active shell, and the active core/active shell/inert shell nanoparticles, allowing an accurate quantification of the effect of varying shell on the luminescence intensity of the core nanoparticle. The luminescence intensity (integrated over all the measured spectroscopic range) of the active core/inert shell ($NaYF_4$: 30%Yb^{3+},

$0.5\%Tm^{3+})/NaYF_4$ nanoparticles is found to be about 23 times higher than that of the core $NaYF_4$: $30\%Yb^{3+}$, $0.5\%Tm^{3+}$ nanoparticles, consistent with previously published results [23]. Remarkably, the luminescence intensity of the active core/active shell ($NaYF_4$: $30\%Yb^{3+}$, $0.5\%Tm^{3+}$)/$NaYbF_4$ is about 121 times higher than that of the core $NaYF_4$: $30\%Yb^{3+}$, $0.5\%Tm^{3+}$nanoparticles, and about 5.3 times higher than that of the active core/inert shell ($NaYF_4$: $30\%Yb^{3+}$, $0.5\%Tm^{3+}$)/$NaYF_4$ nanoparticles. Moreover, compared to the luminescence of the active core/inert shell ($NaYF_4$: $30\%Yb^{3+}$, $0.5\%Tm^{3+}$)/$NaYF_4$ nanoparticles, the NIR emission at 800 nm from the $^3H_4$ state, the blue/red emission at 475/650 nm from the $^1G_4$ state, the UV/blue emission at 360/450 nm from the $^1D_2$state, and the UV emission at 345 nm from the $^1I_6$ state, is enhanced by about 3.2, 3.5, 21, and 57 times for the active core/active shell ($NaYF_4$: $30\%Yb^{3+}$, $0.5\%Tm^{3+}$)/$NaYbF_4$ UCNPs, respectively. The much higher enhancement times for luminescence involving higher-order multiphoton processes indicate that Yb^{3+} ions in the active shell can expedite upconverting processes in the core nanoparticles due to efficient migration of excitation energy from the active shell to the core nanoparticles.

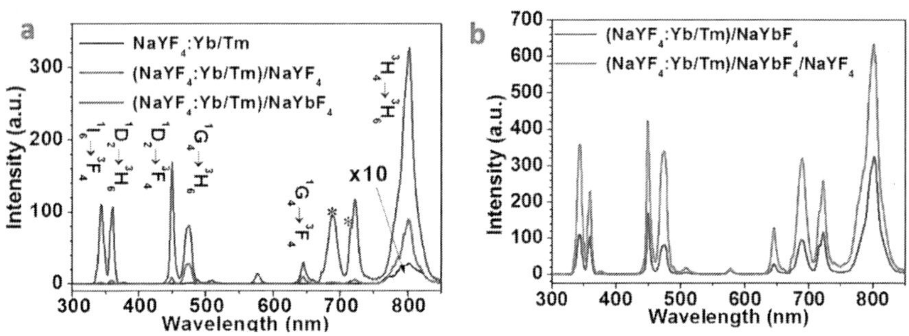

Figure 3. (**a**) Compared upconversion luminescence of colloidal nanoparticles (hexane dispersion, 1 wt.%) of the core $NaYF_4$: $30\%Yb^{3+}$, $0.5\%Tm^{3+}$, the active core/inert shell ($NaYF_4$: $30\%Yb^{3+}$, $0.5\%Tm^{3+}$)/$NaYF_4$, and the active core/active shell ($NaYF_4$: $30\%Yb^{3+}$, $0.5\%Tm^{3+}$)/$NaYbF_4$. The peaks at ~690 and ~720 nm marked by asterisk correspond to the second order of the peaks at 345 and 360 nm; (**b**) Compared upconversion luminescence of colloidal nanoparticles (hexane dispersion, 1 wt.%) of the active core/active shell/inert shell ($NaYF_4$: $30\%Yb^{3+}$, $0.5\%Tm^{3+}$)/$NaYbF_4$/$NaYF_4$ and the active core/active shell ($NaYF_4$: $30\%Yb^{3+}$, $0.5\%Tm^{3+}$)/$NaYbF_4$. The same batch of the core nanoparticles is utilized to grow the active core/inert shell, the active core/active shell, and the active core/active shell/inert shell nanoparticles. The excitation is performed with a diode laser at ~980 nm of about 50 W/cm^2. All the spectra have been calibrated by the spectral sensitivity of the utilized spectrophotometer system.

Figure 3b shows the luminescence comparison result of the active core/active shell/inert shell (NaYF$_4$: 30%Yb^{3+}, 0.5%Tm^{3+})/NaYbF$_4$/NaYF$_4$ UCNPs and the active core/active shell (NaYF$_4$: 30%Yb^{3+}, 0.5%Tm^{3+})/NaYbF$_4$ UCNPs. Luminescence at 800, 476/650, 365/450, and 345 nm from the $^3H_4$, $^1G_4$, $^1D_2$, and $^1I_6$ state in the active core/active shell nanoparticles are all enhanced by about two times. This luminescence enhancement, apparently, is a direct influence of the inert shell on the energy migrations inside the active shell, as the outmost inert shell is spatially isolated from the core nanoparticles. Combining results in Figure 3a,b, it can be easily found that the luminescence of the active core/active shell/inert shell (NaYF$_4$: 30%Yb^{3+}, 0.5%Tm^{3+})/NaYbF$_4$/NaYF$_4$ UCNPs is about ~240 times higher than that of the typically used NaYF$_4$: Yb^{3+}30%/Tm^{3+}0.5% UCNPs, and about 11 times higher than that of the active core/inert shell (NaYF$_4$: Yb^{3+}30%/Tm^{3+}0.5%)/NaYF$_4$ UCNPs. It is noted that the luminescence comparison in Figure 3 is based on the same weight concentration of UCNPs dispersed in hexane. Thus, different number of the core, the active core/inert shell, the active core/active shell, the active core/active shell/inert shell UCNPs is involved for the luminescence comparison. If compared based on the same number of nanoparticles, higher enhancement times can be anticipated.

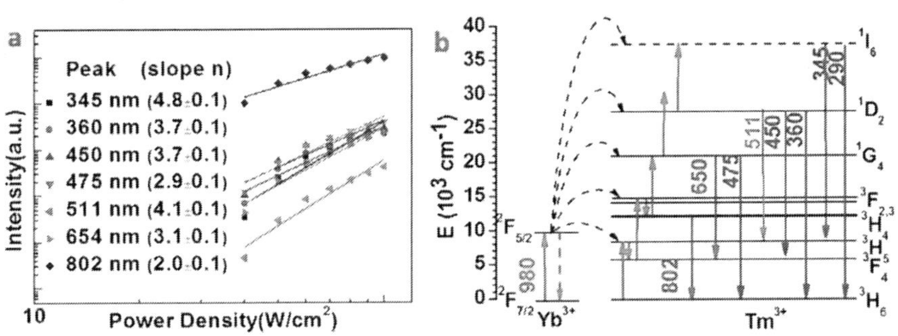

Figure 4. (**a**) A log-log plot of the dependence of various luminescence intensities from the active core/active shell/inert shell (NaYF$_4$: 30%Yb^{3+}, 0.5%Tm^{3+})/NaYbF$_4$/NaYF$_4$ nanoparticles on the excitation density; (**b**) Energy level diagrams of Yb^{3+} and Tm^{3+} ions as well as the involved mechanisms for upconversion luminescence from different energy states of Tm^{3+} ions.

2.3. Upconversion Mechanism

Figure 4a displays a log-log plot of the dependence of the intensities of various emissions on the excitation density for the active core/active shell/inert shell UCNPs. In general, the number of photons which are required to populate the upper emitting state under unsaturated condition can be obtained by the relation [19]:

$$I_L \propto P^n \tag{1}$$

where I_L is the luminescence intensity, *P* is the pump laser power, and *n* is the number of the laser photons required. Hence, the slope values in Figure 4a can provide the information on the number of photons absorbed for upconversion. Slope values of 4.8, 3.7, 3.7, 2.9, 4.1, 3.1, and 2.0 are observed for UC emissions peaked at 345, 360, 450, 475, 511, 654, and 800 nm, respectively. This result illustrates that five-, four-, three-, and two-photon processes are involved to generate the UC emission at 345 (from the $^1I_6$ state), 360/450/511 (from the $^1D_2$ state), 475/645 (from the $^1G_4$ state), and 800 nm (from the $^3H_4$ state), respectively. These observations agree well with previous results of Yb^{3+}/Tm^{3+}-codoped $NaYF_4$materials [24]. Moreover, similar slope values were observed for the active core/active shell ($NaYF_4$: 30%Yb^{3+}, 0.5%Tm^{3+})/$NaYbF_4$, and the active core/inert shell ($NaYF_4$: 30%Yb^{3+}, 0.5%Tm^{3+})/$NaYF_4$UCNPs (data not shown). The similar results indicate similar UC mechanisms for UV, visible, and NIR luminescence generation. Figure 4b shows the energy level diagrams of Yb^{3+} and Tm^{3+} ions as well as the proposed mechanisms for different luminescence bands. The Yb^{3+} ions firstly absorbs laser photons at ~980 nm, and are excited from the ground state $^2F_{7/2}$ to the $^2F_{5/2}$ state. Then, the Yb^{3+} ions in their excited states successively transfer the absorbed energy to the Tm^{3+} ion, and promote it from the ground sate to the $^3H_5$ (first step), $^3F_{2,3}$ (second step), $^1G_4$ (third step), and $^1D_2$ (fourth step) states, respectively. Multi-phonons from the $NaYF_4$ host lattice are involved to assist the energy transfer from Yb^{3+} to Tm^{3+}ions. Nonradiative relaxations from the $^3F_{2,3}$ state can populate the $^3H_4$ state, from which the NIR emission at 800 nm is produced. In addition, the UV emission at 345 nm can be generated by radiative relaxation from the $^1I_6$ state to the first excited state of $^3F_4$. The UC luminescence at 360, 450, and 511 nm arise from radiative relaxations from the $^1D_2$ state to the ground $^3H_6$, the first excited state $^3F_4$, and the second excited state $^3H_5$, of Tm^{3+} ions, respectively. The blue emission at 475 nm and the red emission at 645 nm stems from radiative relaxations from the $^1G_4$ state to the ground $^3H_6$ state, and the first excited state $^3F_4$, of Tm^{3+} ions, respectively.

2.4. Decay of Upconversion Luminescence

Figure 5 displays the decays of upconversion luminescence at 802 nm from the core $NaYF_4$: 30%Yb^{3+}, 0.5%Tm^{3+}, the active core/inert shell ($NaYF_4$: 30%Yb^{3+}, 0.5%Tm^{3+})/$NaYF_4$, the active core/active shell ($NaYF_4$: 30%Yb^{3+}, 0.5%Tm^{3+})/$NaYbF_4$ nanoparticles, and the active core/active shell/inert shell ($NaYF_4$: 30%Yb^{3+}, 0.5%Tm^{3+})/$NaYbF_4$/$NaYF_4$ nanoparticles.

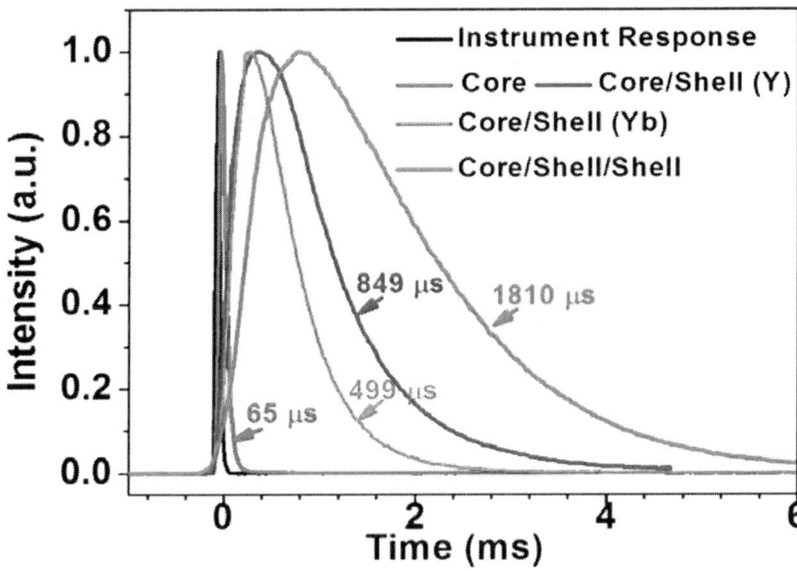

Figure 5. Decays of upconversion luminescence at 802 nm from the core $NaYF_4$: 30%Yb^{3+}, 0.5%Tm^{3+}, the active core/inert shell ($NaYF_4$: 30%Yb^{3+}, 0.5%Tm^{3+})/$NaYF_4$, the active core/active shell ($NaYF_4$: 30%Yb^{3+}, 0.5%Tm^{3+})/$NaYbF_4$, and the active core/active shell/inert shell ($NaYF_4$: 30%Yb^{3+}, 0.5%Tm^{3+})/$NaYbF_4$/$NaYF_4$ nanoparticles.

The average luminescence lifetime is determined to be about 65, 849, 499, 1810 μs, for the core, the active core/inert shell, the active core/active shell, and the active core/active shell/inert shell UCNPs, respectively. The significantly longer average lifetime of the active core/inert shell and the active core/active shell UCNPs than that of the core nanoparticles, demonstrate that surface-quenching effects in both type of core/shell structure have been significantly suppressed [25,26,27]. Note that despite having a thicker shell, the average lifetime of the active core/active shell nanoparticles is shorter than that of the active core/inert shell nanoparticles. This might be caused by the quenching effect produced by surface defects existed at the core surface. Due to a larger lattice mismatch between $NaYbF_4$ and $NaYF_4$ than between $NaYF_4$ and $NaYF_4$, a larger number of surface defects are anticipated at the core surface for the active core/active shell nanoparticle than for the active core/inert nanoparticle, thus leading to a shorter lifetime for the active core/active shell UCNPs. The growth of an inert shell on the active core/active shell nanoparticles can significantly increase the average lifetime of $^3H_4$ state of Tm^{3+}located in the core nanoparticles. This might arises from a better protection of the core nanoparticles due to the larger spacing between the surface of the core and the surrounding environment. Moreover, due to the inert shell coating on the active core/active shell UCNPs, the excitation energy migration within the active shell is relatively confined (Consult Figure 6d), and, thus, become much stronger. The only way to dissipate the excitation energy in the active shell is to migrate it to the core nanoparticle surface (Consult Figure 6d). The stronger energy migration in the active core/active shell/inert shell nanoparticle results in the use of longer time to build up the population in the $^3H_4$ state. This is in good agreement with the experimental observation in Figure 5 where a rising time of ~1 ms for the active core/active shell UCNPs is prolonged to be ~2.5 ms for the active core/active shell/inert shell UCNPs.

2.5. Quenching and Enhancing Mechanisms

Figure 6 illustrates the luminescence quenching mechanism in (a) the core $NaYF_4$: 30%Yb^{3+}, 0.5%Tm^{3+}nanoparticles, as well as the luminescence enhancement mechanism in (b) the active core/inert shell ($NaYF_4$: 30%Yb^{3+}, 0.5%Tm^{3+})/$NaYF_4$ nanoparticles; (c) the active core/active shell ($NaYF_4$: 30%Yb^{3+}, 0.5%Tm^{3+})/$NaYbF_4$ nanoparticles; and (d) the active core/active shell/inert shell ($NaYF_4$: 30%Yb^{3+}, 0.5%Tm^{3+})/$NaYbF_4$/$NaYF_4$ nanoparticles. For the core UCNPs without any shell coating (Figure 6a), numerous lanthanide dopants (sensitizer Yb^{3+} or activator Tm^{3+}) are exposed to surface deactivations (caused by surface defects, lattice strains, as well as ligands and solvents that possess high phonon energy) owing to the high surface-to-volume ratio at nanometer dimension, thus yielding UC luminescence at low efficiency [4,26]. This surface-related quenching mechanism can be minimized by coating with an inert shell (Figure 6b), which can spatially isolate the core nanoparticles from the surface activations, yielding a higher

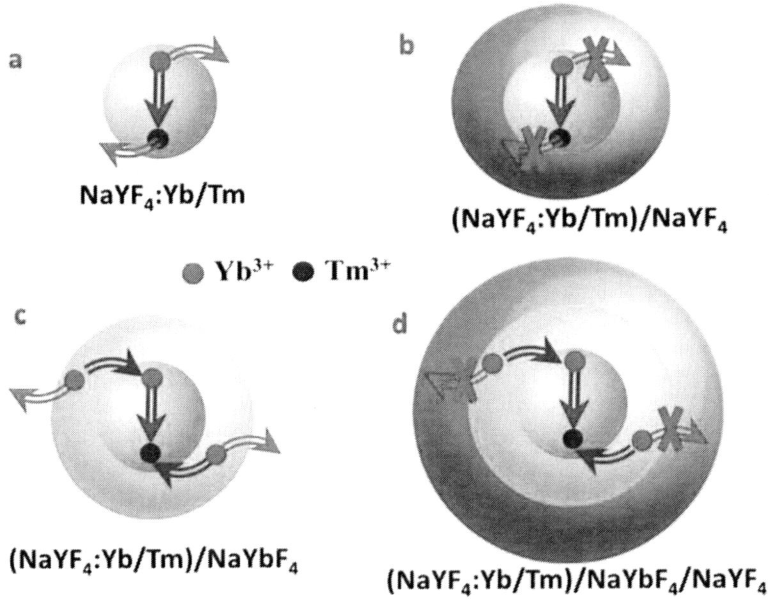

Figure 6. Schematic illustration of the luminescence quenching mechanism in: (**a**) The core $NaYF_4$: 30%Yb^{3+}, 0.5%Tm^{3+}nanoparticles, as well as the luminescence enhancement mechanism in (**b**) The active core/inert shell ($NaYF_4$: 30%Yb^{3+}, 0.5%Tm^{3+})/$NaYF_4$ nanoparticles; (**c**) The active core/active shell ($NaYF_4$: 30%Yb^{3+}, 0.5%Tm^{3+})/$NaYbF_4$ nanoparticles; and (**d**) The active core/active shell/inert shell ($NaYF_4$: 30%Yb^{3+}, 0.5%Tm^{3+}) /$NaYbF_4$/$NaYF_4$ nanoparticles.

UC efficiency. This is as the case of active core/inert shell ($NaYF_4$: 30%Yb^{3+}, 0.5%Tm^{3+})/$NaYF_4$ nanoparticles here which is about 23 times more efficient than the core $NaYF_4$: 30%Yb^{3+}, 0.5%Tm^{3+} nanoparticles. In addition to minimization of surface-related quenching mechanism, in the active core/active shell nanoparticles, a large amount of sensitizer Yb^{3+} ions in the shell can further enhance the UC luminescence and boost population of high-lying energy levels through transferring its absorbed excitation energy to lanthanide dopants located in the core nanoparticles [16]. Indeed, the luminescence of the active core/active shell ($NaYF_4$: 30%Yb^{3+}, 0.5%Tm^{3+})/$NaYbF_4$ is about 5.3 times higher in magnitude than that of the active core/inert shell ($NaYF_4$: 30%Yb^{3+}, 0.5%Tm^{3+})/$NaYF_4$ nanoparticles, and exhibits much higher luminescence intensity ratio of the UV emission at 345 nm from the $^1I_6$ state to the NIR emission at 800 nm from the $^3H_4$ state (consult Figure 3a). However, the excitation energy harvested in the active shell can be dissipated by random energy migration in two ways; one way is to the core nanoparticle, and the other way is to surface deactivations at the outer active shell surface (Figure 6c). The way of random energy migration to the outer surface of the active shell can be cut off by further growing an inert shell (Figure 6d), directing energy migrations solely to the core nanoparticles and thus increasing the luminescence output, in good agreement with the experimental result in Figure 3b.

3. EXPERIMENTAL SECTION

3.1. Synthesis of $NaYF_4$: Yb^{3+}30%/Tm^{3+}0.5% Core Nanoparticles

$NaYF_4$ doped with 0.5%Tm^{3+} and 30%Yb^{3+} nanoparticles were synthesized using a modified co-thermolysis method [6]. All chemicals used in the synthesis were purchased from Sigma-Aldrich (Milwaukee, WI, USA) and used as received. One millimole amounts of Tm_2O_3, Yb_2O_3, and Y_2O_3 were mixed and dissolved in 10 mL 50% concentrated trifluoroacetic acid at 95 °C in a three-necked 100 mL flask. The molar compositions of the cationic lanthanide ions in the solution were [Y/(Tm + Yb + Y)] = 0.695, Tm/(Tm + Yb + Y)] = 0.005, and [Yb/(Tm + Yb + Y)] = 0.3. Then, the solution was evaporated to dryness under the argon gas purge. Next, 2 mmol CF_3COONa, 6 mL oleic acid (90%, technical grade), 10 mL octadecene (90%, technical grade), and 6 mL oleylamine (70%, technical grade) were added into the three-necked flask. The resulting solution was then heated at 120 °C with magnetic stirring for 30 min to remove water and oxygen. The solution was then heated to 300 °C at a rate of about 12 °C per minute under argon gas protection, and kept at this temperature under vigorous stirring for about 0.5 h. A needle was used to let the argon gas out during the synthesis. The mixture was cooled to room temperature and precipitated by ethanol in an ultrasonic bath and collected by centrifugation at 7000 rpm for 5 min. The precipitate was then washed with ethanol for several times, and the nanoparticles were dispersed in 10 mL of hexane for further characterizations.

3.2. Synthesis of ($NaYF_4$: Yb^{3+}30%/Tm^{3+}0.5%)/$NaYbF_4$ Active Core/Active Shell Nanoparticles

The ($NaYF_4$: Yb^{3+}30%/Tm^{3+}0.5%)/$NaYbF_4$ active core/active shell nanoparticle were synthesized following a similar preparation procedure as the core. A 0.5 mmol amount of Yb_2O_3 dissolved in 10 mL 50% concentrated trifluoroacetic acid at 95 °C in a three-necked 100 mL flask. Then, the solution was evaporated to dryness under the argon gas purge. Next, 1 mmol CF_3COONa, 0.5 mmol $NaYF_4$: Yb^{3+}30%/Tm^{3+}0.5% core, 10 mL oleic acid, and 10 mL octadecene were added into the three-necked flask. The resulting solution was heated at heated at 120 °C with magnetic stirring for 30 min to remove water and oxygen. The solution was then heated to 300 °C at a rate of about 12 °C per minute under argon gas protection, and kept at this temperature, under vigorous stirring, for about 0.5 h. The mixture was cooled to room temperature and precipitated by ethanol in an ultrasonic bath and collected by centrifugation at 7000 rpm for 5 min. The precipitate was then washed with ethanol several times, and the nanocrystals were dispersed in 10 mL of hexane for further characterizations.

3.3. Synthesis of ($NaYF_4$: Yb^{3+}30%/Tm^{3+}0.5%)/$NaYF_4$ Active Core/Inert Shell Nanoparticles

The ($NaYF_4$: Yb^{3+}30%/Tm^{3+}0.5%)/$NaYF_4$ active core/inert shell nanoparticles were synthesized following a similar preparation procedure as described in Section 3.2. A 0.5 mmol amount of Y_2O_3 was dissolved in 10 mL 50% concentrated trifluoroacetic acid at 95 °C in a three-necked 100 mL flask. Then, the solutions were evaporated to dryness under the argon gas purge. Next, 1 mmol CF_3COONa, 0.5 mmol $NaYF_4$:Yb^{3+}30%/Tm^{3+}0.5% core, 10 mL oleic acid, and 10 mL octadecene were added into the three-necked flask. The resulting solution was then heated at heated at 120 °C with magnetic stirring for 30 min to remove water and oxygen. The solution was then heated to 300 °C at a rate of about 12 °C per minute under argon gas protection, and kept at this temperature under vigorous stirring for about 0.5 h. The mixture was cooled to room temperature and precipitated by acetone in an ultrasonic bath and collected by centrifugation at 7000 rpm for 5 min. The precipitate was then washed with ethanol for several times, and the nanocrystals were dispersed in 10 mL of hexane for further characterizations.

3.4. Synthesis of ($NaYF_4$: Yb^{3+}30%/Tm^{3+}0.5%)/ $NaYbF_4$/$NaYF_4$ Active Core/Active Shell/Inert Shell Nanoparticles

The ($NaYF_4$: Yb^{3+}30%/Tm^{3+}0.5%)/$NaYbF_4$/$NaYF_4$ active core/active shell/inert shell nanoparticles were synthesized following a similar procedure as the core. A 0.25 mmol amount of Y_2O_3 dissolved in 50% concentrated trifluoroacetic acid at 95 °C in a three-necked 100 mL flask. Then, the solutions were evaporated to dryness under the argon gas purge. Next, 0.5 mmol CF_3COONa, 0.25 mmol ($NaYF_4$: Yb^{3+}30%/Tm^{3+}0.5%)/$NaYbF_4$ active core/active shell nanoparticles, 5 mL oleic acid, and 5 mL octadecene were added into the three-necked flask. The resulting solution was then heated at heated at 120 °C with magnetic stirring for 30 min to remove water and oxygen. The solution was then heated to 300 °C at a rate of about 12 °C per minute under argon gas protection, and kept at this temperature under vigorous stirring for about 0.5 h. The mixture was cooled to room temperature and precipitated by ethanol in an ultrasonic bath and collected by centrifugation at 7000 rpm for 5 min. The precipitate was then washed with ethanol for several times, and the nanocrystals were dispersed in 10 mL of hexane for further characterizations.

3.5. Instruments

The size and morphology of UCNPs were characterized by transmission electron microscopy (TEM) using a JEM-2010 microscope (JEOL USA, Inc., Peabody, MA, USA) at an acceleration voltage of 200 KV. The powder X-ray diffraction (XRD) patterns were recorded by a Siemens D500 diffractometer using Cu Kα radiation (λ = 0.15418 nm). The 2θ angle of the XRD spectra was recorded at a scanning rate of 5°/min. UC luminescence spectra were recorded using a Fluorolog-3.11 Jobin Yvon spectrofluorometer (Horiba Jobin Yvon, Paris, France) with a slit width defining spectral resolution of 2 nm. The luminescence was excited at 975 nm using a fiber-coupled laser diode (Q-Photonics). All UC luminescence spectra have been corrected for the spectral sensitivity of the system. Decays of luminescence at 802 nm for UCNPs of the core $NaYF_4$: Yb^{3+}30%/Tm^{3+}0.5%, the active core/active shell ($NaYF_4$: Yb^{3+}30%/Tm^{3+}0.5%)/$NaYbF_4$, the active core/inert shell ($NaYF_4$: Yb^{3+}30%/Tm^{3+}0.5%)/$NaYF_4$, the active core/active shell/inert shell ($NaYF_4$: Yb^{3+}30%/Tm^{3+}0.5%)/$NaYbF_4$/$NaYF_4$ were recorded at the Infinium oscilloscope (Hewlett-Packard, Palo Alto, CA, USA) coupled to the photomultiplied tube (PMT) of the Fluorolog-3.11 Jobin Yvon spectrofluorimeter, using excitation at 975 nm from laser diode (Q-Photonics) operating in pulsed mode.

4. CONCLUSIONS

The intensity of UC luminescence of $NaYF_4$: Yb^{3+}30%/Tm^{3+}0.5% UCNPs has been enhanced by ~240 times through utilization of the active core/active shell/inert shell ($NaYF_4$: Yb^{3+}30%/Tm^{3+}0.5%)/$NaYbF_4$/$NaYF_4$ structure. Moreover, the UC luminescence of the active core/active shell/inert shell ($NaYF_4$: Yb^{3+}30%/Tm^{3+}0.5%)/$NaYbF_4$/$NaYF_4$ UCNPs is determined to be about 11 times higher than that of well-investigated ($NaYF_4$:

Yb^{3+}30%/Tm^{3+}0.5%)/$NaYF_4$ active core/inert shell UCNPs. Mechanistic investigation reveals that the significant UC enhancement in the active core/active shell/inert shell UCNPs arises from directed energy migration within the second active shell layer solely to the surface of the core nanoparticle. The guidance of energy migration is provided by the outmost inert shell which blocks the way of energy migration to surface quenching sites. The enhancement UC strategy demonstrated here in Yb^{3+}/Tm^{3+}-doped $NaYF_4$ nanoparticles has implications to be extended to other type of lanthanide-doped UCNPs.

ACKNOWLEDGMENTS

This work is supported by Natural Science Foundation of China (No. 51102066), Program for Basic Research Excellent Talents in Harbin Institute of Technology (BRETIII 2012018), and National Science Fund for Distinguished Young Scholars (No.51325201).

CONFLICTS OF INTEREST

The authors declare no conflict of interest.

REFERENCES

1. Auzel, F. Upconversion and anti-stokes processes with f and d ions in solids. *Chem. Rev.* **2004**, *104*, 139–174.

2. Wang, F.; Liu, X.G. Recent advances in the chemistry of lanthanide-doped upconversion nanocrystals. *Chem. Soc. Rev.* **2009**, *38*, 976–989.

3. Naccache, R.; Rodríguez, E.M.; Bogdan, N.; Sanz-Rodríguez, F.; Cruz, M.C.I.; Fuente, J.; Vetrone, F.; Jaue, D.; Solé, J.G.; Capobianco, J.A. High resolution fluorescence imaging of cancers using lanthanide ion-doped upconverting nanoparticles. *Cancers* **2012**, *4*, 1067–1105.

4. Chen, G.; Yang, C.; Prasad, P.N. Nanophotonics and nanochemistry: Controlling the excitation dynamics for frequency upconversion and downconversion in lanthanide-doped nanoparticles.*Acc. Chem. Res.* **2013**, *46*, 1474–1486.

5. Chen, G.; Seo, J.W.; Yang, C.; Prasad, P.N. Nanochemistry and nanomaterials for photovoltaics.*Chem. Soc. Rev.* **2013**, *42*, 8304–8338.

6. Mai, H.X.; Zhang, Y.W.; Si, R.; Yan, Z.G.; Sun, L.D.; You, L.P.; Yan, C.H. High-quality sodium rare-earth fluoride nanocrystals: Controlled synthesis and optical properties. *J. Am. Chem. Soc.* **2006**,*128*, 6426–6436.

7. Boyer, J.C.; Cuccia, L.A.; Capobianco, J.A. Syntheis of colloidal upconverting $NaYF_4:Er^{3+}/Yb^{3+}$ and Tm^{3+}/Yb^{3+} monodisperse nanocrystals. *Nano Lett.* **2007**, *7*, 847–852.

8. Ehlert, O.; Thomann, R.; Darbandi, M.; Nann, T. A four-color colloidal multiplexing nanoparticle system. *ACS Nano* **2008**, *2*, 120–124.

9. Wang, F.; Liu, X.G. Upconversion multicolor fine-tuning: Visible to near-infrared emission from lanthanide-doped $NaYF_4$ nanocrystals. *J. Am. Chem. Soc.* **2008**, *130*, 5642–5643.

10. Chen, G.; Qiu, H.; Fan, R.; Hao, S.; Tan, S.; Yang, C.; Han, G. Lanthanide-doped ultrasmall yttrium fluoride nanoparticles with enhanced multicolor upconversion photoluminescence. *J. Mater. Chem.***2012**, *22*, 20190–20196.

11. Krämer, K.W.; Biner, D.; Frei, G.; Güdel, H.U.; Hehlen, M.P.; Lüthi, S.R. Hexagonal sodium yttrium fluoride based green and blue emitting upconversion phosphors. *Chem. Mater.* **2004**, *16*, 1244–1251.

12. Zhao, J.; Jin, D.; Schartner, E.P.; Lu, Y.; Liu, Y.; Zvyagin, A.; Zhang, L.; Dawes, J.M.; Xi, P.; Piper, J.A.; *et al.* Single-nanocrystal sensitivity achieved by enhanced upconversion luminescence. *Nat. Nanotechnol.* **2013**, *8*, 729–734.

13. Chen, G.; Liu, H.; Liang, H.; Somesfalean, G.; Zhang, Z. Upconversion emission enhancement in Yb^{3+}/Er^{3+}-codoped Y_2O_3 nanocrystals by tridoping with Li^+ ions. *J. Phys. Chem. C* **2008**, *112*, 12030–12036.

14. Chen, G.; Liu, H.; Liang, H.; Somesfalean, G.; Zhang, Z. Enhanced multiphoton ultraviolet and blue upconversion emission in $Y_2O_3:Er^{3+}$ nanocrystals by codoping with Li^+ ions. *Solid State Commun.* **2008**, *148*, 96–100.

15. Schietinger, S.; Aichele, T.; Wang, H.; Nann, T.; Benson, O. Plasmon-enhanced upconversion in Single $NaYF_4:Yb^{3+}/Er^{3+}$ codoped nanocrystals. *Nano Lett.* **2010**, *10*, 134–138.

16. Vetrone, F.; Naccache, R.; Mahalingam, V.; Morgan, C.G.; Capobianco, J.A. The active-core/active-shell approach: A strategy to enhance the upconversion luminescence in lanthanide-doped nanoparticles. *Adv. Funct. Mater.* **2009**, *19*, 2924–2929.

17. Yang, D.; Li, C.; Li, G.; Shang, M.; Kang, X.; Lin, J. Colloidal synthesis and remarkable enhancement of the upconversion luminescence of $BaGdF_5:Yb^{3+}/Er^{3+}$ nanoparticles by active-shell modification. *J. Mater. Chem.* **2011**, *21*, 5923–5927.

18. Wu, F.; Liu, X.; Kong, X.; Zhang, Y.; Tu, L.; Liu, K.; Song, S.; Zhang, H. The real role of active-shell in enhancing the luminescence of lanthanides doped nanomaterials. *Appl. Phys. Lett.* **2013**, *102*, 243104:1–243104:4.

19. Chen, G.; Ohulchanskyy, T.Y.; Kumar, R.; Ågren, H.; Prasad, P.N. Ultrasmall monodisperse $NaYF_4:Yb^{3+}/Tm^{3+}$ nanocrystals with enhanced near-infrared to near-infrared upconversion photoluminescence. *ACS Nano* **2010**, *4*, 3163–3168.

20. Chen, G.; Shen, J.; Ohulchanskyy, T.Y.; Patel, N.J.; Kuikov, A.; Li, Z.; Song, J.; Pandey, R.K.; Ågren, H.; Prasad, P.N. (α-$NaYbF_4:Tm^{3+}$)/CaF_2 core/shell nanoparticles with efficient near-infrared to near-infrared upconversion for high-contrast deep tissue bioimaging. *ACS Nano* **2012**, *6*, 8280–8287.

21. Shen, J.; Chen, G.; Ohulchanskyy, T.Y.; Kesseli, S.J.; Buchholz, S.; Li, Z.; Ågren, H.; Prasad, P.N.; Han, G. Tunable near infrared to ultraviolet upconversion luminescence enhancement in (α-$NaYF_4$:Yb,Tm)/CaF_2 core/shell nanoparticles for *in situ* real-time recorded biocompatible photoactivation. *Small* **2013**, *9*, 3213–3217.

22. Chen, X.; Zhao, Z.; Jiang, M.; Que, D.; Shi, S.; Zheng, N. Preparation and photodynamic therapy application of $NaYF_4$:Yb,Tm-$NaYF_4$:Yb,Er multifunctional upconverting nanoparticles. *New J. Chem.* **2013**, *37*, 1782–1788.

23. Yi, G.S.; Chow, G.M. Water-soluble $NaYF_4$:Yb,Er(Tm)/$NaYF_4$/polymer core/shell/shell nanoparticles with significant enhancement of upconversion fluorescence. *Chem. Mater.* **2007**, *19*, 341–343.

24. Wang, G.; Qin, W.; Wang, L.; Wei, G.; Zhu, P.; Kim, R. Intense ultraviolet upconversion luminescence from hexagonal $NaYF_4:Yb^{3+}/Tm^{3+}$ microcrystals. *Opt. Express* **2008**, *16*, 11907–11914.

25. Chen, G.; Ohulchanskyy, T.Y.; Law, W.C.; Ågren, H.; Prasad, P.N. Monodisperse $NaYbF_4:Tm^{3+}/NaGdF_4$ core/shell nanocrystals with near-infrared to near-infrared upconversion photoluminescence and magnetic resonance properties. *Nanoscale* **2011**, *3*, 2003–2008.

26. Wang, F.; Liu, X. Direct evidence of a surface quenching effect on size-dependent luminescence of upconversion nanoparticles. *Angew. Chem.* **2010**, *49*, 7456–7460.

27. Zhang, F.; Che, R.; Li, X.; Yao, C.; Yang, J.; Shen, D.; Hu, P.; Li, W.; Zhao, D. Direct imaging the upconversion nanocrystal core/shell structure at the subnanometer level: Shell thickness dependence in upconverting optical properties. *Nano Lett.* **2012**, *12*, 2852–2858.

Index

S

U